46
48
56
C000201405
70
72
76
78
80
82
94
96
98
100
102
104
120
122
124
126
128
130
144
146
148
150
154
156
170
172
174
176
178
180
200
202
204
208

# Saunders' Field Guide to GLADIOLI of South Africa

Rod & Rachel Saunders

With Fiona C Ross

Published by Struik Nature
(an imprint of Penguin Random House South Africa (Pty) Ltd)
Reg. No. 1953/000441/07
The Estuaries No. 4, Oxbow Crescent,
Century Avenue, Century City 7441
PO Box 1144, Cape Town, 8000 South Africa

Visit www.penguinrandomhouse.co.za and join the Struik Nature Club for updates, news, events and special offers.

First published in 2021

10 9 8 7 6 5 4 3 2 1

ISBN 978 1 77584 761 8 (Print)
ISBN 978 1 77584 762 5 (ePub)



Publisher: Pippa Parker
Managing editor: Roelien Theron
Editor: Natalie Bell
Concept and cover design: Janice Evans
Designer: Gillian Black
Cartographer: Neil Bester
Proofreader: Emsie du Plessis

Reproduction by Studio Repro
Printed and bound in China by C&C Offset Printing Co., Ltd.

**Front cover:** *Gladiolus pavonia*, Abel Erasmus Pass.
**Back cover (top to bottom):** Rod and Rachel Saunders; *G. carneus*, Stettynsberg; *G. cruentus*, Krantzkloof.
**Title page:** *G. loteniensis.*
**Contents page:** *G. saxatilis*, God's Window, Mpumalanga.

# Acknowledgements

Pippa Parker, publisher of Struik Nature, saw Rod and Rachel's vision and committed immediately to seeing it through. I am more grateful than words can say to Natalie Bell, whose clarity, patience and gentleness have made something beautiful of both the process and the product. Gillian Black was responsible for the elegant design, Neil Bester generated the many maps, Colette Stott assisted with picture research for the Introduction and Roelien Theron guided the process with excellent feedback.

Renette Hendriks and staff at Personal Trust set up the Saunders Guide Trust to secure Rod and Rachel's archive, manage donations and enable the publication of this book.

My thanks to Ondine Schrick for her work as a trustee and more importantly, for her support. At times she had to vacate her office to allow me the space to trawl through Rachel's computer in search of elusive files. These sessions were always accompanied by copious cups of tea, much laughter and the sharing of memories of Rod and Rachel.

I am grateful to John Manning who fielded all my queries – whether significant or laughable – with grace. He has played a tremendously important consulting role in the production of this book, and has given permission for the use of his beautiful line drawings. My thanks also to John Manning and Peter Goldblatt for their lovely foreword.

Matthew Wolfe created an archive from back-ups of Rachel's computers, a challenging job. He also created and maintained our website.

Thanks are due to the Botanical Society of South Africa for their endorsement, to the Mapula Trust for a generous donation, and to the University of Cape Town for giving me leave to complete the book.

To the many friends and supporters of this project – a heartfelt thanks. I hope you feel that Rod and Rachel's vision has been accomplished.

And finally, to Andy Hackland and our daughters Sarah and Hannah, who have been there every step of the way, my love.

**Fiona C Ross**

## Publisher's acknowledgements

The publisher would like to thank Fiona Ross for drawing together the Saunders' notes and many photographs to prepare a full submission for this publication. Our thanks to Saunders Guide Trust and the individual funders for their generous support, and to John Manning, for playing a consulting role during the production of this book, and for the kind use of his line drawings. We are also grateful to Mike Picker for assisting with the identification of insects in some of the Saunders' photographs.

*In memory of*
*Rod and Rachel Saunders –*
*whose wish it was*
*to dedicate this book*
*to the tortoises they saved*
*along the way*

# DONORS

After Rod and Rachel Saunders' deaths, many people gave generously to ensure their *Gladiolus* book could be published. The Saunders Guide Trust received donations from Rod and Rachel's family, friends and customers from around the world of their company, Silverhill Seeds – from Japan, China, Europe, the United Kingdom, the United States of America, Canada and, of course, South Africa. Many donations were made anonymously. We extend our deep gratitude to those named and unnamed people who have helped us to bring *Saunders' Field Guide to Gladioli of South Africa* to completion.

Anonymous donors
Alisdair Aird
Anne Betschart
AP Hamilton
Arnelia Farms
Audrey Cain
Barbara Knox-Shaw
C Grobbelaar
Cecilia JingJing Zhu
Charles Gorenstein
Christell Bosveld and Peter Knippels
Dell C Sherk
Doctor Flower
Elizabeth Biffie Odendaal
Ernie de Marie
Francis J Hartnell
Gastil Buhl
Geoff Crowhurst
George Elder
Gerald and Pixie Lewis
Hiroshi Nakatani
Indigenous Bulb Association of South Africa (IBSA)
Jan van Dijk
Jeremy Spon
JP Potgieter
Kathy and Ron Barker
Kirstin, Dave and Islay Hutton
Living Cycads (Paul and Karen Sternberg)
Mapula Trust
Marianne Wustenhoff and Staf van Opstal
Marie Stobie
Mark Perry, Simon Perry, Justine Brown, Lisa Field and families
Ondine, Thomas, Bradley and Erica Schrick
Peggy Reynolds
Peter Kohn
Random Harvest
Richard and Hanneke Jamieson
Richard White
Robert Rutemoeller and Mary Sue Ittner
Rudi and Naomi Koster
Sachin Doarsamy
Shelley Brown
Silverhill Seeds
Simon Politzer
William Squire

# FOREWORD

'*Not your Grandma's gladioli*' is the saucy title that an American horticultural journal gave to an article on wild gladioli that we wrote for them some years ago. The piece was a commission from the editor, who clearly knew that the blowsy, matronly gladioli that convene in florists' shops were a world away from their varied and sophisticated wild relatives. Garden gladioli are in fact the products of extensive hybridisation and selection from *Gladiolus dalenii* and *G. oppositiflorus*, two of the larger, more ostentatious species that grow wild in southern Africa. But the subcontinent is home to many more kinds of wild gladioli than just this couple, and they produce flowers in an astonishing array of colours and shapes.

Southern Africa, comprising the countries of South Africa, Namibia, Botswana, Lesotho and Eswatini (formerly Swaziland), is the centre of diversity for gladioli; 169 kinds are scattered across its grasslands and savannas, succulent shrublands and fynbos, from the seashore to the highest peaks. They range from the sturdy and upright to the delicate and unassuming, and all have a charm and attraction of their own. Indeed, gladioli exert a fascination among plant fanciers that sometimes defies explanation.

Local bulb specialists cultivate several genera of Iridaceae, mostly somewhat erratically. Their monthly shows in the spring always include an assortment of pots containing sparaxis, tritonias, ixias and often romuleas but among them will always be some gladioli. And somehow the gladioli have a poise that sets them apart from the rest of their family, an indefinable elegance or an eccentricity that is captivating. Take the eager, bright orange and lime-green faces of the *Kalkoentjie* ('little turkey'), *G. alatus*, for instance, radiating a joyousness that is impossible to resist. It would be difficult to imagine a greater contrast to the cool elegance of the Painted lady, *G. debilis*, whose pale, almost shell-like flowers are precisely made-up like a geisha, or to the fluttering, brilliant scarlet flags of the New Year lily, *G. cardinalis*.

Like many of us, Rod and Rachel Saunders fell under the gladiolus spell. They combined their deep and intimate knowledge of plants with a love of exploration. It was this spirit of adventure that prompted them, initially purely as a challenge and a focus, to try to find every species in the wild, a feat that had eluded us while we were working on our monograph *Gladiolus in southern Africa*. It was only later that they conceived the idea of formalising their fun into a field guide. In a remarkably short space of time they succeeded in locating and photographing the great majority of the species, sometimes under extreme and challenging conditions. On one extended hike high into the Drakensberg in search of the elusive *G. symonsii* they were overtaken on the exposed slopes by a massive electrical storm and were lucky to escape with only a drenching.

We always looked forward to their return from an expedition, both for the vicarious enjoyment in their travels and for the intellectual stimulation that stemmed from their careful observations of the plants in the field. Many were the times that we puzzled over the identity of some atypical specimen or explored the epistemology of taxonomic knowledge. Or just revelled in Rod's roguish sense of humour as he gleefully described a particularly garish species as 'a real tart of a plant'.

Each trip they made uncovered something interesting or novel. This guide is a testament to them both and to a group of plants that brought them great joy, which it will surely do for all who page through it.

**John Manning and Peter Goldblatt**

# CONTENTS

# PREFACE

In about 2012, Rod and Rachel Saunders began what was to be the last project of their extraordinary botanical lives – a quest to find and photograph every known species of *Gladiolus* in South Africa. The search took them to archives and books; to experts, gardeners, guides and researchers; and to mountains and plains throughout the sub-region. Sometimes the plants were easy to find, well known, prolific and easy to identify. Other times, they were more elusive. Occasionally they had to wait long periods – until after a fire perhaps, or until they happened to be in a place when conditions were just right. At times, plants were no longer to be found in the locations in which they had previously been recorded. Owing to human impacts such as agriculture, mining, grazing and urbanisation, some locations had changed beyond recognition.

Extracts from some of the emails we found in Rachel's notes are telling. In one emailed exchange she comments: 'The problem with these plants is not only do we have to find them, but then they need to be in flower. And of course, the flowering time depends on rain, weather, fires etc. But I don't need to tell you all this as you know! Last year we went to Mpumalanga and Limpopo province about six times and still missed some of them. With fuel at R12 a litre, it becomes rather expensive as well as time-consuming.'

*Rod and Rachel Saunders were inveterate hikers; seen here in the foothills of the Drakensberg.*

More than time-consuming, the Saunders' passion fuelled expeditions and enthusiasms, a wide web of fellowship and a deep commitment to further recording this elegant genus of plants.

In February 2018, returning from a film-making trip in KwaZulu-Natal, Rod and Rachel were abducted and murdered. By then, they had found and photographed all but one of the known *Gladiolus* species they had been seeking. Although their laptop and notes have never been recovered, there is a partial record of their search, including a photographic collection, email correspondence, handwritten notes, jotted notes about locations or the right person to consult about specific species, scrawled directions and phone numbers.

Offering these remnants, the Saunders' family asked close friends of the couple to complete their field guide project. Matt Wolfe compiled an archive from various electronic back-ups; John Manning and Peter Goldblatt generously answered my questions and those of the editorial team, as they had done for Rod and Rachel; Pippa Parker of Struik Nature enthusiastically supported publication; and Natalie Bell and the design team at Struik carefully crafted this book. Drawing from the Saunders' archive, Goldblatt and Manning's magisterial and exquisite monograph *Gladiolus in Southern Africa*, and subsequent publications, and through monies raised from donations across the world, we have completed the work Rod and Rachel began. The book you hold is the result.

Rod and Rachel's passion for southern Africa's wild spaces, for botany, and for the *Gladiolus* genus in particular, has left the world with an extraordinary legacy. This book is part of it, but by no means all of it. Their expeditions generated thousands of photographs of gladioli alone, including species that had not been seen in generations. Some of the images help to fill the gaps in published knowledge and to demonstrate the extraordinary diversity of this beautiful genus. Their work may assist scientists to figure out the relationships between gladioli and

*Rod and Rachel Saunders photographing* Ornithogalum, *Devil's Peak, Western Cape.*

their environments – other plants, soils, climatic conditions and pollinators, and the ways that these have interacted with human incursions. To protect what remains of the Saunders' data, we have formed a Trust that will be responsible for safeguarding this material for future research.

We acknowledge here the incredible connections and relationships that must have informed Rod and Rachel's project. We do not have a full list of names of people who helped, so wish simply to recognise that knowledge comes not only through individuals' hard work and exploration, but also through dense networks spread across time and space. As is clear from even a cursory examination of the extensive email correspondence relating to this project that was generated, such connections are marked by extraordinary generosity in the sharing of locations, data, social networks, expertise and passion. In that sense, the project has already touched many lives and here – in tangible book form – represents a tracery of the connections that made it possible.

We are tremendously grateful to the many donors who gave generously to make this publication possible. A list of donors is included, but many donations were made anonymously, and we wish to record our deep gratitude to all who have contributed. In addition, I thank the University of Cape Town for granting me sabbatical leave in 2020 to work on this project.

Rod and Rachel always intended to dedicate the book to the tortoises they saved from road deaths. We do not know what they would have said in their dedication, but to honour their intentions, this book recalls the tortoises.

Historically, gladioli symbolised courage. In contemporary floral lore, they also represent perseverance and remembrance, a fitting tribute to Rod and Rachel's lives and work.

**Fiona C Ross**
Professor of Anthropology
University of Cape Town

# INTRODUCTION

The genus *Gladiolus* in the family Iridaceae has fascinated and delighted floral enthusiasts for centuries. Beloved of poets and artists as much as of gardeners, and included in some of Carl Linnaeus' earliest botanical descriptions, gladiolus – the sword lily – is widely acknowledged as a symbol of beauty. There are almost 270 recognised species, distributed across Africa, Europe, the Middle East and Madagascar. Southern Africa is the centre of diversity for the genus, home to more than half of the species. Within South Africa, the Western Cape is a critical location and speciation zone.

## THE *GLADIOLUS* IN HISTORY

Gladioli have not only been treasured for their appearance but also for their medicinal qualities. In medieval Europe, gladiolus corms were used to make poultices and treatments for colic and stomach disorders. Their value to precolonial populations in Africa is not fully recorded, but alongside the pleasure of their beauty, some species were and continue to be used for a range of medicinal purposes and sometimes as a food source across the continent – a tantalising indication of their significance.

European travellers, collectors, scientists and colonial authorities were entranced by the beauty and in some cases floriferous nature, unusual forms or delicate colouring of African species. As early as the late 1600s, gladioli had been sent from southern Africa to Europe and were described in the key botanical texts of the time. *Gladiolus alatus* for example, the much-loved, bright orange gladiolus common in the winter-rainfall region, had been recorded in 1739 by Breyne and described by Linnaeus in 1760.

*bonæ ſpei*
Veronica africana, floribus ad genicula pedicellis biuncialibus. *Herm. afr.* 783 *Pluk. phyt.* 320 *f.* 5.
*Herba* ſtatura Pedicularis, *Caules* pedales, proſtrati, læves.
*Folia* inferna terna (ſuperiora ad flores ſæpe alterna), petiolata, lanceolata, obtuſa, pinnatifida. *Flores* axillares, alterni: *Pedunculo* longo, unifloro; purpurei lineis albentibus. *Fructificatio* proxima Veronicæ, ſed Calyx 5-partitus.

2. GLADGOLUS *alatus* foliis enſiformibus, petalis lateralibus latisſimis.
Siſyrinchium viperarum *Pluk. phyt.* 124: *f* 6.
Bulbus. *Caulis* ſpithamæus, craſsiuſculus, inter flores flexuoſus.
*Folia* enſiformia, ſtriata, lævia, obtuſiuſcula: floralibus brevioribus diſtiche digeſtis. *Corollarum* Galea falcata, anguſta. Alæ ejusdem longitudinis, rhombeæ, latisſimæ. Labium inferius tripartitum: folioilis ovatis, æqualis longitudinis.

3. LEUCADENDRON *hirtum* foliis lanceolatis apice calloſis, caule hirſuto, floribus ſparſis axillaribus.
Lepidocarpodendrum foliis ſericeis brevibus confertisſimis, fructu gracili longo. *Boerh. lugdb.* 2. *p.* 194. *f.* 194.
Frutex. *Folia* breviter lanceolata, vix pubeſcentia, conferta, apice callo ſimplici. *Calyces* glabri: ſquamis acuminatis, ſolitarii, axillares foliorum, eorumque longitudine.

4. SCABIOSA *rigida* corollis quadrifidis ſubradiantibus: ſquamis calycinis obtuſis, foliis lanceolatis ſerratis.
Scabioſa africana fruteſcens, foliis rigidis ſplendentibus & ſerratis, flore albicante. *Comm. hort.* 1. *p.* 185. *f.* 93. *Raj. ſuppl.* 237.
Simillima *Scab. leucanthemæ*. *Caulis* fruticoſus, nudus. *Folia* ſerrata, nuda: inferiora obovata, in petiolis deſinentia; reliqua lanceolata, ſesſilia. Ex alis *Stipulæ* quaſi duæ lineares. *Flos* longe pedunculatus. *Calyx* ſquamis ſubrotundis *Corollulæ* quadrifidæ, extus tomentoſæ: Lacinia extima paullo majore; unde parum radiatus flos.

5. CHI-





**Top:** Gladiolus undulatus, *Curtis's Botanical Magazine, 1801.* ***Above:*** *The plant we know as* G. alatus *was described by Carl Linnaeus in 1760.* ***Opposite:*** G. gracilis.

***Above:*** *Early advertising; a field of cultivated gladioli illustrated for a 1929 seed catalogue.*

Gladiolus merianellus *(previously* G. bonaspei*) was named as a diminutive of the similar* Watsonia meriana, *which was named for Maria Sibylla Merian.*

Jacob Marrel, Public domain, via Wikimedia Commons

*Maria Sibylla Merian (1647–1717), a German-born artist, naturalist and botanist. She recorded, in exquisite artistic detail, every stage in the life cycle of caterpillars from egg to butterfly, at a time when it was believed that insects emerged from the mud. Merian travelled around Suriname for two years, pursuing her interest in insects and other exotic fauna and flora.*

By the beginning of the 1700s, gladioli were flowering in botanical collections in the United Kingdom, the Netherlands and Sweden, and by the beginning of the 1800s, they were beginning to be hybridised.

Some of the early history of scientific collection and description is beautifully related in Fraser and Fraser's book *The Smallest Kingdom*, and in Lewis and Obermeyer's *Gladiolus: A Revision of the South African Species*. T. Barnard describes the history of hybridisation. Peter Goldblatt and John Manning's exquisite and encyclopedic monograph *Gladiolus in Southern Africa* brings together a comprehensive review of the scientifically recorded history of each species, with fascinating details about their collection, description and classification.

The scientific names and descriptions bear traces of southern Africa's colonial histories. Many species are named in recognition of collectors, colonial authorities and landowners, for instance. Lacking from these accounts is the backstory of colonisation and the scientific impulse that accompanied and was made possible by it. Early expeditions made use of enslaved, 'mixed heritage' and indigenous people as labourers here, as they did elsewhere in colonial territories. For the most part, though, indigenous people's knowledge and enjoyment of plants is absent from the literature – except for descriptions of plants that are of medicinal use.

The European sensorium shaped scientific description, too. Take for example, descriptions of the sweet scent of a number of plants, especially that of species from the winter-rainfall regions. Flowers were described as being scented like vanilla, cloves and other spices that were familiar to elites in Europe through the spice trade – the same trade that was to play so significant a part in the establishment of mercantile capital and European rule in southern Africa.

Of course, names and descriptions need not be set in stone. Sometimes they change, usually because plants have been reclassified. In one example, the orange-scarlet *G. bonaspei*, endemic to the Cape Peninsula, was recently reverted to its older name, *G. merianellus*, which derives from an earlier name given by Linnaeus. In a scientific domain dominated by men, this name is significant, for it recognises the contributions of Maria Sibylla Merian, a seventeenth-century naturalist and illustrator who, among other things, was the first European to document the life cycle of butterflies. Until her close observations of metamorphosis, Europeans had believed that butterflies and other insects emerged from the mud, and eschewed them as being 'of the devil'. It is not known whether Merian ever encountered the gladiolus that now bears her name, but perhaps, having been somewhat critical of slavery elsewhere, she would have been equally so of the history of enslavement and colonial conquest that is the foundation of modern South Africa.

## About hybrids

A natural hybrid is created when two species cross-pollinate successfully and set viable seed. The resultant plant will have features of both parent generations, but its seeds may be sterile. In the commercial flower sector, hybrids are created to improve or develop certain characteristics such as productivity, appearance or disease resistance. Gladioli have been extensively hybridised to develop brighter, more colourful flowers for the cut-flower industry. The first European experimentation with hybrids using southern African gladioli began two centuries ago, and several species have become central in global hybridisation programmes.

*A natural hybrid of* Gladiolus saundersii *and* G. oppositiflorus.

Experimentation with hybridisation of gladioli was pursued with vigour in European centres. By the beginning of the nineteenth century, William Herbert reported hybrids of *G. angustus, G. carinatus, G. carneus, G. caryophyllaceus, G. liliaceus* and *G. tristis.* Successful hybridisation of gladioli led to their extensive production as cut flowers and garden stock, and with that came a growing familiarity with these plants across the globe. The fascination with gladioli has not ceased. *G. dalenii* and *G. oppositiflorus* are important elements in modern hybrids. Hybrid gladioli continue to account for a large proportion of the global flower trade, especially in the Netherlands, North America and China. South African scholars regularly note the untapped commercial potential of *Gladiolus* species. The genus may have become a commodity, but it has lost nothing of its ability to be an ordinary delight.

Delight is something familiar to both amateur and professional botanists. As Delpierre and Du Plessis put it in *The Winter-growing Gladioli of South Africa*: 'There are few more pleasant surprises for the lover of the veld than to stumble suddenly across a *pypie* in its lonely grace' (1974:7). *Pypie* is the Afrikaans common name for gladioli. And the surprises continue; in the decade after the publication of Goldblatt and Manning's monograph (1998), five new species of *Gladiolus* were identified and described in the scientific literature; some rare species were found in new sites; and some species not seen in decades were also recorded. At the same time, however, habitats have continued to come under pressure from urbanisation, agriculture, mining and climate change, rendering plant and insect life vulnerable.

Heinz Jacobi / Shutterstock.com

*Gladioli are cultivated on a large scale as they are popular cut flowers in the florist trade.*

G. reginae *was described in 2009.*

## *GLADIOLUS* MORPHOLOGY

It's one thing to take delight in a beautiful flower, another to put a name to the species. In order to identify a plant correctly, all parts of the plant, not just the flower, need to be examined, as different species may have remarkably similar flowers, or even no flowers if the plant is dormant – usually the case with bulbs during the dry season.

Below are descriptions and illustrations of the key gladiolus plant parts needed for identification.

*Bilaterally symmetrical flowers.*

*Radially symmetrical flower.*

### Flowers

Gladiolus flowers are generally zygomorphic (bilaterally symmetrical), with six tepals. Two upper lateral tepals flank the dorsal (uppermost) tepal, which is frequently hooded, giving the gladiolus its characteristic shape. The lower lateral and median tepals are often of a contrasting colour or bear different markings. A few species have actinomorphic (star-shaped) flowers. The sepals and petals of gladioli flowers are not differentiated as they are in many other flowers, where they often have different colouration and shape; instead they are joined and together are referred to as the tepals. The perianth tube is the lower portion of the tepals, where they are joined into a slender tube.

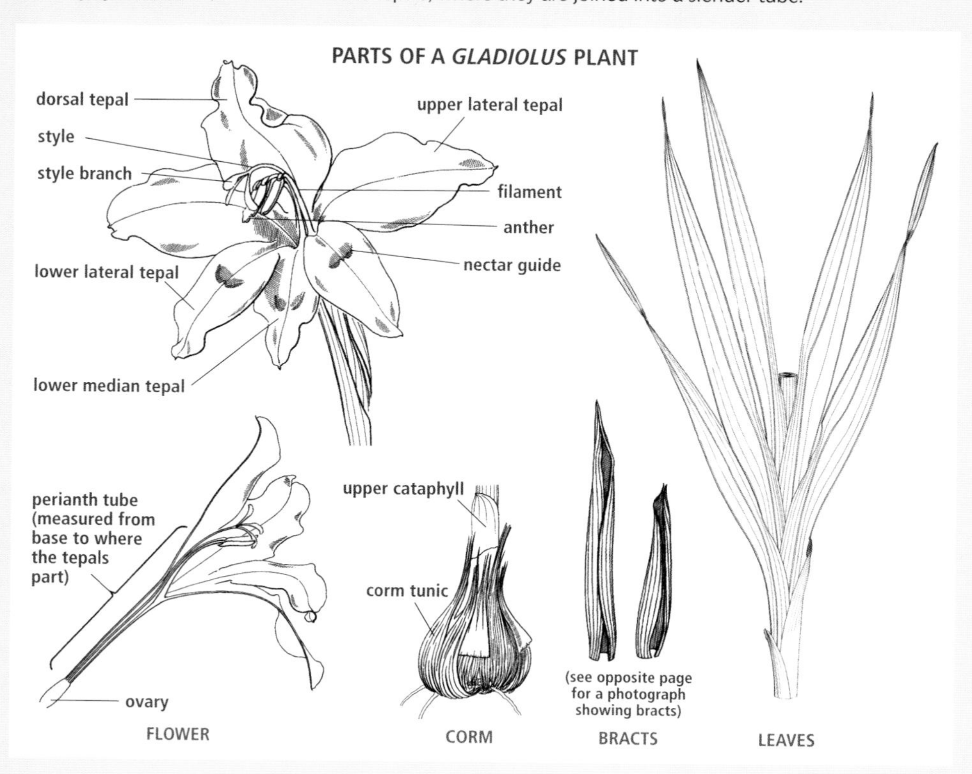

Most species open in the day and close in the late afternoon or evening, but some – those pollinated by moths – are closed by day and open in the evenings. The anthers and the pollen that they bear can both be of a surprising colour. Some species produce nectar as a reward for pollinators; many attract pollinators with their delightful scent.

## Spikes

Spikes bear the plant's flowers. Plants may have a simple flowering spike (described as unbranched) or they may have a branched spike. A spike may be floriferous (bearing many flowers), have only a few flowers, or even a single flower.

*Spikes;* G. abbreviatus.

## Leaves

The shape and form of gladiolus leaves are key in classifying a species. The genus *Gladiolus* gets its Latin name from the shape of its leaves, *gladiolus* meaning 'sword'. However, not all gladioli have sword-shaped leaves, and this is especially true of those in winter-rainfall areas. Some species produce different leaves in flowering and non-flowering seasons; the latter are called 'foliage leaves' and may look quite different from the leaves of flowering plants.

## Cataphylls

Cataphylls also play an important part in identification. They are tubular structures at the base of the plant. Some appear above ground, some wrap around the stems, some change colour along their length and some may seem dried. They do not photosynthesise.

*White-mottled cataphylls;* G. viridiflorus.

## Bracts

Bracts are modified leaves, and are useful in gladiolus identification. For the purposes of this book, only outer bracts are described, but experts will know that gladioli also have inner bracts.

*Dark bracts;* G. cunonius.

## Corms

Gladioli grow from corms, which are underground storage organs. They are the source of food for the plant's growth and are important in vegetative (asexual) reproduction. Corms are covered with tunics, which may be coriaceous (leathery) in appearance, or fibrous, papery or woody. When they decay, they often leave a distinctive mesh shape, which sometimes forms a fibrous neck around the base of the stem. Some species have long stolons (horizontal stems) ending in cormlets; others have large cormlets with thick tunics, and yet others produce cormlets around the base, on short stolons.

## Size

Species vary in size, from the tall *G. dalenii*, *G. geardii* and *G. undulatus*, which can reach 150cm in height, to the tiny *G. uysiae*, which is only 7–15cm high.

A *visiting honeybee on* G. appendiculatus.

*The Mountain pride butterfly is known to pollinate* G. stokoei.

## REPRODUCTION AND POLLINATION

All gladioli species are able to reproduce both sexually (through pollination and seed production) and vegetatively (through regeneration from the corm). Although the plants have both female and male sexual organs, most are self-incompatible and rely on pollination for sexual reproduction. *G. gueinzii* is an important exception; dwelling on the coast, where insect pollinators are not common, this species is autogamous (self-pollinating).

Bees are the most common pollinators and are thought to have been the ancestral pollinators. Over time, and dependent on climatic conditions and insect diversity, specialisation has occurred and plants have adapted to different pollinators. These include long-tongued flies, butterflies, moths and even sunbirds. Pollinators may be attracted by colour and sometimes also scent. It is possible to predict the pollinator from close observation of specific adaptations of the flower; for example, plants that are strongly scented at night are likely to be pollinated by moths. In summer-rainfall areas, flowers pollinated by bees are small, pale and usually scentless, whereas in winter-rainfall areas they are brightly coloured and strongly scented, presumably to attract pollinators in the short flowering season.
In general, flowers pollinated by flies are scentless, as are those pollinated by butterflies. Fly-pollinated flowers tend to have pale pink or cream-coloured flowers, while those pollinated by butterflies are generally red with a large supply of nectar. Bird-pollinated flowers tend to be orange-red or greenish, scentless, and with strong stems

### The role of fire and soil

Two factors are particularly important in the ecology of gladioli, namely fire and soil type. Where important, these factors are mentioned in the **Ecology & notes** paragraph for a species.

**Fire** Many gladioli species flower vigorously after fire, though some may only bloom for a year or two after a burn before going dormant for extended periods. Species in *Blandi,* namely *G. oreocharis*, *G. crispulatus* and *G. phoenix*, show particular propensities for this, as does *G. uitenhagensis*. Intense flowering in the seasons after fire and diminished flowering thereafter seems to be a characteristic of many winter-rainfall species. This is not necessarily true of summer-rainfall species; *G. pubiger*, for example, flowers the year after fire, but *G. microcarpus* may not flower at all.

G. emiliae *flowering after fire.*

**Soil types** Many species, particularly in the winter-rainfall areas, display edaphic fidelity; this means that they prefer one soil type over another. In these areas, where soils are either nutrient-poor quartzitic sandstone derivatives or shale clays, the effect is marked, and soil type can be used as a fairly reliable means to species identification.

and firm bracts to prevent damage; they have abundant glucose- and fructose-heavy nectar (except in section *Homoglossum*). Only one species, *G. meliusculus*, is known to be pollinated by monkey beetles, although the latter may also be opportunistic pollinators of plant species whose usual pollinator is a different insect. Monkey beetles also pollinate a similar-looking romulea that flowers at the same time. This is an example of a pollination guild, where a single pollinator is responsible for pollinating a variety of species.

*Common reed frog between capsules on* G. densiflorus.

Some plants have adapted their flowering time as a pollination strategy; they may flower late in the season when there is less competition, or flower at the same time as a different species to take advantage of a specific pollinator. So important are these multiple pollination strategies that they are considered to be one of the major ways that speciation in the genus has occurred.

*Ants on* G. pole-evansii.

### Capsules and seeds

Once pollinated, the gladiolus plant produces fruit called capsules. When these dry, they split open and the seeds are dispersed. The seeds are adapted for wind dispersal, those of most species having wings that allow them to be carried.

*Dehiscent capsules of* G. ecklonii *have split open to expel the seeds.*

G. viridiflorus *seeds awaiting wind dispersal.*

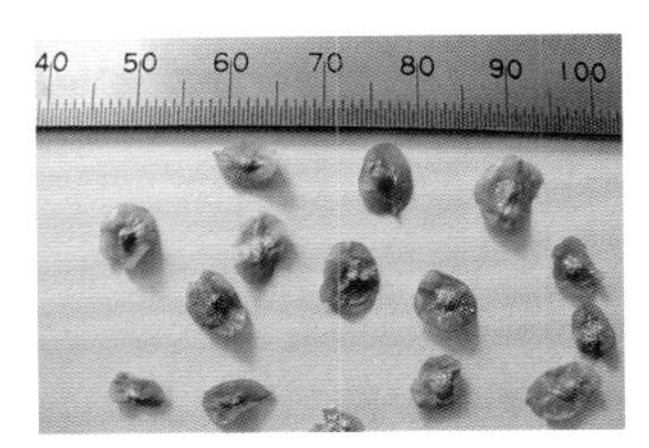



*Typical gladiolus seeds have broad wings that enable wind dispersal.*

### Corms and stolons

Gladioli reproduce vegetatively by producing cormlets, either directly from the base of the corm or on stolons, which are horizontal, stem-like structures that grow from the corm. The stolons enable the plant to create new corms and so begin the cycle again. A few species in *Hebea* are also able to grow corms from leaf axils below the ground.

G. phoenix *bears small cormlets on short stolons.*

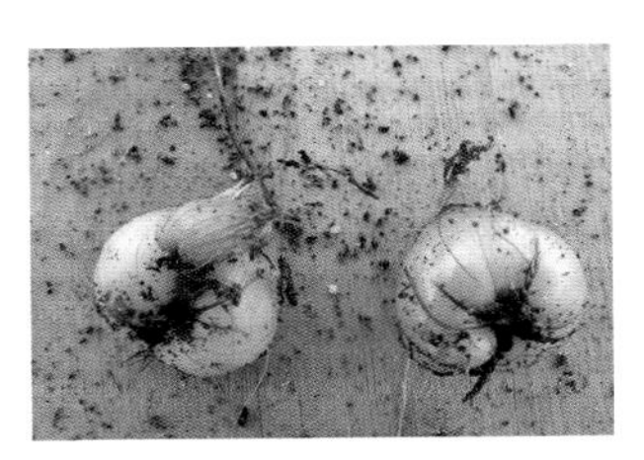

*The corms of* G. virescens *are small with papery tunics.*

*Corms of* G. equitans *have fibrous, leathery tunics.*

## ABOUT THIS FIELD GUIDE

This book aims to assist readers with identifying gladioli in the field, an activity requiring good research skills, strong legs, eagle eyes and a love of the search. It offers a photographic record of all scientifically described gladioli species in South Africa. Of course, plants are no respecters of political boundaries and the notion that there exists a South African account of the genus is nonsensical; plants grow where their dispersal and growing conditions permit, not where state boundaries insist. Some species are highly localised or endemic: *G. uitenhagensis*, scientifically observed only three times in two tiny populations, is an extreme example, but there are many Rare or Vulnerable plants that have restricted ranges. Other species are more widespread: *G. dalenii*, for example, grows across the African continent. Nevertheless, and for reasons to do with the state – such as national borders, permits, collection regimes and permissions – this book focuses on gladioli in South Africa.

It may come as a surprise to the amateur botanist that gladioli are divided into different sections based on a range of distinguishing features, which include leaves, corms and seeds; and certainly not solely – or even necessarily – based on their flower shape or colour. Scientific plant descriptions are made from a type specimen and do not account for variation in a species; this is especially true of flower colour. For example, *G. patersoniae* is scientifically described as having blue, grey or cream flowers, but, as seen on page 341, in some populations they may be white, mauve or pink. Populations of a given species may lack typical markings or exhibit other variations. Another example is *G. carinatus* on page 284, which has three colour forms: blue-mauve, yellow and pink, each of which may appear darker, paler, brighter or more intense, depending on the circumstances. The Saunders' archive records 24 distinct shades for *G. carinatus*! (In the endpaper thumbnails, we show a typical blue form of *G. carinatus*.)

## *GLADIOLUS* TAXONOMY

Currently, 166 species of *Gladiolus* are recorded in South Africa, three with recognised subspecies. Making sense of so large a genus with heterogeneous characteristics and a huge distribution is not straightforward, evidenced by the considerable revision of scientific descriptions of species and the genus over time, most recently in Peter Goldblatt and John Manning's *Gladiolus in Southern Africa*, published in 1998.

In their monograph, Goldblatt and Manning divided the genus *Gladiolus* into seven sections that share similar morphological characteristics. The sections, which give taxonomic order to the genus and serve as the structure for this guide, are as follows:

- Two sections, ***Densiflori*** (page 32) and ***Ophiolyza*** (page 74), occur solely in summer-rainfall areas.
- ***Blandi*** (page 106) is concentrated mostly in the Western Cape (the winter-rainfall area), with a few species occurring along the coastal region in summer-rainfall areas.
- ***Linearifolii*** (page 152) has a disjunct distribution across the winter- and summer-rainfall areas.
- Section ***Heterocolon*** (page 188) falls almost entirely in the summer-rainfall area, barring three species (*G. kamiesbergensis*, *G. marlothii* and *G. mostertiae*), which occur in a highly restricted area in the bulb-rich region of the southern parts of the Northern Cape.
- ***Hebea*** (page 206) is widespread throughout South Africa and spreads northwards into Namibia, Zimbabwe and Botswana.
- ***Homoglossum*** (page 274) is also widespread, occurring throughout South Africa (but apparently not beyond its borders) except in the drylands of the Karoo.

The descriptions that follow summarise the key characteristics of each section in the genus. A map is provided for each section, showing the distribution range of, and species richness within, that section. Of course, section distributions overlap considerably and readers should be alert to this, particularly in areas of high species diversification such as the Western Cape.

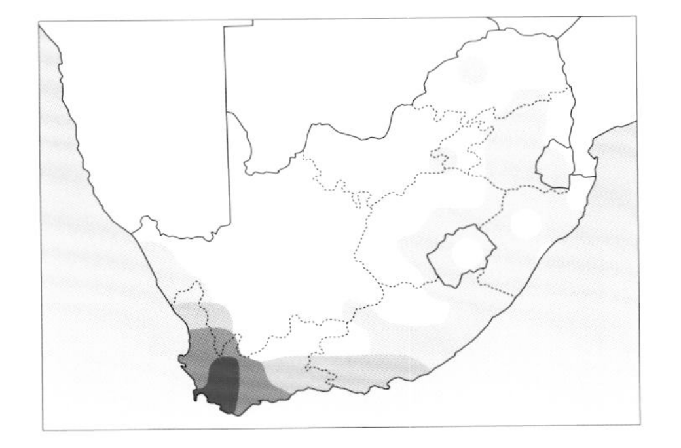

***Right:*** *Map showing distribution range of section* Homoglossum; *the intensity of shading on the section maps indicates species richness, with darker tones signifying more species in an area.*

*Unusual anthers extend into long spurs;* G. appendiculatus.

*Thickened leaf margins and midribs;* G. crassifolius.

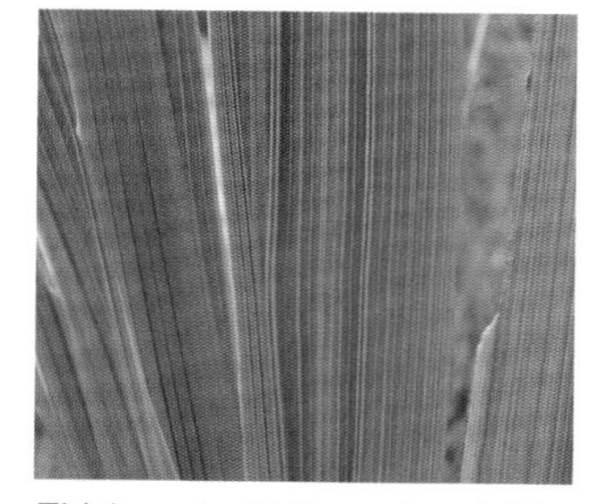

*Thickened midribs and scabrid leaves;* G. scabridus.

## DENSIFLORI

Section *Densiflori* is widespread in summer-rainfall areas of southern Africa and does not occur in winter-rainfall areas. The least specialised of all the gladioli, plants in this section are generally of medium size to large, with flat, sword-shaped leaves that form a fan. Their corms are globose, with papery tunics that decay into irregular fragments. Cataphylls are hairless. Stems are firm, with branched or unbranched spikes carrying numerous small flowers without claws. The lower tepals are generally the same length as, but narrower than, the upper tepals, with the dorsal arched over the stamens. The perianth tubes are long – ranging from two-thirds to one-and-a-half times the length of the dorsal tepals. The flowers of some species have nectar guides, but all are unscented. The flowers are mostly adapted for pollination by long-tongued bees, although a few are pollinated by long-tongued flies.

The section *Densiflori* contains 20 species. Two, *G. paludosus* and *G. papilio*, occur in marshy areas, where their well-developed stolons enable vegetative reproduction, often producing dense colonies.

*G. crassifolius*, *G. hollandii*, *G. serpenticola*, *G. exiguus*, *G. densiflorus*, *G. ferrugineus*, *G. saxatilis* and *G. varius* have thickened leaf margins and midribs, and small flowers with short bracts and perianth tubes (except in *G. saxatilis and G. varius*).

Three species, *G. appendiculatus*, *G. calcaratus* and *G. macneilii,* are restricted to the Drakensberg escarpment. Their unusual anthers, which are extended into long spurs, make identification comparatively easy.

The seven remaining species, *G. ochroleucus*, *G. mortonius*, *G. microcarpus*, *G. scabridus*, *G. cataractarum*, *G. pavonia* and *G. brachyphyllus,* have large pink flowers and long perianth tubes (except *G. ochroleucus*). Most have scabrid leaves.

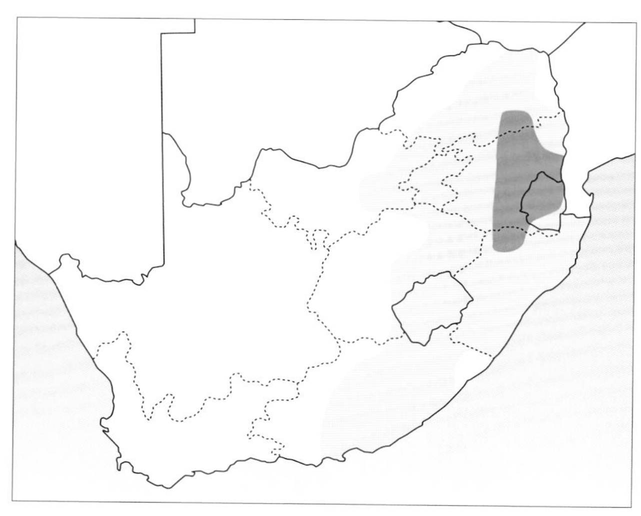

## OPHIOLYZA

Section *Ophiolyza* occurs throughout South Africa's summer-rainfall areas. Plants are usually of medium size to large with lanceolate leaves forming a distichous basal fan. Corms are leathery, decaying into fibres or fragments. Medium-sized to large flowers with long bracts are borne on erect spikes. The flowers have large dorsal tepals, with much smaller lower tepals marked with blotchy nectar guides. The unscented flowers are often pale and speckled. These plants are adapted for pollination by long-tongued bees, but there is evidence of pollination by sunbirds and butterflies in some species too.

The 15 species in section *Ophiolyza* are spread throughout the summer-rainfall areas. Two species, *G. dalenii* and *G. oppositiflorus*, have been widely used in breeding the showy spikes familiar in cut-flower gladioli.

*G. reginae*, *G. dolomiticus*, *G. pole-evansii*, *G. oppositiflorus*, *G. sericeovillosus* and two subspecies, and *G. elliotii* have distinctive flowers that form in distichous (opposed) ranks on the spike. They have fairly long bracts, with the outer bract being inflated and the inner having two tips.

*G. ecklonii*, *G. vinosomaculatus* and *G. rehmannii* are characterised by long bracts that exceed the flowers. Additionally, *G. ecklonii* and *G. vinosomaculatus* have striking dark speckles.

*G. antholyzoides*, *G. aurantiacus*, *G. dalenii*, *G. flanaganii*, *G. saundersii* and *G. cruentus* have large flowers with long perianth tubes. The vegetative and reproductive phases of their growth cycle occur at different times of the year in the first two species.

*Long perianth tubes and bracts;* G. antholyzoides.

*Two-ranked (distichous) flowers;* G. elliotii.

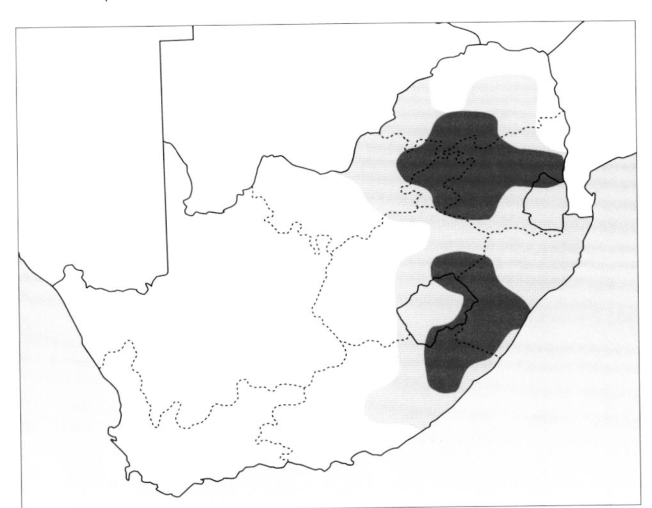

*Long bracts;* G. rehmannii.

## BLANDI

The species in section *Blandi* are concentrated in winter-rainfall areas and adapted to the sandstone soils generally found there. Medium-sized to large plants have several sword-shaped, markedly veined leaves in a basal fan. Corms are papery to leathery, becoming fibrous or fragmented with age. The spikes are usually branched; erect spikes bear a few medium-sized to large cream, pink or red flowers, which have pale nectar guides with dark outlines, and large bracts. The perianth tubes are narrow and the tepals are subequal. Flowers are adapted for pollination by bees and long-tongued flies, with some species pollinated by butterflies or sunbirds. Some species have developed adaptations for late flowering.

Three of the 22 species in section *Blandi* are adapted to fire, flowering only in the first few years immediately after a burn and not thereafter. They are *G. oreocharis*, *G. crispulatus* and *G. phoenix*.

One unusual member of the section is *G. gueinzii*, a coastal sands dweller that, unlike others in *Blandi*, has actinomorphic (star-shaped) flowers, a form thought to be an adaptation to enable self-pollination.

Several other species are adapted for pollination by butterflies: *G. carneus*, *G. pappei*, *G. geardii*, *G. aquamontanus*, *G. undulatus*, *G. angustus*, *G. buckerveldii*, *G. bilineatus*, *G. dolichosiphon*, *G. insolens*, *G. cardinalis*, *G. sempervirens*, *G. stephaniae* and *G. carmineus*.

Four species have short stems and unusual fibrous corms, which form a neck around the base of the stem. They have different pollinators, with *G. rudis* being pollinated by bees, *G. grandiflorus* by long-tongued bees, *G. floribundus* by long-tongued flies and *G. miniatus* by sunbirds.

*Fibrous corms form a thick neck around the stem;* G. miniatus.

*Flowering after fire;* G. phoenix.

*Sword-shaped leaves;* G. cardinalis.

## LINEARIFOLII

Species in section *Linearifolii* occur in both summer- and winter-rainfall areas. Plants are usually small to medium in size, with sword-shaped or linear leaves arranged in basal fans or sheathing the stem. The leaves are usually hairy. Corms are leathery, becoming fibrous with age. Unbranched stems end in spikes crowded with small flowers. The flowers of some species are scented. Most *Linearifolii* species flower aseasonally, thus maximising pollination at a time when there are fewer flowers and there is less competition for visits from pollinators.

There are 17 species of *Linearifolii* in South Africa. Five, *G. woodii*, *G. malvinus*, *G. pardalinus*, *G. pubiger* and *G. parvulus*, occur in the summer-rainfall regions, flower in spring and are adapted for pollination by bees.

Twelve other species grow on sandstone-derived soils. Barring two species, *G. caryophyllaceus* and *G. hirsutus*, they flower in summer or autumn, unlike most gladioli in the winter-rainfall regions, which flower from late winter to early summer. The plants are variously adapted to pollination by bees, moths, long-tongued flies, butterflies and sunbirds: *G. hirsutus*, *G. caryophyllaceus*, *G. rhodanthus*, *G. guthriei*, *G. emiliae*, *G. overbergensis*, *G. merianellus* (formerly *G. bonaspei*); *G. aureus*, *G. brevifolius*, *G. monticola*, *G. nerineoides* and *G. stokoei*.

*Flowers appear in spring;* G. merianellus.

*Unbranched stems of* G. stokoei *end in spikes of large flowers; Riviersonderend Mountains, Western Cape.*

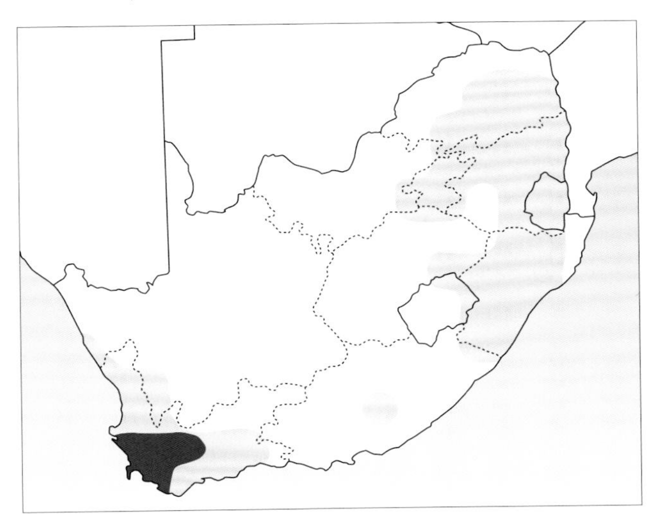

*Hairy leaves and stems;* G. woodii.

## HETEROCOLON

The adaptations of species in section *Heterocolon* to dry habitats in both summer- and winter-rainfall areas are indicated by thickly tunicked corms and fibrous leaves. Plants are medium sized, with small flowers on straight spikes and a few narrow leaves. Long lower tepals have darkly outlined nectar guides and are narrowed to claws. Flowers are usually unscented, and most are adapted for pollination by long-tongued bees. Some species flower in the dry season when the leaves are withered.

There are eight species in *Heterocolon* in South Africa. *G. oatesii* is found in bushveld in North West, Limpopo and Gauteng. It has unusually short and smooth leaves, and flowers before the rainy season.

*G. pretoriensis* and *G. filiformis* occur north of the Vaal River and can be identified by their distinctive terete leaves, flowers with bands of colour and indistinct nectar guide markings, and unusual seeds.

*G. rufomarginatus* and *G. vernus* occur in the summer-rainfall areas, from northern KwaZulu-Natal to Limpopo, while *G. kamiesbergensis*, *G. marlothii* and *G. mostertiae* occur in the Namaqualand–Bokkeveld escarpment in the Western Cape. They have spathulate, twisted lower tepals and all but those of *G. vernus* are minutely speckled.

*Speckled and with twisted lower tepals;* G. rufomarginatus.

*Spathulate lower tepals;* G. mostertiae.

*Tepals narrow to claws;* G. marlothii.

***Left:*** *Short, smooth leaves;* G. oatesii. ***Right:*** *Terete leaves (oval in cross section);* G. pretoriensis.

*Unusual wingless seeds;* G. filiformis.

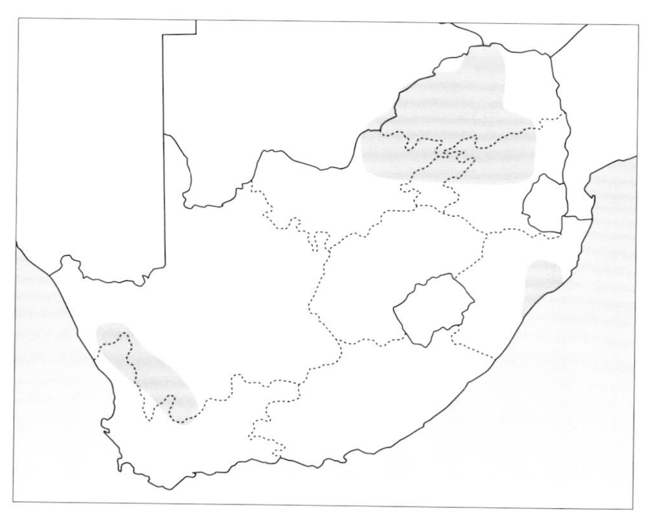

## HEBEA

This section of 33 species is widespread across southern Africa, with great species richness in the drier parts of the winter-rainfall area. Plants in this section are usually small, with linear, narrow leaves and margins that are not thickened. Corms vary, with some being fibrous and others papery or with woody claws. Several have slender stolons that produce cormlets. Branched stems end in inclined spikes. Plants have short, soft bracts and small to medium-sized, diversely coloured and usually sweetly scented flowers with unusual nectar guides and clawed lower tepals. Pollination is quite diverse, with species pollinated by bees, sunbirds or moths. Ellipsoid capsules and dark seed bodies are characteristic.

Seven species in the southern part of the Western Cape lack the sweet scent and clawed upper tepals of other species in *Hebea*. They are adapted for pollination by bees or sunbirds: *G. leptosiphon*, *G. loteniensis*, *G. involutus*, *G. vandermerwei*, *G. cunonius*, *G. splendens* and *G. saccatus*.

*G. permeabilis* and two subspecies, *G. sekukuniensis*, *G. uitenhagensis*, *G. acuminatus*, *G. karooicus*, *G. wilsonii*, *G. inandensis* and *G. robertsoniae* have arched dorsal tepals forming 'windows' between the dorsal and lateral tepals when viewed in profile. *G. stellatus* is star-shaped. Two are pollinated by moths.

Seven species, *G. arcuatus*, *G. viridiflorus*, *G. deserticola*, *G. scullyi*, *G. venustus*, *G. salteri* and *G. lapeirousioides*, occur in the Northern Cape. They have hard, coriaceous corm tunics and other adaptations to aridity.

*G. orchidiflorus*, *G. watermeyeri*, *G. virescens*, *G. ceresianus*, *G. uysiae*, *G. equitans*, *G. speciosus*, *G. pulcherrimus*, *G. alatus* and *G. meliusculus* are highly specialised. Small plants with bright, strongly scented flowers, they occur in the west of the winter-rainfall area in the Western Cape. Several are easily confused, especially in areas north of Cape Town where they overlap. *G. meliusculus* is pollinated by monkey beetles.

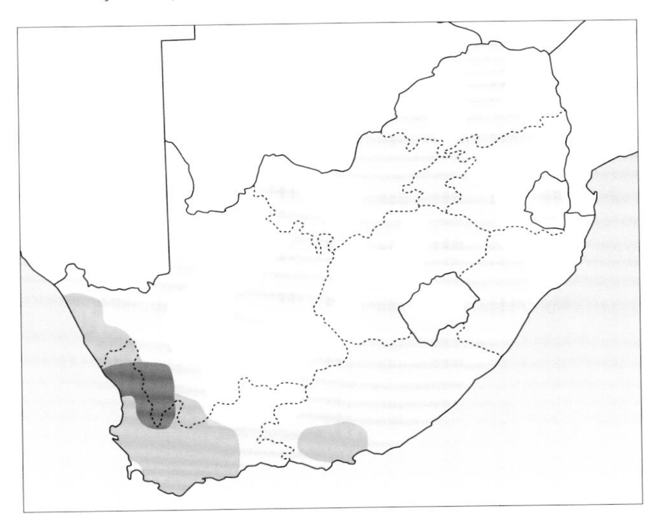

*Fibrous corms;* G. venustus.

*Visible 'window';* G. permeabilis *subsp.* permeabilis.

*Adapted for pollination by sunbirds;* G. splendens.

*Monkey beetles are known to pollinate* G. meliusculus.

## HOMOGLOSSUM

Section *Homoglossum* is concentrated in the winter-rainfall region, with a high degree of species richness in the southern part of the Western Cape. Species restricted to nutrient- and water-poor sandstone soils have fewer, narrower leaves and a wider range of pollinators than those in other sections. Plants are usually small to medium in size, with two to four narrow, superposed leaves. Some species have unusual leaves that appear cross shaped in cross section, while others have thick leathery leaves.

The corms are often woody, but in some species they are leathery or fibrous. Unbranched spikes carry medium-sized to large flowers.

The flowers in this section display a great deal of colour diversity, and all have nectar guides. Many species have delicate, somewhat inflated flowers and a number have boldly delineated tepal patternings of dots, chevrons, spears or lozenges. Some have elongate lower tepals, giving them an unusual flower shape and profile.

*Delicate, inflated flowers;* G. brevitubus.

Most are adapted for pollination by long-tongued bees, with long-tongued flies, moths and sunbirds pollinating others. Some species attract moth pollinators by opening in the evenings, changing colour or becoming more scented; the flowers of these species are also usually large. One species, *G. brevifolius*, does not resemble a *Gladiolus* so much as a *Geissorhiza* or a *Hesperantha*, making identification very confusing.

This is the largest section of the genus, with 51 species spread across South Africa: *G. atropictus*, *G. violaceolineatus*, *G. comptonii*, *G. roseovenosus*,

*Flowers of* G. liliaceus *change colour and become strongly scented at night.*

*Inflated flowers and elongate lower tepals;* G. patersoniae.

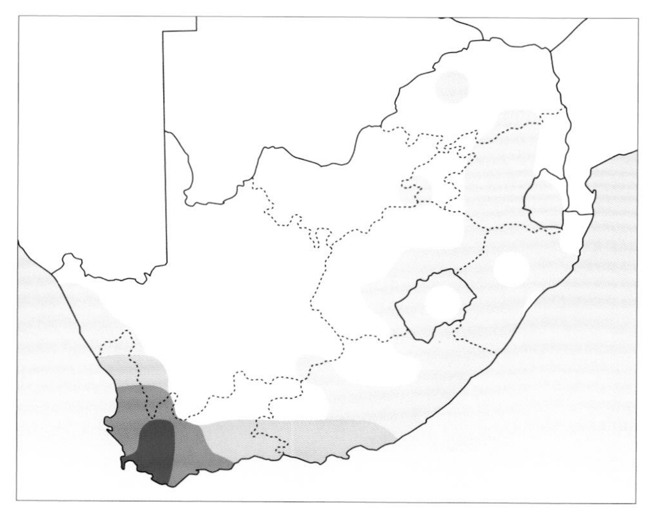

Homoglossum *species have narrow and sometimes unusual leaves* ***(left to right):*** G. priorii; G. mutabilis; G. gracilis; G. delpierrei; G. sufflavus.

*G. carinatus*, *G. griseus*, *G. quadrangulus*, *G. mutabilis*, *G. exilis*, *G. vaginatus*, *G. maculatus*, *G. albens*, *G. meridionalis*, *G. priorii*, *G. brevitubus*, *G. rogersii*, *G. bullatus*, *G. blommesteinii*, *G. virgatus*, *G. debilis*, *G variegatus*, *G. vigilans*, *G. ornatus*, *G. inflexus*, *G. taubertianus*, *G. gracilis*, *G. caeruleus*, *G. recurvus*, *G. inflatus*, *G. cylindraceus*, *G. nigromontanus*, *G. engysiphon*, *G. patersoniae*, *G. subcaeruleus*, *G. martleyi*, *G. jonquilliodorus*, *G. trichonemifolius*, *G. sufflavus*, *G. pritzelii*, *G. delpierrei*, *G. hyalinus*, *G. liliaceus*, *G. tristis*, *G. longicollis* and two subspecies, *G. symonsii*, *G. watsonius*, *G. teretifolius*, *G. quadrangularis*, *G. huttonii*, *G. fourcadei* and *G. abbreviatus*.

*Many species in* Homoglossum *have bright nectar guides; here,* G. debilis.

*Markedly reduced lower tepals;* G. abbreviatus.

*Mottled cataphylls;* G. griseus.

# HOW TO USE THIS BOOK

## Find a flower in the gallery

Finding gladioli in the field is largely dependent on their being in flower, and because most readers are likely to rely on the flowers to initiate species identification, the book offers a gallery of close-up photographs of all 166 species' flowers on the front and back endpapers. These photographs have been specially selected to portray a typical example of each species, and are arranged sequentially to correspond with the presentation of species in the book.

## Observe a flower in the field

To start the process, observe closely the shape characteristics of the flower in question, including the dorsal (uppermost) tepals, the habits of lateral and lower tepals, and the presence of nectar guides and additional colours or patterning.

Note, however, that the flower colour of members of a single *Gladiolus* species may vary considerably, depending on a range of factors, including genetics, geology and weather conditions. For this reason, it is important to examine *all* the various plant features as clues to make an identification.

Scientific name of species

Meaning of scientific name

Current IUCN Red List status

Ecology notes draw from published sources and the Saunders' own records

Similar species text helps to distinguish confusing species

Comparative table draws attention to finer details of similar gladioli, helping readers to sidestep the more technical descriptions

Section name

Flowering months

Distribution map

*Gladiolus taubertianus*

Named after Paul Taubert (1867–1892), a German botanist.

*Gladiolus taubertianus* has sweetly scented, pale mauve to blue flowers that appear in August and September. It is endemic to the mountainous regions of the Western Cape. Flowers are similar to *G. inflexus* but differ in markings.

J F M A M J J A S O N D

STATUS Rare

**DESCRIPTION** Plant 20–50cm. Corm globose with hard layered tunics. Cataphylls uppermost purple and membranous. Leaves 3, narrow (up to 3mm wide), the lowermost as long as the spike or exceeding it; midrib raised and prominent. Spike flexed outward above sheathing leaves, unbranched, 1–3 flowers. Bracts grey-green, inner bract twisted so that it lies under the outer bract. Flowers pale pinky mauve to blue with streaks of purple; lower tepals with pale yellow to cream transverse band outlined with darker mauve streaks. Perianth tube up to 14mm. Anthers pale lilac. Pollen cream. Capsules unknown. Seeds unknown. Scent lightly scented.

**DISTRIBUTION** Known only from the eastern and southern Cederberg in the southwestern Cape.

**ECOLOGY & NOTES** Grows among dry mountain fynbos in sandstone-derived soils, in areas that receive ±400–500mm of rain per annum. Often found among rocks, which presumably give them protection from porcupines and mole-rats. Although it is possible that they flower better after fires, we have seen them flowering well in mature fynbos on several occasions.

**POLLINATORS** Long-tongued bees.

**SIMILAR SPECIES** Closely related to *G. gracilis* and *G. inflatus* but in flower is unlikely to be confused for either.

*There are streaks on all the tepals, with darker streaks on the lower ones.*

| | Leaves | Flowers | Markings |
|---|---|---|---|
| ***G. taubertianus*** | blade linear, 2–3mm wide, midrib prominent | not bell-shaped, pinky mauve | yellow transverse band with purple markings |
| ***G. inflexus*** | blade linear, 1–2mm wide, midrib raised | not bell-shaped, shades of purple | purple streaks on a cream background |
| ***G. inflatus*** | blade oval to terete with 4 narrow longitudinal grooves | slightly bell-shaped, pink to purple | spade- or spear-shaped markings on lower tepals |

324 • HOMOGLOSSUM

## Turn to the relevant description

Once you have pinpointed a likely species match on the endpaper galleries, note the page reference and turn to the relevant species description. The species are arranged according to taxonomic order within their sections.

## Study the photographs

All species descriptions are accompanied by photographs selected to show the greatest detail or the most typical aspects of the species, such as markings, perianth tubes, bracts, etc., and, in some instances, leaves and seeds.

## Read the species accounts

Armed with detailed species information and multiple photographs, users can check their found plant against the various species features – flowers, spikes, number of flowers, scent, leaves, corms, seeds, size of plant, etc. – all important factors that could either confirm identification, or send the user back to try another match.

## Key terminology

A detailed explanation of the parts of a gladiolus plant on page 16 is supplemented by a short glossary of terms on page 378.

Photographs of details such as leaves, spikes, seeds or capsules have been provided wherever possible

Where a species is diverse, multiple close-up images are included to demonstrate the variation recorded; here we see a variety of colour forms

At least one representative habitat image per species is included

*Rachel Saunders photographs* G. saxatilis, *God's Window, Mpumalanga.*

# Gladiolus paludosus

***paludosus* = of marshes, referring to its habitat.**

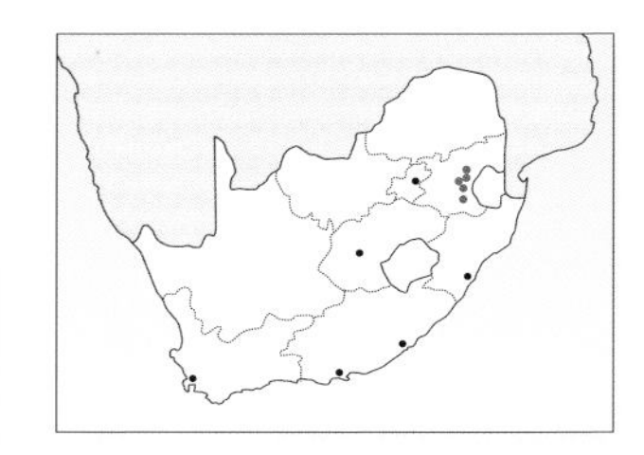

J F M A M J J A S O N D

**STATUS** Vulnerable

*Gladiolus paludosus* is found in marshy areas in Mpumalanga. Its mauve to pink flowers appear from mid-October to mid-November. It is sometimes confused with *G. crassifolius*, which has smaller flowers, appearing late in summer.

**DESCRIPTION** **Plant** 35–55cm. **Corm** globose, papery, with cormlets on short stolons at base. **Cataphylls** pale green and membranous, firm textured. **Leaves** 4–6, 7–10mm wide, soft textured, lanceolate and short, reaching midway up stem; moderately thickened margins and hyaline midrib lightly raised. **Spike** erect, sometimes branched, with 5–9 flowers but occasionally up to 20. **Bracts** up to 20mm, pale to grey-green. **Flowers** mauve to pink to reddish; lower lateral tepals with dark mauve or purple diamond-shaped or semicircular band across the midline. Perianth tube 10–14mm. **Anthers** yellow or light mauve. **Pollen** yellow. **Capsules** almost round. **Seeds** yellowish brown, winged. **Scent** unscented.

*Mauve or purple markings can be seen on the lower tepals.*

**DISTRIBUTION** Mpumalanga. The centre of distribution seems to be Middelburg and eMakhazeni, south to eMkhondo and Wakkerstroom, and north to Dullstroom.

**ECOLOGY & NOTES** Grows in or near marshes and vleis that are wet all year round or dry out for only a few months. Plants flower after the first rains in October and November or later, blooming before the surrounding flora has grown to its full height. This species is not often seen, seemingly becoming rare. Many of the areas where it was once common are now forested or grazed by stock. Human activities are changing waterways. In the summer of 2015, a year of late rains, we found two populations in flower in mid-December. In 2017 a population at Verlorenvlei had plants that were still in flower in late January, in wet grassland but not in marshes.

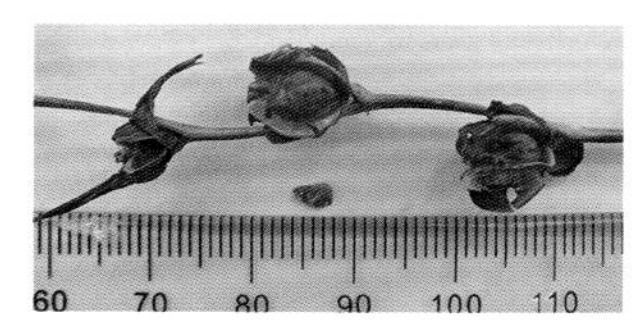


*Rounded capsules release yellowish-brown, unevenly winged seeds.*

**POLLINATORS** Probably adapted for pollination by long-tongued bees.

**SIMILAR SPECIES** If rains are late, *G. paludosus* may flower at the same time as the similar-looking *G. crassifolius*.

| | Flowers | Leaves | Habitat |
|---|---|---|---|
| ***G. paludosus*** | early summer, upper tepal 20–25mm long | margins moderately thickened | marshy |
| ***G. crassifolius*** | late summer, upper tepal 18–22mm long | margins and veins heavily thickened | grassland often at higher altitudes, also serpentine soils |

*Short pale bracts support medium-sized flowers.*

*Spikes are erect, the flowers with short perianth tubes.*

*The flowers are secund and alternating.*

*An unspecified beetle exits the perianth. Note the dark markings on the lower tepals.*

G. paludosus *grows in or near vleis and marshlands, here flowering near Chrissiesmeer, Mpumalanga.*

# Gladiolus papilio

***papilio* = butterfly-like, probably referring to markings on the lower tepals.**

*Gladiolus papilio* is widespread in summer-rainfall areas. Flowers vary in colour from cream to greenish, pink or light purple, with dark blotches on the lower tepals. The large nodding flowers that appear between November and February are unusual, making this species easy to identify.

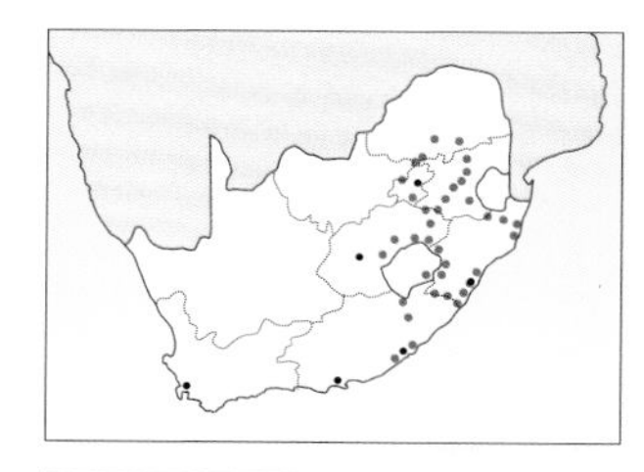

| J | F | M | A | M | J | J | A | S | O | N | D |
|---|---|---|---|---|---|---|---|---|---|---|---|

*The nodding flowers of* G. papilio *are distinctive.*

**DESCRIPTION** **Plant** 50–75cm. **Corm** hardly developed at flowering time; fine stolens produce cormlets. **Cataphylls** leathery. **Leaves** 6 or more, narrow, 9–14mm wide; margins and midrib lightly thickened. **Spike** slightly inclined, unbranched, 4–8 flowers. **Bracts** green or flushed grey-purple, held at an angle from the axis. **Flowers** nodding, variable, usually mauve to pink, but also cream, yellow to greenish; translucent pink or purple on inside of upper tepals; lower tepals with nectar guides of dark green or purple blotches. Perianth tube 18–20mm, curved, emerging from middle of bracts. **Anthers** purplish above, cream below. **Pollen** cream. **Capsules** ovate-oblong, large. **Seeds** rusty brown, broadly winged. **Scent** unscented.

**DISTRIBUTION** From near Butterworth in the Eastern Cape, north through KwaZulu-Natal, Lesotho, eastern Free State, Eswatini, Mpumalanga and Limpopo.

**ECOLOGY & NOTES** Usually in marshes and seeps, sometimes in damp grassland, from sea level to 2,200m. Sometimes forms dense colonies, presumably because of vegetative reproduction. The flowers are very variable in colour. The species is one of the parents used in *Gladiolus* hybridisation.

**POLLINATORS** Probably miner bees such as *Amegilla capensis* of family Anthophoridae, which pollinate other *Gladiolus* species.

**SIMILAR SPECIES** These unique, large, nodding flowers cannot be confused with any other.

*Narrow, sword-shaped leaves form a fan.*

*Bracts are green or flushed with grey-purple.*

*The curved perianths are shorter than the bracts.*

*Recurved perianths direct tepals downwards into nodding flowers.*

*Anthers are purplish with cream pollen.*

*Lower tepal markings may resemble butterflies.*

*The mauve form is the most common of all.*

G. papilio *flowering en masse near Memel, Free State.*

# Gladiolus crassifolius

***crassifolius* = thick-leaved.**

*Gladiolus crassifolius* is widespread across eastern southern Africa, Lesotho and Eswatini. Its pink to light purple flowers bloom from February to March.

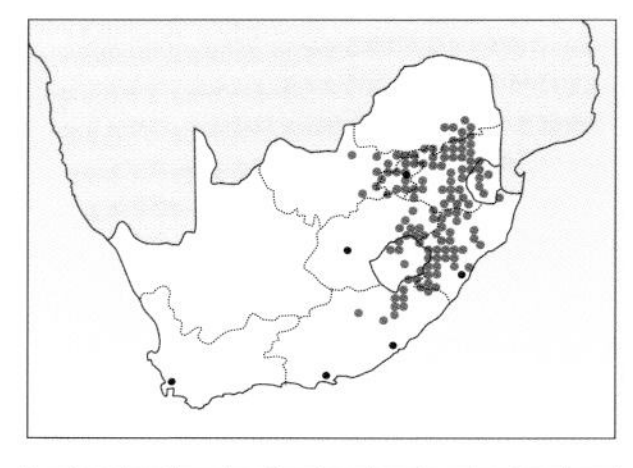

| J | F | M | A | M | J | J | A | S | O | N | D |
|---|---|---|---|---|---|---|---|---|---|---|---|

**DESCRIPTION** **Plant** 35–90cm. **Corm** leathery with coarse tunics. **Cataphylls** leathery and red-brown. **Leaves** 4–8, in basal fan reaching or exceeding spike, 5–12mm wide; the midrib, margins and two or more pairs of secondary veins are hyaline and thickened. **Spike** weakly inclined, may be branched, with 16–22 flowers. **Bracts** ±25mm long, pale green or flushed purple. **Flowers** pink or light purple, occasionally orange-red or cream; lower tepals each with a dark band of colour, the dark band often edged with cream. Perianth tube curving, 9–17mm. **Anthers** pale lilac to purple. **Pollen** white to pale yellow. **Capsules** obovoid with apex slightly sunken. **Seeds** broadly winged, wing often transparent. **Scent** unscented.

**DISTRIBUTION** Widespread in eastern South Africa from Khowa in the Eastern Cape, through Lesotho and Eswatini to Limpopo, as far west as Rustenburg. (Also found in tropical Africa as far north as southern Tanzania.)

**ECOLOGY & NOTES** Found in hilly areas in well-drained rocky habitats, usually in grassland in full sun. Flowers in summer after fires.

**POLLINATORS** *Amegilla* species of long-tongued bees.

G. crassifolius *has long leaves and spikes.*

**SIMILAR SPECIES** May be confused with *G. paludosus*, *G. exiguus* or *G. serpenticola* but is most likely to be mistaken for *G. densiflorus*, particularly in Mpumalanga where they grow in close proximity. There, *G. crassifolius* populations may have orange, red or bright pink flowers, many without typical nectar guides, and leaf venation is not consistent within populations. While *G. crassifolius* is usually found at higher altitudes, both species flower within a kilometre of each other in the area around Barberton.

| | Flowers | Leaves | Habitat |
|---|---|---|---|
| ***G. crassifolius*** | late summer, dark band on lower tepals | margins and veins heavily thickened | grassland, at higher altitudes; also on serpentine soils |
| ***G. paludosus*** | early summer | margins moderately thickened | marshy |
| ***G. serpenticola*** | late summer, cream band on lower tepals | grey-green, margins and midrib moderately thickened | serpentine soils |
| ***G. densiflorus*** | midsummer | margins and midrib lightly thickened, other veins fine | grassland, usually lowlands |

*Narrow leaves form a fan, partly sheathing the stem.*

*This floriferous species bears as many as 22 flowers on a spike.*

*The coral-coloured form is endemic to Mpumalanga.*

*The cream form with faint nectar guides is less common.*

*Flower colours vary; in KwaZulu-Natal they are pink-purple.*

*Small flowers are adapted for pollination by long-tongued bees.*

*Note the dark purple anthers.*

*Yellow outlines on nectar guides.*

*Near Harrismith, Free State.*

# *Gladiolus hollandii*

**Named after FH Holland (1873–1955) who collected the plants that were successfully grown at Kirstenbosch and later used as the type specimen.**

*Gladiolus hollandii* is restricted to the northeastern Lowveld. It has striking, large, spotted, pale pink flowers borne on densely flowered spikes from February to April.

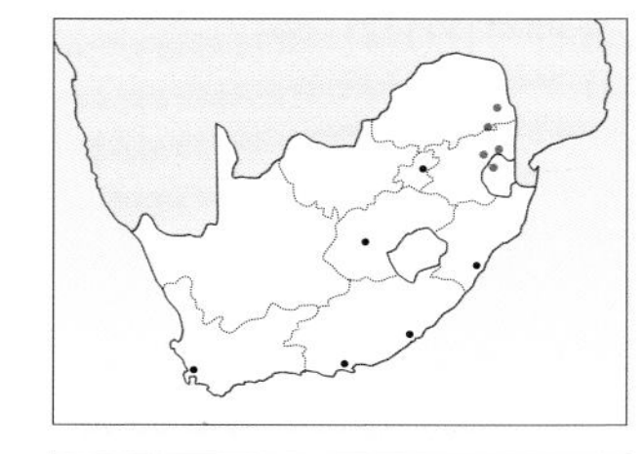

J F M A M J J A S O N D

**DESCRIPTION** **Plant** 65–120cm. **Corm** depressed-globose, with a few cormlets; tunics leathery, fragmenting with age. **Cataphylls** up to 10cm above ground, green or flushed with purple. **Leaves** up to 10, long and narrow, 8–18mm wide; midrib and margins lightly thickened. **Spike** erect, may be branched, with 18–30+ flowers. **Bracts** green with tapered apex, soon turning brown. **Flowers** pale pink with minute, dark pink spots, closely packed; all tepals with dark pink to reddish median streak running entire length. Perianth tube 18–30mm. **Anthers** dark purple. **Pollen** cream. **Capsules** narrow and truncate. **Seeds** poorly and unevenly winged. **Scent** unscented.

*Flowers with typical minute dots and dark median streaks.*

**DISTRIBUTION** Eastern South African Lowveld, from Barberton in Mpumalanga to Limpopo. (Also found in Eswatini and extending into southern Mozambique.)

**ECOLOGY & NOTES** Found on lower slopes of mountains (at altitudes of ±800m) and on granite outcrops, in sun and light to moderate shade. As we photographed the plants, we noticed many ants on the flowers. It was not apparent what they were doing on the plants, but they spent some time investigating the insides of the flowers.

*The dorsal tepal is arched.*

**POLLINATORS** The flowers have longer tubes than those of their closest relatives, suggesting that they probably have a different, more specialised pollinator. It is thought that the pollinators may be tangle-veined flies (Nemestrinidae).

**SIMILAR SPECIES** When in flower, this species is not likely to be confused with any other. The red stripes on the white to pale pink tepals, together with the relatively long perianth tube and the significant height of the plants, make it unmistakable. When not in flower, the leaves might be mistaken for those of *G. serpenticola*.

*Dense spots make the pale pink flowers appear darker.*

*Narrow leaves form a fan, partly sheathing the stem.*

*Spikes may have up to 30 short-bracted flowers.*

*The flowers have long styles and dark purple anthers.*

G. hollandii, *near Mbombela, Mpumalanga. Note how tall these plants are.*

# Gladiolus serpenticola

**serpenticola = referring to the serpentine soil type on which it grows.**

*Gladiolus serpenticola* is endemic to the Barberton area of Mpumalanga. A very tall species, it bears long, floriferous spikes of large, pale pink to white flowers between late February and March. It is sometimes confused with *G. crassifolius* and *G. densiflorus*.

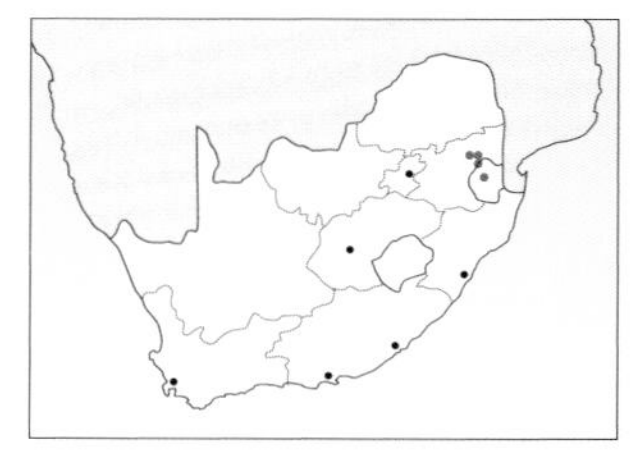

J F M A M J J A S O N D

**STATUS** Rare

**DESCRIPTION** **Plant** 75–150cm. **Corm** depressed-globose, with leathery tunics becoming fibrous with age. **Cataphylls** 10–12cm above ground, green or purple. **Leaves** ±7–10mm wide, sword shaped to linear, grey-green; margins and midrib moderately thickened. **Spike** inclined, usually branched, with 18–30 flowers. **Bracts** 12–16mm, pale green in bud and dry at flowering; apices tapering and twisted. **Flowers** pale pink to white; lower lateral tepals each with a central yellow streak surrounded by a light pink to purple blotch. **Anthers** purple. **Pollen** cream. **Capsules** ovoid, small. **Seeds** ovate with narrow wings, often on one side only. **Scent** unscented.

*Flowers are usually pale pink with purple anthers.*

**DISTRIBUTION** Endemic to the Barberton area of Mpumalanga and extending into Eswatini. Plants are found in the Kaap Valley where there are extensive outcrops of verdite, a dark green form of serpentine.

**ECOLOGY & NOTES** Grows in eponymous serpentine soils. Serpentine-rich rock has a mottled, olive-green colour with a waxy feel. Derived from igneous rocks, it has a very low silica content and is high in iron and magnesium. This rock type (called ultramafic) makes up the Earth's mantle. If metamorphosed and hydrated, it becomes serpentinite, which breaks down to form serpentine soils. They are deficient in nitrogen, phosphorus, potassium and sulphur, have low calcium to magnesium ratios, and have elevated levels of the potentially toxic minerals nickel, cobalt and chromium. Serpentine soils are challenging habitats for plants, because of the high magnesium and low calcium levels, and the high levels of heavy metals. This has resulted in speciation and endemism. Plants are often found in open, steep landscapes that have reduced moisture retention.

**POLLINATORS** Probably long-tongued bees.

**SIMILAR SPECIES** May be confused with *G. densiflorus* but is most likely to be mistaken for *G. crassifolius*.

| | Leaves | Flowers | Soil |
|---|---|---|---|
| ***G. serpenticola*** | narrow, glaucous, moderately thickened midrib | yellow markings on lower tepals | serpentine soils only |
| ***G. crassifolius*** | coarsely ribbed, thickened midrib, veins and margins | dark markings on lower tepals | most soil types |

*Green bracts are dry and twisted at flowering.*

*A stem usually has one or two branches but may have more.*

*Flowers are almost two ranked.*

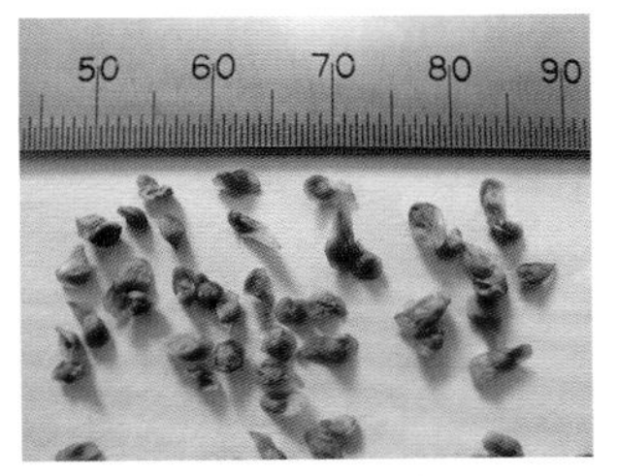


*Seeds are often winged on one side only.*

*Pale-flowered form.*

G. serpenticola *can reach a height of 150cm.*

# Gladiolus exiguus

***exiguus*** **= small, slender, referring to the plants and flowers.**

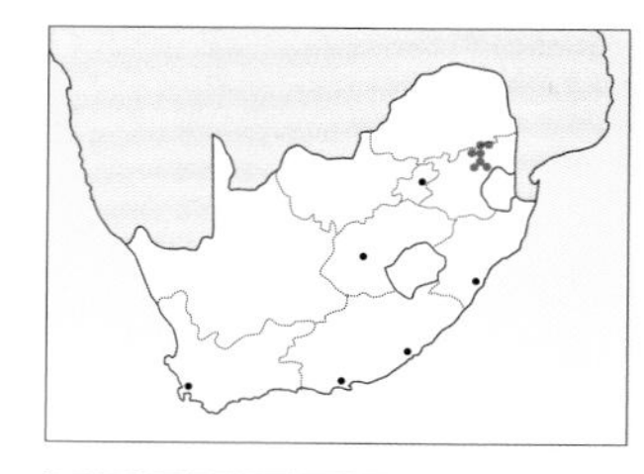

J F M A M J J A S O N D

*Gladiolus exiguus* is endemic to the Drakensberg escarpment of Mpumalanga. The flowers range in colour from pale to deep pink, appearing between late January and March.

**DESCRIPTION** **Plant** small, 30–40cm. **Corm** depressed-globose, with tunics accumulating to form a thick neck of fibres around stem base. **Cataphylls** membranous. **Leaves** 7 or 8, fanned, 15mm wide; margins moderately thickened, midrib less so, and blades slightly twisted. **Spike** inclined, unbranched, with 12–20 flowers. **Bracts** dull green to grey-purple, folded along the midline. **Flowers** pale creamy pink to deep pink, occasionally white; no nectar guides. Perianth tubes short. **Anthers** brownish purple. **Pollen** cream. **Capsules** broadly ovoid with rounded apex. **Seeds** small, ovate to triangular, unevenly winged, sometimes winged only at one end. **Scent** unscented.

**DISTRIBUTION** At high elevations on the northern Drakensberg escarpment of Mpumalanga, usually between 1,600 and 2,300m above sea level, extending from Mount Sheba southwards to Emgwenya.

**ECOLOGY & NOTES** Plants grow in open grassland in well-drained sites on rocky slopes, usually in quartzite and sandstone. Often forming fairly dense colonies, they are conspicuous on the drier western slopes of Long Tom Pass.

**POLLINATORS** Long-tongued bees pollinate the short-tubed flowers.

**SIMILAR SPECIES** Can be difficult to identify because of its lack of distinguishing features. Most likely to be confused with *G. densiflorus* and perhaps *G. crassifolius*.

*Creamy pink flowers have dark anthers and no nectar guides.*

| | Plant size | Leaves | Flowers |
|---|---|---|---|
| ***G. exiguus*** | 40cm | thickened margins and midrib, other veins fine | inclined spikes, no markings on flowers, larger than *G. densiflorus* |
| ***G. densiflorus*** | up to 120cm | similar to *G. exiguus* | weakly inclined, often speckled, bands of colour on lower tepals |
| ***G. crassifolius*** | 90cm | thickened margins, midribs and secondary veins | weakly inclined spikes, flowers with bands of colour on lower tepals |

Robust leaves sheath the stem in a tight fan.

Bracts are dark with a clear midline fold.

Flowers are occasionally white.

Darker pink form.

Note arching upper tepal.

Plants flowering en masse between Mbombela and Barberton, Mpumalanga.

# *Gladiolus densiflorus*

***densiflorus*** **= densely flowered.**

*Gladiolus densiflorus* is widespread throughout the Lowveld and coastal regions of eastern South Africa. It has dense spikes with up to 45 flowers, in various colours. It generally flowers between December and March, except on the northern coastal plains of KwaZulu-Natal, where it flowers at any time of the year.

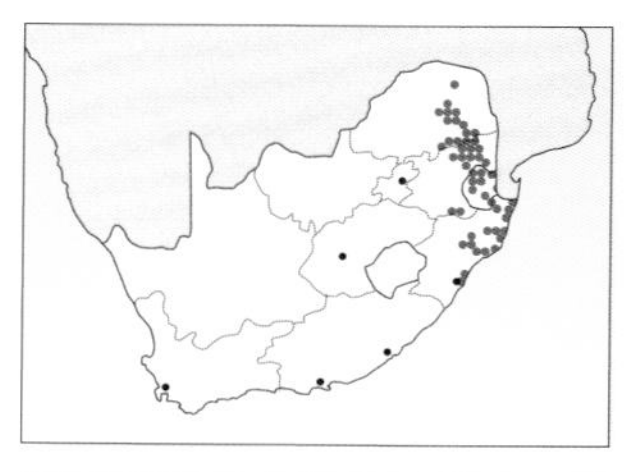

| J | F | M | A | M | J | J | A | S | O | N | D |
|---|---|---|---|---|---|---|---|---|---|---|---|

**DESCRIPTION** **Plant** 65–120cm, stem usually unbranched. **Corm** depressed-globose, tunics fragmenting with age. **Cataphylls** leathery and pale, green or reddish above ground. **Leaves** 8 or 9, in a fan, 5–24mm wide; margins and midrib thickened and raised, remaining veins poorly developed and closely set. **Spike** inclined, with 25 or more smallish flowers. **Flowers** variable; cream, greenish, pink, mauve, slate-grey, occasionally orange and reddish, usually minutely speckled with pink to purple spots; either without nectar guides or the lower 3 tepals with a poorly defined pale yellow zone along the midline. Perianth tubes short. **Anthers** dark purple or black. **Pollen** white to cream. **Capsules** obovoid, small. **Seeds** ovate, poorly and unevenly winged. **Scent** unscented.

G. densiflorus, *cream form.*

**DISTRIBUTION** A widespread species of the Lowveld and coast of eastern South Africa. It extends from Limpopo through the Mpumalanga Lowveld and Eswatini to coastal southern Mozambique and KwaZulu-Natal, as far south as Durban.

**ECOLOGY & NOTES** Usually found below 300m in altitude, but also occurs near Graskop and Sabie, and near Barberton at elevations above 1,200m. Found in grassland in full sun or partial shade and usually in areas of fairly high rainfall. This species is extremely variable in size and flower colour. On the Mpumalanga escarpment, plants are short and slender, whereas in other areas plants are much more robust. The flowers are most commonly light pinkish mauve with purple speckles, but they can vary from reddish to orange to cream. The presence of speckles is also variable, with speckled and unspeckled plants present in the same population.

**POLLINATORS** Long-tongued bees.

**SIMILAR SPECIES** Most often confused with *G. crassifolius*, which generally occurs on the Highveld; *G. densiflorus* grows mainly in the lowlands, but there are areas of overlap on the Mpumalanga escarpment. Key differences are leaf venation and flower markings, including speckles. However, these characteristics are not stable within populations and the species can be extremely difficult to tell apart.

| | Plant size | Leaves | Flowers |
|---|---|---|---|
| ***G. densiflorus*** | up to 120cm | thickened margins and midrib, other veins fine | weakly inclined, often speckled, bands of colour on lower tepals |
| ***G. crassifolius*** | up to 90cm | thickened margins, midribs and secondary veins | weakly inclined spikes, flowers with bands of colour on lower tepals |

Leaves, varying in width, make a dense fan.

Bracts are smooth.

Up to 45 small flowers are densely stacked on spikes.

Reddish-orange form.

Mauve form.

Anthers are dark with white-cream pollen.

Deep pink dots make the cream flowers appear darker.

Note the speckles and yellow marking in the midline.

Plants may reach 120cm in height.

# *Gladiolus ferrugineus*

***ferrugineus*** **= rust-brown, referring to the colour of the floral bracts.**

*Gladiolus ferrugineus*, which occurs along the eastern escarpment of southern Africa, has pale flowers that flush pink or blue on fading. It usually flowers between January and March but sometimes flowering begins earlier and/or ends later.

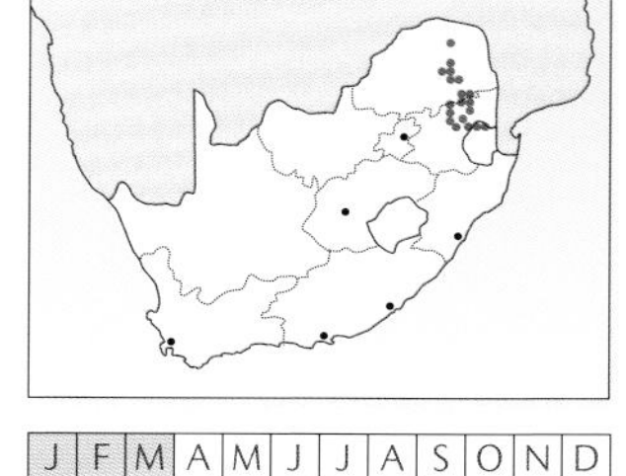


**DESCRIPTION** **Plant** 35–60cm. **Corm** globose, tunics decaying to vertical fibres. **Cataphylls** rust coloured. **Leaves** 6–8, reaching or exceeding spike, lanceolate to linear and twisted, ±8mm wide; margins and midrib thickened and raised, remaining veins not raised. **Spike** inclined, unbranched, with 8–15 flowers. **Bracts** greenish at first, becoming rust coloured and dry at time of flowering. **Flowers** white, cream, pearl-grey to pale pink, fading to pink or blue; with cream to pale yellow blotches on the lower tepals; small with short perianth tubes, usually up to 20mm long. **Anthers** lilac to purple. **Pollen** yellow. **Capsules** obovoid with a flat to sunken apex. **Seeds** small and brown with pale, uneven wings. **Scent** unscented.

**DISTRIBUTION** Along the escarpment of eastern South Africa, from Piggs Peak in northern Eswatini and Barberton in Mpumalanga, to the Wolkberg and Haenertsburg in northern Limpopo.

**ECOLOGY & NOTES** Plants grow in marshy grassland or on well-watered stony slopes, but we know of at least two populations of plants in rocky, well-drained habitats. There appear to be three distinct forms. We have photographed a white form north of Graskop in short, marshy grassland overlying rock sheets at an altitude of about 1,400m. This form, with short, slender plants, flowers in late November and December. In the Steenkampsberg and in the mountains above Barberton, there is a more robust dark pink form that grows in rock crevices, flowering in February and March, and also a white form. The Steenkampsberg site is extremely well drained and very cold in winter, at an altitude of about 2,200m. The third form comes from Long Tom Pass, where we have found plants with pale pink and whitish flowers in late April. These plants were growing on a grassy, well-drained slope at an altitude of about 1,750m. Hybrids of this species and *G. varius* have been recorded on Long Tom Pass. These are intermediate in size, between the two parents, and the pink flowers have dark streaks on the lower tepals.

**POLLINATORS** The short-tubed pale flowers are adapted for pollination by large long-tongued bees, which visit the flowers in search of nectar.

**SIMILAR SPECIES** Similar to *G. saxatilis* and *G. varius*. Most likely to be confused with *G. varius* as both species occur at high elevations on the eastern escarpment in Mpumalanga, where we have seen them in close proximity to one another.

| | Flower size | Flower colour |
|---|---|---|
| ***G. ferrugineus*** | small, with short tubes ≤20mm | white to pale pink |
| ***G. varius*** | larger flowers with tubes 30–50mm long | pink with reddish nectar guides on lower tepals |

*Pale pink flowers fade to blue.*

*Bracts become rapidly dry as flowers open.*

*The rust colour of the floral bracts is distinctive.*

*White form.*

G. ferrugineus *has lance-shaped, twisted leaves, seen here at Steenkampsberg, Mpumalanga.*

# *Gladiolus saxatilis*

(= *Gladiolus lithicola*)

***saxatilis*** **= growing among rocks.**

*Gladiolus saxatilis* is known only from the lower Drakensberg escarpment in Mpumalanga. Its pale flowers, which appear in mid-March, darken to mauve or pink with age.

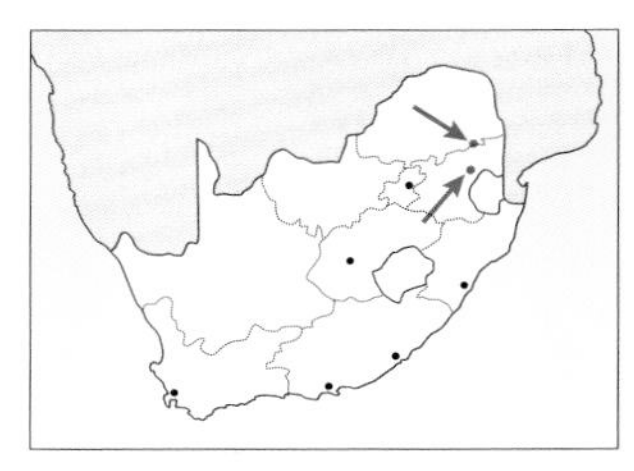

| J | F | M | A | M | J | J | A | S | O | N | D |
|---|---|---|---|---|---|---|---|---|---|---|---|

**STATUS** Rare

*G. saxatilis may have pinkish flowers with mauve streaks.*

**DESCRIPTION** **Plant** 45–80cm. **Corm** depressed-globose, papery tunics become partly fibrous with age. **Cataphylls** dry and dark brown above the ground. **Leaves** 6–10, in a fan, narrowly sword shaped, 18–35mm wide; margins and midrib not thickened, midrib lying to one side. **Spike** erect or inclined, unbranched, with 9–16 flowers. **Bracts** purplish green-grey and soft textured, becoming rust coloured, slightly folded along the midline. **Flowers** can be white flushed with pale mauve, or pinkish; each lower tepal marked with a slightly darker streak. Perianth tube 30–40mm. **Anthers** blueish mauve. **Pollen** cream. **Capsules** globose. **Seeds** pale translucent brown with uneven wings. **Scent** unscented.

**DISTRIBUTION** A rare endemic found along the lower Drakensberg escarpment between Graskop and Mariepskop, a small area spanning ±40km.

**ECOLOGY & NOTES** Found in two distinct habitats: either on steep cliffs, hanging from rock cracks, or in shady places among sandstone rocks where the plants have an upright growth habit. The plants grow either on shallow soil in leaf litter or in peaty sand on rocky pavements. Near Graskop, we saw both growth forms, but obviously the upright plants on rocky pavements were far easier to photograph than those hanging from cliff faces!

**POLLINATORS** Probably long-tongued flies, which tend to pollinate long-tubed, pale pinkish-mauve flowers.

**SIMILAR SPECIES** *G. saxatilis* could be confused with *G. ferrugineus* and *G. varius*; all three species have similar flowers and bracts, and grow in the same habitats.

| | Leaves | Perianth tube | Flowers |
|---|---|---|---|
| ***G. saxatilis*** | margins and midribs lightly thickened, soft textured, up to 35mm wide, | 30–40mm | large, white to pale pink, weakly defined nectar guides |
| ***G. ferrugineus*** | 5–8mm, twisted, margins and midrib thickened | 12–20mm | white to pale pink, faint cream markings on lower tepals |
| ***G. varius*** | 4–9mm, margins and midrib strongly thickened | 30–50mm | deep pink with darker nectar guides |

*Wide, sword-shaped leaves form a loose fan.*

*Seeds are unevenly winged.*

*Slightly folded bracts become rust coloured.*

*White form.*

*Pink form flushed with mauve.*

G. saxatilis *on rocky pavement, God's Window, Mpumalanga.*

# Gladiolus varius

**_varius_ = variable, as the original species included three varieties.**

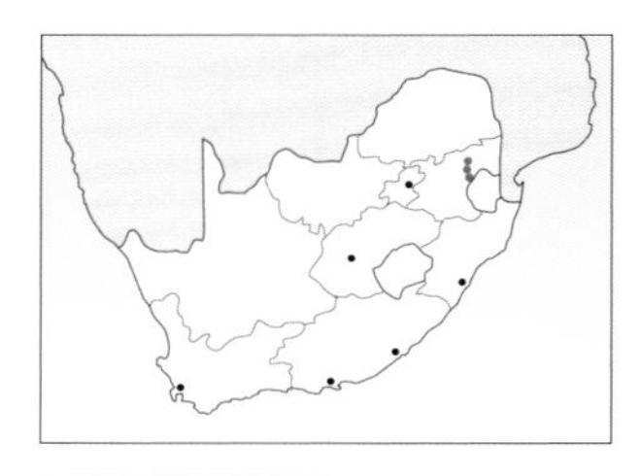

J F M A M J J A S O N D

*Gladiolus varius* is endemic to the high mountains of Mpumalanga. Its pink flowers appear between February and March.

**DESCRIPTION** **Plant** 45–70cm. **Corm** depressed-globose, with leathery tunics becoming fibrous with age. **Cataphylls** accumulate around the stem to form a fibrous neck. **Leaves** 7, in a fan reaching base of spike, 4–9mm wide; margins and midrib thickened. **Spike** lightly inclined, unbranched, with 8–12 flowers. **Bracts** greenish brown below, rust coloured above. **Flowers** deep pink, occasionally lighter; 3 lower tepals with darker median streak. Perianth tube long, 37–50mm. **Anthers** lilac. **Pollen** cream. **Capsules** oblong. **Seeds** discoid to angular with dark seed body, small irregular wing translucent brown. **Scent** unscented.

**DISTRIBUTION** Endemic to Mpumalanga and restricted in range to the high mountains from Barberton to Mount Sheba.

**ECOLOGY & NOTES** Always associated with rocky ridges; the corms are found in rock cracks. We photographed them on Long Tom Pass in quartzite. Shorter-tubed plants, with perianth tubes shorter than 37mm, occur in the Barberton mountains; these flowers are almost white. Goldblatt & Manning (2000) suggest that these are perhaps hybrids with the similar *G. ferrugineus*.

**POLLINATORS** Long-tongued flies in search of the large amounts of nectar produced by the flowers. This species belongs to a pollination guild including *Disa amoena*, *Nerine angulata*, *Radinosiphon leptostachya* and *Watsonia wilmsii* – all pollinated by the long-tongued fly *Prosoeca ganglbaueri*.

G. varius *has large pink flowers.*

**SIMILAR SPECIES** Can be confused with both *G. ferrugineus* and *G. saxatilis*. All three species have rust-coloured bracts, lack any distinctive markings on the pink to white flowers, flower at similar times and grow in almost identical habitats. *G. varius* and *G. ferrugineus* may hybridise.

| | Leaves | Perianth tube | Flowers |
|---|---|---|---|
| ***G. varius*** | 4–9mm, margins and midrib strongly thickened | 30–50mm | deep pink with darker nectar guides |
| ***G. ferrugineus*** | 5–8mm, twisted, margins and midrib thickened | 12–20mm | white to pale pink, faint cream markings on lower tepals |
| ***G. saxatilis*** | soft textured, up to 35mm wide, margins and midribs lightly thickened | 30–40mm | large, mauve to pale pink, nectar guides weakly defined |

*Leaf margins and midribs are thickened; plant grows in cracks.*

*Perianth tubes are long; bracts are green below and rust above.*

*Note lilac anthers.*

*Spikes are lightly inclined, bearing eight to 12 flowers.*

*A paler form of* G. varius.

G. varius *grows on rocky ridges and is endemic to Mpumalanga.*

# Gladiolus appendiculatus

***appendiculatus* = with small appendages, referring to the long sterile spurs or 'tails' on the anthers.**

*Gladiolus appendiculatus* is found in northern KwaZulu-Natal through Mpumalanga and Eswatini to southern Limpopo. Its pale flowers appear on floriferous spikes in April and May. This species forms a clade with *G. calcaratus* and *G. macneilii*, both of which share its distinctive tailed anthers.

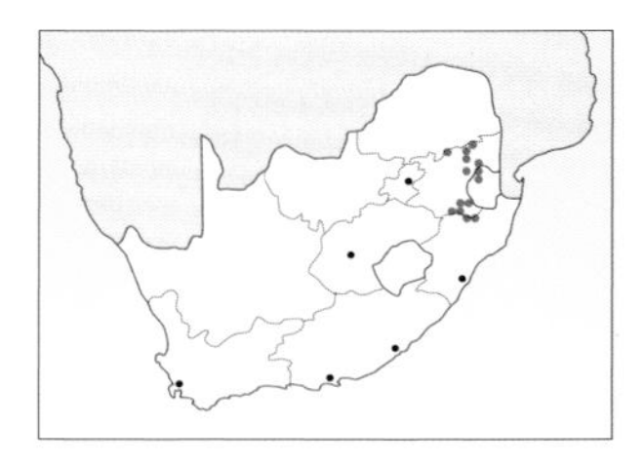

| J | F | M | A | M | J | J | A | S | O | N | D |
|---|---|---|---|---|---|---|---|---|---|---|---|

**DESCRIPTION** **Plant** 35–60cm. **Corm** globose, with papery tunics becoming fibrous. **Cataphylls** pale. **Leaves** either narrow and reaching the base of the spike, or broad and much shorter; margins thickened, midrib lightly thickened. **Spike** erect, unbranched, with 6–14 flowers, sometimes more. **Bracts** green to greyish purple, brown tipped at time of flowering, each overlapping the next. **Flowers** small, white to pink; lower tepals with a small dark purplish spot, throat yellowish. Perianth tube 15mm. **Anthers** pale yellow fading to mauve, with distinctively spurred sterile white tails. **Pollen** white. **Capsules** obovoid. **Seeds** ovate to pear shaped, narrowly winged. **Scent** unscented.

**DISTRIBUTION** Along the eastern escarpment from Vryheid in KwaZulu-Natal, across Eswatini to Pilgrim's Rest and Mariepskop, close to the Mpumalanga–Limpopo border.

**ECOLOGY & NOTES** Found in rocky grassland usually above 1,000m. There are two forms: one, more robust, has narrow leaves and up to 27 flowers per spike; the other is shorter, has broader leaves and only up to 12 flowers per spike.

**POLLINATORS** The short-tubed flowers are pollinated by anthophorid bees: as a bee moves into the flowers in search of nectar, it pushes against the angled spurs of the anthers, causing the tips to move down and deposit pollen on its upper body.

**SIMILAR SPECIES** Only three species have sterile spurs on the anthers: *G. appendiculatus*, *G. calcaratus* and *G. macneilii*. Presumably because of this characteristic, the three species are thought to be closely related, but they are dissimilar in appearance and are unlikely to be confused.

G. appendiculatus *is pollinated by African honeybees* (Apis mellifera scutellata).

G. appendiculatus *has distinctively spurred anthers.*

*Greenish-purple bracts.*

*Densely floriferous spikes.*

*Anthers fade to mauve.*

*Pale pink form.*

*Dark pink form.*

*Note yellowish throat.*

G. appendiculatus *is found in rocky grassland near Msauli, Mpumalanga.*

# *Gladiolus calcaratus*

***calcaratus*** **= spurred, referring to the long sterile tails on the anthers.**

*Gladiolus calcaratus* grows in Mpumalanga, along the escarpment. The pale flowers, which bloom from mid-February to early April, fade to lilac or pink. It forms a clade with *G. appendiculatus* and *G. macneilii*, both of which share its distinctive tailed anthers, but this species is distinguished by white flowers on a straight spike.

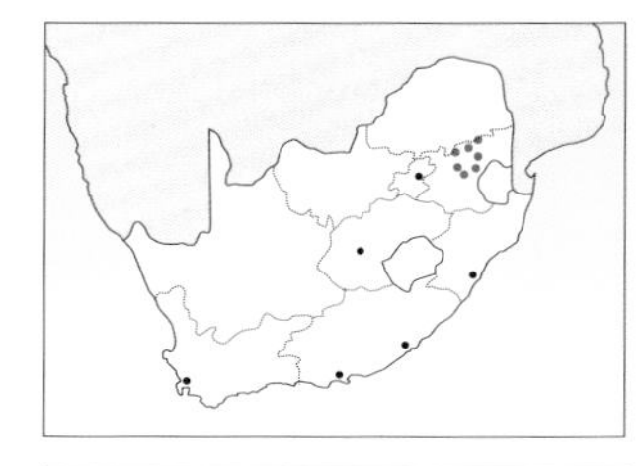

| J | F | M | A | M | J | J | A | S | O | N | D |
|---|---|---|---|---|---|---|---|---|---|---|---|

**DESCRIPTION** **Plant** 30–45cm. **Corm** globose with small cormlets, tunics initially woody, decaying to fibrous. **Cataphylls** up to 3cm above ground, purple or dry and brown. **Leaves** 6–8, sword shaped to linear, 4–8mm wide, reaching middle of stem; midrib and margins moderately thickened. **Spike** erect and unbranched, with 6–10 flowers. **Bracts** dark green, firm textured, folded on the midline. **Flowers** white fading to pink or lilac, yellow in the throat; lower tepals with a pale yellowish streak, upper 3 tepals the largest. Perianth tube 25–40mm long, obliquely funnel shaped. **Anthers** whitish with sterile 'tails'. **Pollen** cream. **Capsules** obovoid. **Seeds** narrowly ovate to pear shaped, unevenly and weakly winged. **Scent** unscented.

*The white flowers of* G. calcaratus *fade to pink.*

**DISTRIBUTION** Found in the higher mountains of Mpumalanga, from Dullstroom in the south to Mount Sheba in the north. Most collections have been from Long Tom Pass.

**ECOLOGY & NOTES** Found in deeper soils and damp areas; may also occur around the edge of damp depressions. Always seen as isolated plants, which is surprising since the corms produce many cormlets around the base. We have only seen one population in flower, north of Emgwenya, where they grew among quartzitic rocks. The plants were in damp grassland together with *Dierama* species, *Agapanthus inapertus* and *Erica woodii*.

*Specialised anthers enable short-tongued insects such as bees to pollinate the flowers.*

**POLLINATORS** The long perianth tube requires an insect with a long tongue, such as *Prosoeca robusta*, to reach the nectar. However, because of the long, sterile tails on the anthers, even an insect with a short tongue is capable of pollinating the flowers. As the insect moves into the flower, it brushes against the sterile tail causing the anthers to tilt and deposit pollen on the insect's body. Bees have also been observed on the flowers.

**SIMILAR SPECIES** Only three *Gladiolus* species have sterile spurs on the anthers: *G. appendiculatus*, *G. calcaratus* and *G. macneilii*. Presumably because of this characteristic, the three species are thought to be closely related, but they are so dissimilar in appearance that they are not likely to be confused.

*Narrow leaves have thickened midribs and margins.*

*Bracts are firm, folded in the midline.*

*Note yellowish throat.*

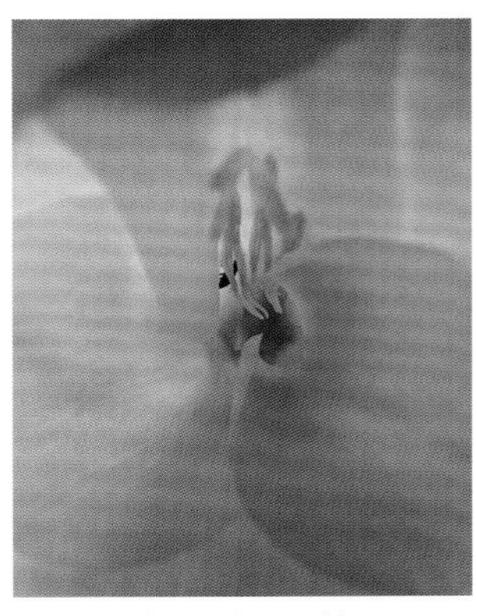

*Long, sterile anthers with distinctive 'tails' aid in pollination.*

G. calcaratus *grows high up in the mountainous region near Mbombela, Mpumalanga.*

# Gladiolus macneilii

**Named after Gordon MacNeil (1908–1986), a farmer who was an enthusiastic naturalist and cultivator of indigenous bulb plants; he sent specimens of the plant to Kirstenbosch in 1967.**

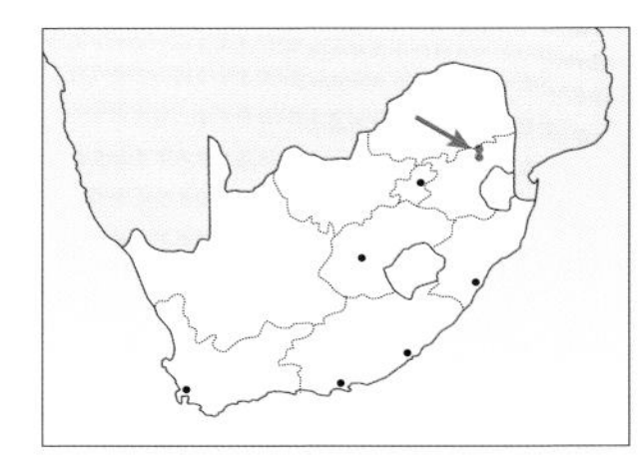

| J | F | M | A | M | J | J | A | S | O | N | D |
|---|---|---|---|---|---|---|---|---|---|---|---|

**STATUS** Critically Endangered

*Gladiolus macneilii* occurs only on the eastern escarpment of Limpopo. The salmon-pink flowers appear on inclined spikes in March and April. Part of a clade with *G. appendiculatus* and *G. calcaratus*, it is easily distinguished by its flower colour.

**DESCRIPTION** **Plant** 70–90cm. **Corm** ovoid, with layered tunics decaying to vertical fibres. **Cataphylls** pale and membranous, dry and brown above ground. **Leaves** 8 or 9, 60cm long, 2.5–4mm wide, reaching the spike; midrib moderately thickened and raised, margins only slightly thickened. **Spike** erect, either unbranched or with a few short branches and up to 17 flowers. **Bracts** pale green in bud, becoming dry at flowering. **Flowers** cream to salmon pink, with dark red streaks on the lower 3 tepals. Perianth tube long and curved, up to 45mm. **Anthers** purple with sterile tails. **Pollen** pale yellow. **Capsules** with brown speckles or reddish. **Seeds** broadly winged. **Scent** unscented.

**DISTRIBUTION** Known only from the slopes of the Abel Erasmus Pass north of Ohrigstad in Limpopo, at an altitude of about 1,300m.

**ECOLOGY & NOTES** Grows in wooded grassland among dolomite outcrops in loamy soils; the area is heavily grazed. Most are found in cracks in rocks in semi-shade under trees and shrubs such *as Kirkia wilmsii*, *Bauhinia tomentosa*, *Diospyros lycioides* and *Dichrostachys cinerea*. We found the plants on east-facing slopes together with *Aloe fosteri*, both in flower.

**POLLINATORS** Long-tongued flies. The sterile-tailed anther formation, which deposits pollen on an insect's back as it passes, may also enable pollination by bees.

**SIMILAR SPECIES** The cream to pale salmon flowers with long tubes and red markings on the lower tepals, together with the tails on the anthers, make this species easy to identify. Only three species have sterile spurs on the anthers: *G. appendiculatus*, *G. calcaratus* and *G. macneilii*. Presumably because of this characteristic, the three species are thought to be closely related, but they are so dissimilar in appearance that they are not likely to be confused. *G. sekukuniensis* (section *Hebea*), which grows in this area, has similar-looking flowers but no sterile anthers.

*Flowers of* G. macneilii *have a distinctive shape and colour.*

*A carpenter bee (of the genus* Xylocopa*) crawls across an unscented salmon-pink flower, robbing the flower of nectar by piercing the perianth tube.*

*Narrow leaves have thickened midribs.*

*Perianth tubes are very long and slightly curved.*

*Viewed from above,* G. macneilii *has a distinctive growth habit.*

*Purple anthers have sterile tails.*

*The dorsal tepal is much larger than the others; lower tepals are narrow.*

*Tall plants growing near Abel Erasmus Pass, Limpopo.*

# Gladiolus ochroleucus

***ochroleucus* = brownish yellow, for the flower colour of the type collection.**

Widespread in the Eastern Cape, *Gladiolus ochroleucus* has a long flowering period, throughout summer. Its flowers are pink, purple, reddish or white. Sometimes confused with *G. mortonius*, which has a much longer perianth tube and thicker leaves.

J F M A M J J A S O N D

**DESCRIPTION** **Plant** 40–80cm. **Corm** globose, with papery tunics decaying to fragments. **Cataphylls** up to 12cm above ground, green. **Leaves** 7–12, sword shaped and in a tight fan reaching the base of the spike, ±20mm wide; margins and all veins lightly to moderately thickened, with margins sometimes minutely hairy. Sometimes evergreen. **Spike** slightly inclined and may be branched, with 4–12 or more flowers. **Bracts** 25–40mm, green. **Flowers** pink, reddish or purple, or white to yellow with brown reverse. Dorsal tepal is hooded and lower lateral tepals are small, giving an almost closed appearance; lower tepals with whitish nectar guides sometimes highlighted with reddish median streaks. Perianth tube short, 15–20mm. **Anthers** 9mm, pale mauve. **Pollen** cream. **Capsules** shorter than bracts. **Seeds** light brown, broadly winged. **Scent** unscented.

G. ochroleucus *tends to have pink flowers when it grows along the coast.*

**DISTRIBUTION** Common in the Eastern Cape. Found from Makhanda (Grahamstown) and King William's Town, moving northwards close to Richmond in KwaZulu-Natal.

**ECOLOGY & NOTES** From sea level to altitudes of 1,200m; grows in coastal sand to light clay. Found in grassland, often in light shade, and occurring in areas where rain falls all year round. Plants in these areas are often evergreen. *G. ochroleucus* was previously recognised as two varieties because it shows such variation over its range. Plants with creamy yellow flowers and a shorter perianth tube tend to occur inland while the pink to reddish form occurs further south and closer to the coast.

**POLLINATORS** Long-tongued bees (of the genera *Amegilla* and *Anthophora*).

**SIMILAR SPECIES** Similar to *G. mortonius*, another pink-flowered species from the Eastern Cape. *G. mortonius* has leaves with thickened margins and longer perianth tubes.

| | Perianth tube | Anthers | Leaves |
|---|---|---|---|
| ***G. ochroleucus*** | 15–20mm | 9mm | fanned, midrib and margins moderately thickened |
| ***G. mortonius*** | 30–45mm | 12–15mm | fanned, midrib and margins strongly thickened |

*Leaves form a tight fan.*

*Soft green bracts are fairly long.*

*Spikes are slightly inclined.*

*This cream-yellow form with brown reverse was found near Matatiele, Eastern Cape.*

*Note the pale mauve anthers in this cream-yellow form.*

*A reddish-orange form found near Stutterheim, Eastern Cape.*

G. ochroleucus, *Morgan Bay, Eastern Cape.*

# Gladiolus mortonius

**The type illustration was drawn from plants sent by a Mr Morton to the United Kingdom in 1838.**

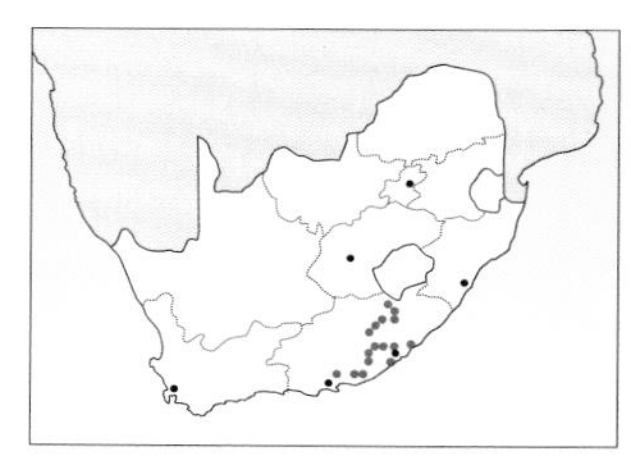

| J | F | M | A | M | J | J | A | S | O | N | D |
|---|---|---|---|---|---|---|---|---|---|---|---|

Growing in the Eastern Cape, *Gladiolus mortonius* has pink, scented flowers on flexed spikes from February to April. Sometimes confused with *G. oppositiflorus*, it can be distinguished by the latter's two-ranked spike and unique bracts.

**DESCRIPTION** **Plant** 40–70cm. **Corm** globose, with papery tunics decaying to fragments. **Cataphylls** green and leaf-like. **Leaves** 7–12, in a tight fan, 15–25mm wide, reaching the base of the spike; margins, midrib and secondary veins strongly thickened. **Spike** erect, rarely with single branch, 8–16 flowers. **Bracts** pale green. **Flowers** pink, fairly large (up to 95mm); tepals unequal (upper 3 largest) with undulate margins, the lower tepals with reddish streaks. Perianth tube 30–45mm. **Anthers** long, usually purple with a short, pointed structure (apiculus) at the end. **Pollen** white. **Capsules** obovoid-oblong, shorter than the bracts. **Seeds** broadly and evenly winged. **Scent** unscented.

G. mortonius *is restricted to the Eastern Cape.*

**DISTRIBUTION** Found only in the Eastern Cape, from Bosberg near Somerset East and the Zuurberg Mountains, through the Amatola Mountains and the Winterberg to Khowa near Lesotho.

**ECOLOGY & NOTES** Plants grow in grassland in stony areas on mountain slopes. This species is evergreen, with new leaves developing beside those of the previous year.

**POLLINATORS** Long-tubed pink flowers with reddish nectar guides and widely spreading tepals are usually pollinated by long-tongued flies.

**SIMILAR SPECIES** Similar to *G. ochroleucus*. It can also be difficult to distinguish this species from *G. oppositiflorus* (section *Ophiolyza*) whose leaves and flower colour may be similar, as well as the habitat.

| | Perianth tube | Anthers | Bracts | Spike |
|---|---|---|---|---|
| ***G. mortonius*** | 30–45mm | 12–15mm | not inflated | secund (flowers face one side only) |
| ***G. ochroleucus*** | 15–20mm | 9mm | not inflated | secund |
| ***G. oppositiflorus*** (sect. *Ophiolyza*) | 40–50mm | 11mm | inflated, inner bracts have margins united around the ovary | distichous (flowers arranged in 2 opposed ranks) |

*Lanceolate leaves form a fan.*

*Perianth tubes are curved.*

*Flowers are usually secund but may appear two ranked.*

*Red median streaks are visible.*

*Tepal margins are undulate.*

*The arched dorsal tepal is shorter than the laterals.*

G. mortonius *flowering near the Otto du Plessis Pass, Eastern Cape.*

# *Gladiolus microcarpus*

***microcarpus* = with a small fruit (capsule).**

This long, drooping species is endemic to the Drakensberg range. *Gladiolus microcarpus* bears bright pink flowers between December and January, and later at higher elevations.

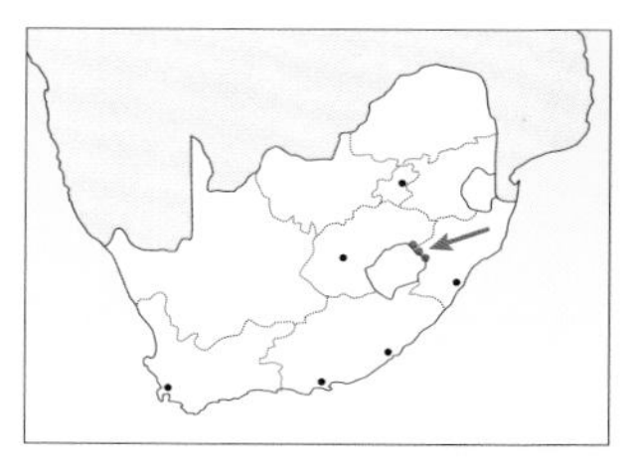

J F M A M J J A S O N D

**DESCRIPTION** **Plant** up to 1m, usually pendulous. **Corm** globose, with tunics decaying into fine fibre. **Cataphylls** pale green with dense short hairs. **Leaves** 5–7, trailing and forming a sheathing 'stem', 6–12mm wide; midrib strongly thickened and scabrid to the touch. **Spike** arching upwards and may be branched, with 8–12 almost distichous flowers in two ranks about 70° apart. **Bracts** pale green, lightly ridged and diverging from the stem. **Flowers** bright pink; with broad white longitudinal area on 3 lower tepals, white zone has a narrow red streak in the middle. Perianth tube is long, 35–50mm. **Anthers** whitish to lilac, with small appendages on the apex. **Pollen** white. **Capsules** oblong, enclosed in bracts. **Seeds** small, broadly and evenly winged. **Scent** unscented.

*The two-ranked arrangement of flowers is clear when viewed from beneath.*

**DISTRIBUTION** Known only from the Drakensberg Mountains of Lesotho, the Free State and KwaZulu-Natal. Its range extends from Cathkin Peak in the south to Royal Natal National Park in the north.

**ECOLOGY & NOTES** Found between 1,800 and 2,700m; more common at higher altitudes. They grow on large boulders and steep cliffs, with corms firmly wedged in crevices in either basalt or sandstone rock. The stems and leaves hang downwards, while the flower spikes either emerge vertically or arch upwards. The plants are continuously moist during the summer, either from mist and rain or from seeps over the rock faces. It appears that this species is very sensitive to fires. When visiting the Sentinel in the northern Drakensberg after a particularly devastating fire in the winter of 2014, we were unable to find any flowering plants in areas where they had been recorded regularly in the past. A few flowering plants were recorded in the area in February 2016, indicating some recovery after the fire.

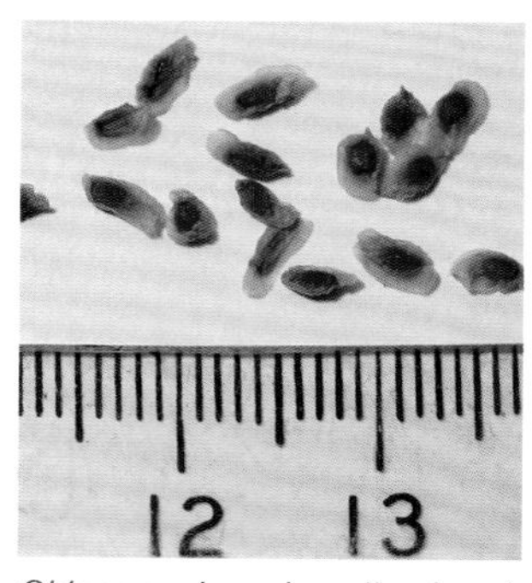


*Oblong seeds are broadly winged.*

**POLLINATORS** A long-tongued fly, *Prosoeca ganglbaueri*, which also pollinates several other pink- to cream-flowered plants, all with long tubes. This group of plants makes up a pollination guild, with all species depending on one pollinator. The guild includes *Hesperantha scopulosa*, *Watsonia wilmsii*, several *Disa* species and *Zaluzianskya microsiphon*.

**SIMILAR SPECIES** Although similar to *G. scabridus*, the distinctive habitat and restricted distribution range make confusion unlikely.

*The trailing, scabrid leaves are distinctive.*

*Stems and leaves hang down, giving a drooping appearance.*

*The unusual appendages on the anthers are clearly visible.*

*Anthers are whitish to lilac with white pollen.*

G. microcarpus *grows high in the Drakensberg. This picture was taken on the Sentinel; note the moisture on the rocks.*

# Gladiolus scabridus

***scabridus* = rough, referring to the feel of the stem and bracts.**

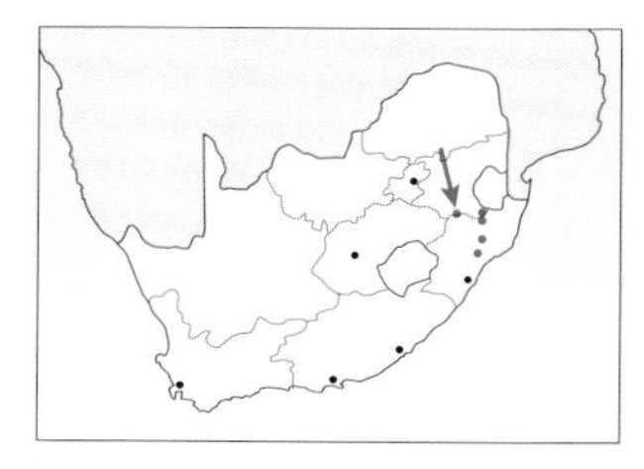

*Gladiolus scabridus* grows at elevations above 1,000m in northern KwaZulu-Natal and Eswatini. Its bright pink, unscented flowers appear in December and January, and sometimes later in cooler sites.

| J | F | M | A | M | J | J | A | S | O | N | D |
|---|---|---|---|---|---|---|---|---|---|---|---|

**DESCRIPTION** **Plant** up to 100cm. **Corm** globose, with papery tunics decaying into fibrous fragments. **Cataphylls** pale green or dull purple, with or without short hairs. **Leaves** 7–9, sword shaped, in a loose fan, up to 20mm wide, reaching the same height as the spike; midrib and secondary veins moderately raised and thickened; scabrid. **Spike** erect, simple or branched, scabrid, 10–16 flowers. **Bracts** dull green, scabrid. **Flowers** bright pink; lower 3 tepals have a narrow, white longitudinal G with a reddish streak in the midline. Perianth tube 35–40mm. **Anthers** pale yellow. **Pollen** white. **Capsules** oblong. **Seeds** large and with ovate brown seed body, evenly or unevenly winged. **Scent** unscented.

*Flowers are bright pink.*

**DISTRIBUTION** Records show that *G. scabridus* is restricted to northern KwaZulu-Natal (south of the Pongola River) and southern Eswatini. Although not recorded from southern Mpumalanga, it probably occurs there as well, on the northern side of the Pongola River.

**ECOLOGY & NOTES** Found between 1,000 and 2,000m above sea level, always among rocks where the corms are safe from porcupines and baboons.

**POLLINATORS** Probably long-tongued flies associated with long-tubed flowers.

*Stems emerge from cracks in rocks.*

**SIMILAR SPECIES** Originally described as a subspecies of *G. microcarpus*, *G. scabridus* was raised to species rank by Goldblatt & Manning (1998). It grows at a higher altitude than *G. microcarpus*.

| | Habitat | Bracts | Stems |
|---|---|---|---|
| ***G. scabridus*** | well-drained rocky habitats at altitudes of 1,000–2,000m, plants are upright | scabrid | scabrid and erect |
| ***G. microcarpus*** | large boulders and steep cliffs in the Drakensberg at altitudes of 1,800–2,700m, plants are pendulous | smooth | smooth, inclined or pendulous |

*Corms are wedged into rocks. Sword-shaped leaves form a loose fan.*

*Spiny sugar ants (*Polyrhachis gagates*) explore the long perianth tube.*

*The lower 3 tepals have distinctive red streaks.*

G. scabridus *has two-ranked flowers, scabrid bracts and an erect stem.*

G. scabridus *grows at higher altitudes; here near Bivane Dam, KwaZulu-Natal.*

# *Gladiolus cataractarum*

***cataracta* = waterfall, named for the waterfall at Lunsklip, Mpumalanga, where it was first scientifically recorded.**

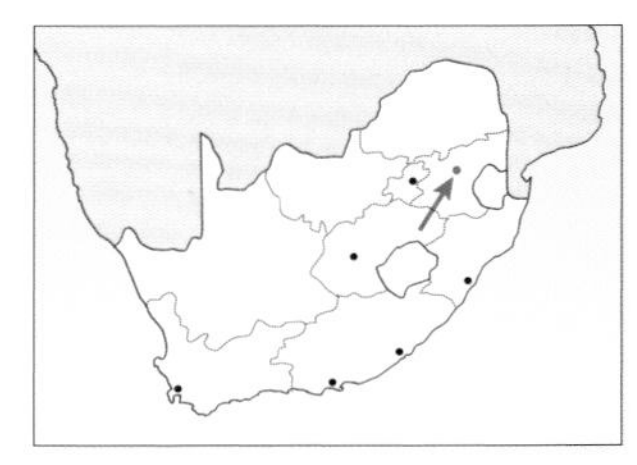

J F M A M J J A S O N D

**STATUS** Endangered

*Gladiolus cataractarum* only grows in shaded areas on cliffs of the eastern escarpment of Mpumalanga. Its large pink flowers streaked with reddish lines, and with long perianth tubes, appear between February and mid-March.

**DESCRIPTION** **Plant** up to 70cm. **Corm** globose, with leathery tunics becoming partly fibrous with age. **Cataphylls** pale green. **Leaves** 8 or 9, in a fan, 35–45mm wide, reaching the spike; margins and midrib thickened and translucent. **Spike** erect and unbranched, with 8–16 flowers. **Bracts** firm textured and green, up to 70mm. **Flowers** large and pink; reddish lines on the lower 3 tepals. Perianth tube long, 40–50mm. **Anthers** yellow. **Pollen** yellow. **Capsules** oblong. **Seeds** broadly and evenly winged, with dark seed body. **Scent** unscented.

**DISTRIBUTION** Restricted to the eastern escarpment of Mpumalanga. Known from Lunsklip Waterfall near Dullstroom and from a few other sites towards Mashishing, always along the escarpment edge.

**ECOLOGY & NOTES** Found on cliffs and in steep rocky areas, often in shade, on south-facing slopes. We found some plants growing among *Begonia sutherlandii* in damp to wet soil, while others were among rocks in grassland on much drier slopes.

**POLLINATORS** Presumed to be long-tongued flies of the family Nemestrinidae.

**SIMILAR SPECIES** Closely related to *G. mortonius, G. microcarpus* and *G. scabridus* but owing to its restricted habitat, it is unlikely to be confused with any other species.

*Erect, unbranched stems bear large flowers.*

*Habitat at Lunsklip Waterfall in Mpumalanga.*

*Broad leaves form a soft fan.*

*Large pink flowers have reddish lines. Anthers and pollen are usually yellow.*

*The species has long perianth tubes and bracts.*

G. cataractarum *grows in the shade near Lunsklip Waterfall, Mpumalanga.*

# Gladiolus pavonia

***pavonia*** **= peacock; named for the eye in the centre of the flower, which resembles a peacock feather.**

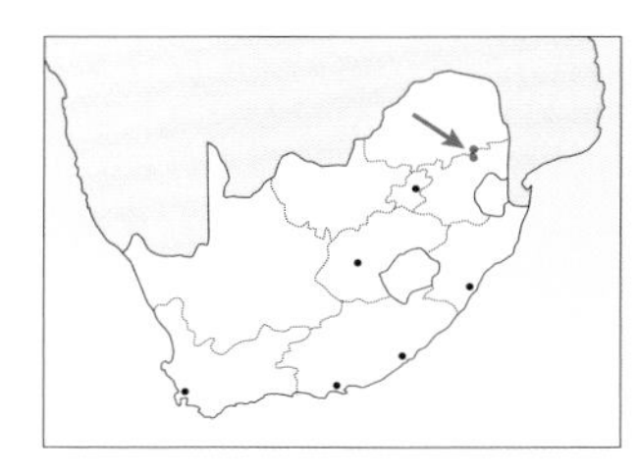

| J | F | M | A | M | J | J | A | S | O | N | D |
|---|---|---|---|---|---|---|---|---|---|---|---|

**STATUS** Critically Endangered

*Gladiolus pavonia* grows on dolomitic soils in mountainous areas of Mpumalanga/Limpopo. It can be distinguished from *G. brachyphyllus* by its pale rather than dark pink flowers, which appear in November and December.

**DESCRIPTION** **Plant** 45–80cm. **Corm** ovoid with stolons at base; papery tunics. **Cataphylls** brown to purple. **Leaves** ±7, soft textured, slightly twisted in the upper half, 8–14mm wide, reaching or exceeding the spike base. **Spike** erect and unbranched, 4–7 flowers in two ranks about 50° apart. **Bracts** pale green and soft textured. **Flowers** pale pink; with red streaks on lower 3 tepals and a red 'eye' in the mouth of the tube. Perianth tube ±16mm long. **Anthers** dark purple. **Pollen** cream. **Seeds** large, 7–10mm, broadly and evenly winged (previously unrecorded). **Scent** unscented.

*Dark purple anthers bear cream-coloured pollen.*

**DISTRIBUTION** An endemic species known from one small area, *G. pavonia* has only been found on the slopes of the Abel Erasmus Pass in Mpumalanga/Limpopo.

**ECOLOGY & NOTES** Plants grow among dolomite rocks with their corms protected in narrow cracks. As this species produces stolons, plants are often found in small clumps; this is unusual behaviour for species found in a rocky environment as the corms are well protected from predators and stolons are not necessary for their survival. We saw plants at an altitude of ±900m above sea level; the area is hot and dry. Plants were in shade and sun, among *Dichrostachys cinerea* and *Vachellia* or *Senegalia* species. Many of the trees are dwarfed by the harsh environment. The habitat is north facing and well drained.

**POLLINATORS** Not known, but the pink flowers with a short tube and a dark pink 'eye' are probably pollinated by long-tongued bees, the section's ancestral pollinators.

**SIMILAR SPECIES** *G. pavonia* is related to *G. brachyphyllus* and has similar flowers and leaves. They are unlikely to be confused because they grow in different soils and flower at different times.

| | Habitat | Flowering plants | Flowers | Flowering season |
|---|---|---|---|---|
| ***G. pavonia*** | dolomite | leaves well developed | pale pink with dark red in mouth of tube | rainy season |
| ***G. brachyphyllus*** | sandy soils, not dolomitic | leaves with very reduced blades | dark pink with white on lower 3 tepals | at the end of the dry season |

*Narrow leaves have raised midribs and tend to twist at the ends.*

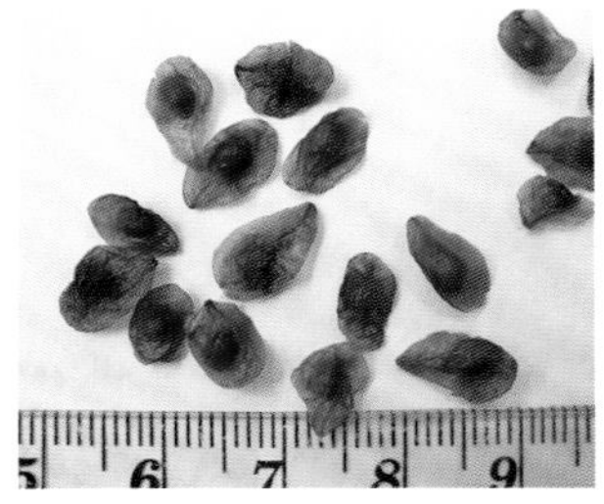


*Mature seeds measure 7–10mm and are evenly winged.*

*Plants grow among dolomite rocks, with their corms wedged in cracks.*

*Short tubes suggest pollination by long-tongued bees, the section's ancestral pollinators.*

*Bracts are green and soft.*

G. pavonia *on the slopes of Abel Erasmus Pass, Mpumalanga/Limpopo.*

# *Gladiolus brachyphyllus*

***brachyphyllus*** **= short-leaved, for the short leaf blades of flowering spikes.**

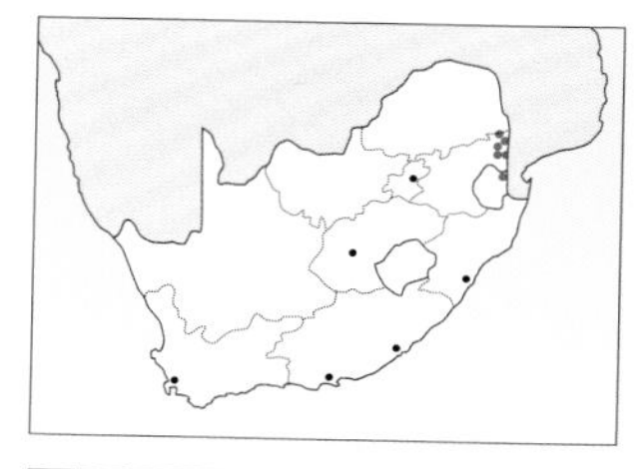

J F M A M J J A S O N D

*Gladiolus brachyphyllus* grows in the dry savanna regions of eastern South Africa, Mozambique and Eswatini. Its flowers are deep pink, making it relatively easy to distinguish from *G. pavonia*. Unusual for species in *Densiflori*, it flowers at the end of the dry season, in October and November.

*Flowers are deep pink; lower 3 tepals have a pale band with a purple streak.*

**DESCRIPTION** **Plant** 55–80cm. **Corm** globose, with leathery tunics becoming partly fibrous with age. **Cataphylls** 5cm above ground, green. **Leaves** present on flowering plants but underdeveloped at flowering, sheathing with short blades; 6–12mm wide on non-flowering plants; midrib and margins lightly thickened, soft textured. **Spike** erect, occasionally branched, 7–9 flowers. **Bracts** pale green, relatively large. **Flowers** dark pink; lower 3 tepals almost equal, each with a broad band of white with a dark purple streak in the centre of the white. Perianth tube ±20mm long. **Anthers** violet. **Pollen** lilac. **Capsules** ovoid, slightly sunken above. **Seeds** body darker than broad wings. **Scent** unscented.

**DISTRIBUTION** A Lowveld species, found from Klaserie in the north, through the Kruger National Park, to northern Eswatini and southern Mozambique.

**ECOLOGY & NOTES** Grows in well-drained sandy soils, among rocks in dry savanna woodland. It has adapted to the harsh conditions prevailing in October when the plants flower: this is the end of the dry winter, and the first flowers are found only a few weeks after the early rains. Within six weeks of flowering, the capsules are ripe and the seeds are released in time to germinate and grow during the same summer. The plants cope with this rapid early growth and flower production by producing reduced sheathing leaves, and by using the stems and leaf sheaths for photosynthesis. Non-flowering plants produce normal leaves with long blades that also emerge in October. The plants we saw all grew in grassland in light shade among *Combretum* species, *Dichrostachys cinerea*, *Sclerocarya birrea* subsp. *caffra* and various *Vachellia* or *Senegalia* species. Although the plants flower better after a burn, we found many flowering plants in older, unburnt grassland.

**POLLINATORS** Likely to be generalists such as bees, flies with longish tongues, and butterflies.

**SIMILAR SPECIES** *G. brachyphyllus* has similar flowers and leaves to *G. pavonia* but is unlikely to be confused as they grow in different soils and flower at different times.

| | Habitat | Flowering plants | Flowers | Flowering season |
|---|---|---|---|---|
| ***G. brachyphyllus*** | sandy soils, not dolomitic | leaves with very reduced blades | dark pink with white on lower 3 tepals | at the end of the dry season |
| ***G. pavonia*** | dolomite | leaves well developed | pale pink with dark red in mouth of tube | rainy season |

*The flowers have short perianth tubes.*

*Bracts are pale green on erect stems.*

*This species grows among rocks in grasslands and woodlands, here near Malelane, Mpumalanga.*

G. brachyphyllus *growing in well-drained soils near Malelane, Mpumalanga.*

*Rod Saunders records* G. cruentus *on the steep slopes of Krantzkloof, KwaZulu-Natal.*

# *Gladiolus reginae*

***reginae*** **= queen. Named after the Queen of Sheba, who according to biblical texts presented King Solomon with gifts of gold; she was said to control mines in Africa. The type locality was on a platinum mine.**

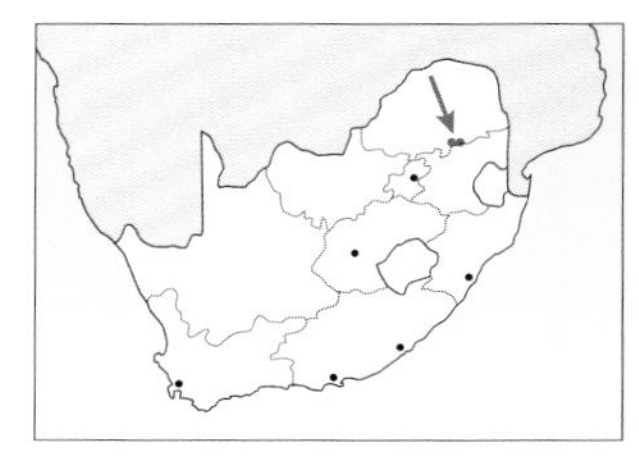

| J | F | M | A | M | J | J | A | S | O | N | D |
|---|---|---|---|---|---|---|---|---|---|---|---|

**STATUS** Critically Endangered

*Gladiolus reginae* is known from a few sites in the Dwarsriver Mountains on the Limpopo–Mpumalanga border. Its large pinkish flowers are streaked with red, appearing in floriferous spikes between mid-March and mid-April.

*This species is adapted to soils rich in heavy metals.*

**DESCRIPTION** **Plant** up to 1.5m. **Corm** globose, with papery tunics. **Cataphylls** above ground, brownish. **Leaves** narrow, up to 10mm wide, sword shaped; midrib raised, blade slightly twisted. **Spike** erect, usually unbranched, 7–16 flowers arranged in two ranks almost opposite. **Bracts** 20–32mm long, pale green, becoming straw coloured and dry. **Flowers** pale salmony pink; lower 3 tepals flushed deep red and marked with deep red streaks. Perianth tube reddish, ±30mm. **Anthers** pale purple. **Pollen** cream. **Capsules** obovoid. **Seeds** translucent, evenly or unevenly winged. **Scent** unscented.

**DISTRIBUTION** Only a few Limpopo populations are known from the hot savanna of the Dwarsrivier Mountains close to Steelpoort.

**ECOLOGY & NOTES** Plants grow in semi-shade and sun among igneous rocks (gabbronorite), where the corms are safe from predators. These rocks contain higher than normal amounts of heavy metals such as chrome and platinum. *G. reginae* shares an intriguing characteristic with *G. pole-evansii*: both species exude a sugary solution (photosynthate) from the tips of the bracts, which attracts ants and other insects, possibly for defensive purposes. Threatened by platinum mining and grazing.

**POLLINATORS** Probably long-proboscid flies including *Stenobasipteron wiedmannii*, which also pollinate *Ocimum tubiforme* (Lamiaceae).

**SIMILAR SPECIES** Flowers look similar to *G. pavonia* (section *Densiflori*) and it is closely related to *G. dolomiticus*, but it is unlikely to be confused with them because its location is specific.

| | Stems & leaves | Flowers | Stolons |
|---|---|---|---|
| ***G. reginae*** | not hairy | flesh-pink with red tube and red streaks (33mm tube) | no |
| ***G. dolomiticus*** | hairy | pale pink with yellow centre (25mm tube) | no |
| ***G. pavonia*** (sect. *Densiflori*) | hairy | pale pink with dark red in mouth of tube (16mm tube) | yes |

*Plants grow among igneous rocks.*

*Plants may reach 1.5m in height.*

*Note red streaks and purple anthers.*

G. reginae *was scientifically described only in 2009.*

# *Gladiolus dolomiticus*

***dolomiticus*** **= of dolomite (the rock on which this species grows).**

*Gladiolus dolomiticus* is found in Limpopo. Unscented, pale pink flowers appear around March and April.

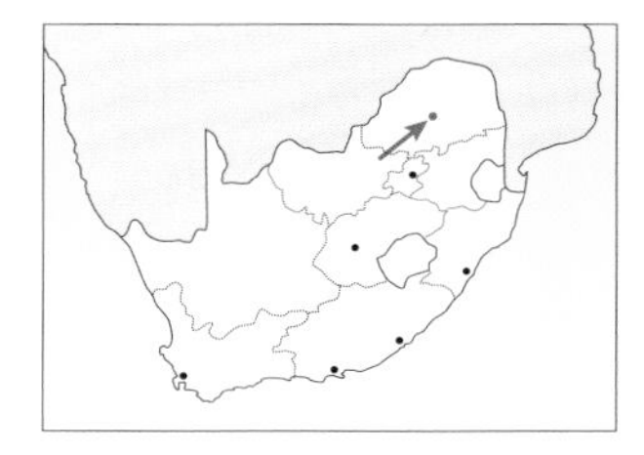

| J | F | M | A | M | J | J | A | S | O | N | D |
|---|---|---|---|---|---|---|---|---|---|---|---|

**STATUS** Rare

**DESCRIPTION** **Plant** 70–100cm. **Corm** depressed-globose, with leathery tunics becoming fibrous. **Cataphylls** up to 10cm above ground, green and hairy. **Leaves** 7 or 8, blades 10–15mm wide, rigid, in a loose fan, pale green, with soft hairs on the veins and surface; margins, midribs and secondary veins are thickened and raised. **Spike** erect, often branched, softly hairy, with 10–17 almost alternating flowers facing in a single direction. **Bracts** up to 30mm long, pale green, softly hairy, becoming dry and brown at tips. **Flowers** pale pink, apricot or cream; with yellow markings on the lower tepals, and a yellow throat. Perianth tube up to 27mm. **Anthers** pale yellow. **Pollen** pale yellow. **Capsules** obovoid-oblong. **Seeds** large, dark brown, broadly winged. **Scent** unscented.

**DISTRIBUTION** Known only from Makapansgat in Limpopo.

**ECOLOGY & NOTES** Many plants have been found on east-facing slopes growing in bands of dolomite, at an altitude of about 1,500m. The corms are invariably wedged into cracks among the rocks. Plants have been found in sun and semi-shade in grassland, along with *Dichrostachys cinerea*, *Kirkia wilmsii*, *Xerophyta retinervis*, *Ziziphus mucronata* and various *Aloe* species. These areas are hot and fairly dry. By May, when the seeds are ripe, the surrounding vegetation is brown and in a state of dormancy. Interestingly, large ants were observed on the stems and in the flowers, similar to those seen on *G. pole-evansii*, a closely related species.

**POLLINATORS** Long-tongued bees that forage for nectar in the flowers.

**SIMILAR SPECIES** Not likely to be confused with any other species. Its choice of habitat (dolomite), late-summer flowering time, pale pink flowers with yellow markings, hairy bracts and cataphylls, and leaves with thickened margins and veins are unlike any other species that could be found in the area.

*The plant has a lax fan of leaves.*

*Short perianth tubes suggest pollination by long-tongued bees.*

*Note the almost alternating arrangement of flowers.*

*Flowers are secund (facing in one direction).*

*This cream-white form has a pale yellow throat.*

G. dolomiticus *is known only from the grasslands of Makapan's Valley, Limpopo.*

# *Gladiolus pole-evansii*

**Named after the Welsh-born South African botanist Illtyd Buller ('IB') Pole-Evans (1879–1968).**

*Gladiolus pole-evansii* is known only from a restricted range in Limpopo and Mpumalanga. A very tall plant, it has floriferous spikes of pinkish to translucent flowers appearing from mid-January to March.

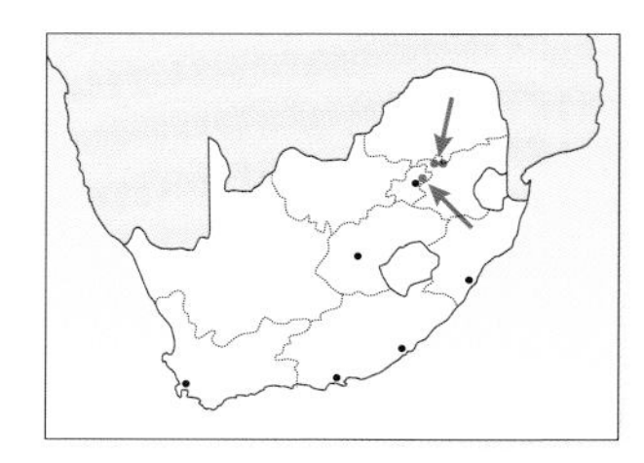

| J | F | M | A | M | J | J | A | S | O | N | D |
|---|---|---|---|---|---|---|---|---|---|---|---|

**STATUS** Rare

**DESCRIPTION** **Plant** 60–110cm. **Corm** globose, with papery tunics becoming fibrous. **Cataphylls** 10cm above ground, green and softly hairy. **Leaves** 10–15mm wide, greyish green in a lax fan; margins lightly thickened, while midrib and a pair of veins are heavily thickened; leaf surface and veins hairy. **Spike** erect, often branched, scabrid, 15–20 flowers. **Bracts** pale green, softly hairy and exuding sugar solution from the tips. **Flowers** translucent pink; the upper lateral tepals have a red streak along the midline, lower tepals with yellow markings in the midline. An unusual form has flowers very speckled with dark pink to mauve spots. **Anthers** light purple. **Pollen** yellow. **Capsules** oblong. **Seeds** oval, large and dark brown, unevenly winged. **Scent** unscented.

*A typical form.*

**DISTRIBUTION** Occurs in the granite hills in the Groblersdal area of Limpopo – the site where IB Pole-Evans first collected them for science. Another large population occurs close to Loskop Dam north of Middelburg, Mpumalanga, in Waterberg quartzite.

**ECOLOGY & NOTES** Found in sun and semi-shade, the corms are always lodged among rocks and in rock cracks. The flowers are usually almost translucent, but the population near Loskop Dam bears heavily spotted flowers. Originally this species was thought to associate with granite specifically, but the Mpumalanga population is on quartzite. The sugary sap produced by the floral bracts attracts several ant species as well as small wasps. When a spike was disturbed, ants on the plant appeared to be on the defensive. Several weeks later capsules and seeds on the same plants were almost completely unparasitised, prompting speculation that this may have been due to the protective role played by the ants. *Gladiolus* capsules and seeds are frequently parasitised; it is unusual to find no insect damage at all.

*Lanceolate leaves form a loose fan.*

**POLLINATORS** Thought to be large long-tongued bees, probably of family Anthophoridae, similar to those that pollinate *G. dolomiticus* and *G. sericeovillosus*.

**SIMILAR SPECIES** Easily distinguished by the tall plants, often over 1m in height, and the branched spikes with many flowers, often covered in ants.

*Ants are attracted to the sugary sap on the bracts.*

*Bract apices exude a sugar solution.*

*Pink flowers have translucent edges and yellow markings.*

*An unusual, richly speckled form.*

*Emerging from rock cracks near Loskop Dam, Mpumalanga.*

# *Gladiolus oppositiflorus*

***oppositiflorus*** **= referring to the arrangement of flowers on the spike.**

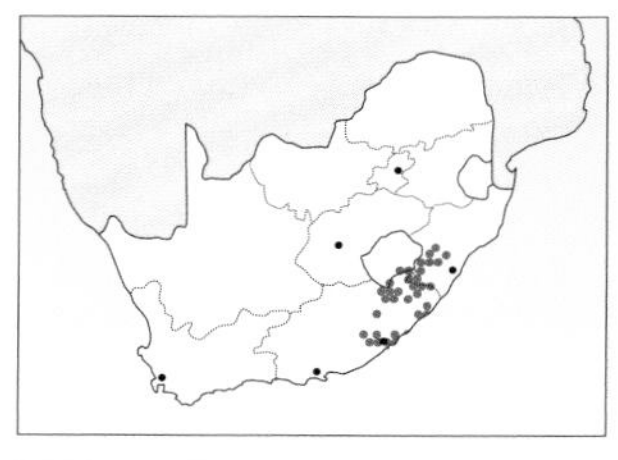

J F M A M J J A S O N D

*Gladiolus oppositiflorus* has distinctive, oppositely ranked flowers. In the Eastern Cape, flowers tend to be pink or mauve, while in KwaZulu-Natal, the flowers are salmon coloured and the plants shorter. Flowering around February to March.

**DESCRIPTION** **Plant** 60–160cm. **Corm** depressed-globose; papery tunics form a neck around the stem. **Cataphylls** up to 10cm above ground, velvety and green. **Leaves** 7 or 8 blades, ±18mm wide, firm; margins and midrib strongly thickened and velvety. **Spike** erect and unbranched, 7–15 or more flowers in two ranks up to 180° apart. **Bracts** green and hairy, inflated, sheathing base of floral tube. **Flowers** large, pale pink to mauve, pink-purple or salmon; lower 3 tepals with reddish streak. Perianth tube 40–50mm. **Anthers** light mauve. **Pollen** cream. **Capsules** narrowly oblong. **Seeds** broadly and evenly winged. **Scent** unscented.

*Mauve-pink form.*

**DISTRIBUTION** Found in southern KwaZulu-Natal and the Eastern Cape. Two variants used to be treated as subspecies: those from coastal areas in the Eastern Cape occur mainly in the vicinity of East London, whereas the other variant occurs further inland, at higher altitudes.

**ECOLOGY & NOTES** Found in full sun in grassland. Plants from coastal areas are often evergreen, whereas the inland form found at higher altitudes is deciduous. The coastal form is taller with flower spikes of 1m or more and flowers 180° apart; the inland form is shorter with flowers 100–150° apart. Intermediates between the two forms have been found. This was a very important species in early *Gladiolus* breeding; many garden hybrids and cut flowers have genes from this plant. Natural hybrids with *G. saundersii* have been found. (See page 15 for more information about natural hybrids.)

**POLLINATORS** Thought to be long-tongued flies, possibly *Prosoeca ganglbaueri*.

**SIMILAR SPECIES** *G. mortonius* (section *Densiflori*) is most likely to cause confusion: it has similar-coloured flowers, also occurs in the Eastern Cape, and flowers at the same time of year. *G. sericeovillosus* and *G. elliotii* also have inflated floral bracts and flowers in two ranks; however, both have much smaller flowers and occur much further north.

| | Perianth tube | Anthers | Bracts | Spike |
|---|---|---|---|---|
| ***G. oppositiflorus*** | 40–50mm | 11mm | inflated, inner bracts have margins united around the ovary | distichous (flowers arranged in two opposed ranks) |
| ***G. mortonius*** (sect. *Densiflori*) | 30–45mm | 12–15mm | not inflated | secund |

*Sword-shaped leaves form a loose fan.*

*Softly pubescent bracts sheath the flowers.*

*The flowers are two ranked up the stem.*

*Note the reddish throat and base of the perianth.*

*Note the distinct red streaks on the lower tepals.*

*Tall coastal pink form, here in the Eastern Cape.*

*Salmon-pink form in grasslands, KwaZulu-Natal.*

# Gladiolus sericeovillosus

***sericeovillosus* = silky-haired, referring to the soft white hairs on leaves and bracts.**

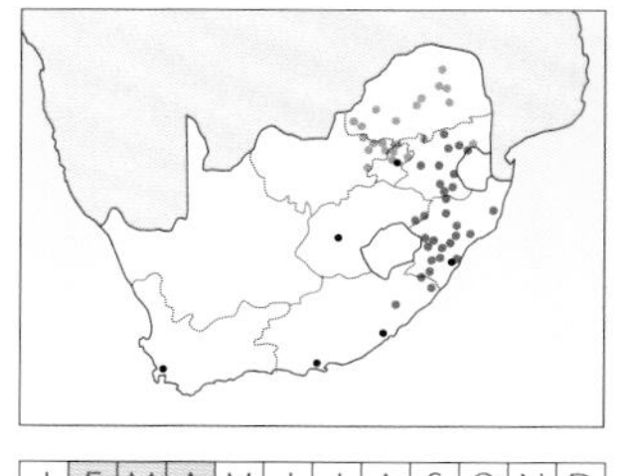

J F M A M J J A S O N D

*Gladiolus sericeovillosus* is widespread in summer-rainfall areas. Small cream, greenish, pink, lilac or dull red flowers grow on erect spikes, appearing from February to April. There are two subspecies.

**DESCRIPTION** **Plant** 35–100cm. **Corm** depressed-globose, tunics papery or leathery, decaying to vertical strips. **Cataphylls** up to 16cm above ground, green with hairs. **Leaves** 5–7, linear, 3–16mm wide; margins, midrib and some secondary veins thickened and translucent; surface is hairy or glabrous; long leaf sheaths form a pseudostem. **Spike** straight and erect, may be branched; 12–20 or more distichous flowers. **Bracts** pale green, hairy or smooth, having an inflated appearance. **Flowers** greenish to cream, pink, or pale lilac to dull red, sometimes dotted with red or maroon spots; lower tepals often with red markings and a yellow centre. Perianth tube 10–16mm. **Anthers** yellow. **Pollen** whitish. **Capsules** oblong with 'ears' on the top. **Seeds** ovate, broadly and evenly winged. **Scent** unscented.

**SUBSPECIES** *G. sericeovillosus* has two subspecies.
***G. sericeovillosus* subsp. *sericeovillosus*** (green on map) Cataphylls, leaves, stems and bracts are pubescent.
***G. sericeovillosus* subsp. *calvatus*** (orange on map) ***calvatus* = bald.** It is distinguished by its hairless bracts and leaves, which give the subspecies its name.

**DISTRIBUTION** This species is found throughout the summer-rainfall areas of South Africa. Subsp. *sericeovillosus* is found in the northern Eastern Cape, KwaZulu-Natal and southern Mpumalanga; it is most common in the KwaZulu-Natal interior. The distribution of subsp. *calvatus* ranges from Mpumalanga, through Limpopo, and into eastern Zimbabwe.

**ECOLOGY & NOTES** Generally found in full sun in open grassland or light woodland.

**POLLINATORS** Anthophorid bees.

**SIMILAR SPECIES** *G. sericeovillosus* is closely related to and can be mistaken for *G. oppositiflorus* and *G. elliotii*.

| | Pseudostem | Leaf venation | Bracts & leaves |
|---|---|---|---|
| ***G. sericeovillosus* subsp. *sericeovillosus*** | yes | midrib, margins and secondary veins thickened | hairy |
| ***G. sericeovillosus* subsp. *calvatus*** | yes | midrib, margins and veins thickened | glabrous |
| ***G. oppositiflorus*** | no | midrib and margins strongly thickened | hairy |
| ***G. elliotii*** | no | margins thickened, midrib and veins not thickened | glabrous, sometimes hairy |

**Top:** *Subsp.* sericeovillosus, *northern KwaZulu-Natal.*
**Above:** *Subsp.* calvatus *near Zeerust, North West.*

*Hairy leaves with thickened veins; subsp.* sericeovillosus.

*A small pink form of subsp.* sericeovillosus.

**Top:** *Hairy bracts on subsp.* sericeovillosus.
**Above:** *Smooth bracts distinguish subsp.* calvatus.

*Subsp.* sericeovillosus, *here growing in northern KwaZulu-Natal.*

*Subsp.* calvatus, *encountered near Zeerust, North West.*

# *Gladiolus elliotii*

**Named after British biologist and plant collector George Scott Elliot (1862–1934), who published on the pollination of South African flora in the late 1800s.**

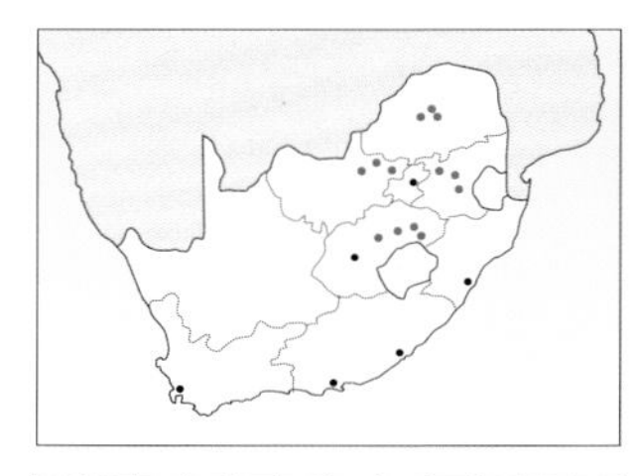

J F M A M J J A S O N D

The range of *Gladiolus elliotii* extends through the summer-rainfall areas of the Highveld. Blueish, grey or cream-coloured flowers appear on distichous spikes between November and February. Can be confused with *G. sericeovillosus*.

**DESCRIPTION** **Plant** 40–60cm. **Corm** depressed-globose, with papery or leathery tunics fragmenting over time. **Cataphylls** up to 6cm above ground, green flushed with purple. **Leaves** 4–6, 14–20mm wide, lanceolate, arranged in a distichous fan; margins strongly thickened and hyaline, midrib and other veins fine. **Spike** erect, occasionally branched, with 14–20 distichous flowers. **Bracts** up to 37mm long, pale green and slightly inflated. Folded along the middle, with the tip folded backwards; lower margins fused around the ovary. **Flowers** blueish grey to cream, with blue to red speckling, sometimes concentrated so the flower appears to have stripes. Perianth tube ±16mm. **Anthers** mauve. **Pollen** cream. **Capsules** oblong with ear-like lobes. **Seeds** reddish brown and almost round, broadly winged. **Scent** unscented.

*This species is widespread on the Highveld.*

**DISTRIBUTION** Found on the Highveld and in the northern Bushveld, from the eastern Free State to the Waterberg in Limpopo, and westwards into Botswana.

**ECOLOGY & NOTES** Grows in moist grassland in full sun, or sometimes in lightly wooded areas. Although it has a wide distribution, it is no longer commonly seen owing to habitat destruction by agricultural practices. All the populations we found were in rail or road reserves, or on the edges of ploughed fields.

**POLLINATORS** Unknown; thought to be long-tongued anthophorid bees.

**SIMILAR SPECIES** *G. elliotii* is closely related to and can be mistaken for *G. oppositiflorus* and *G. sericeovillosus*.

| | Pseudostem | Leaf venation | Bracts & leaves |
|---|---|---|---|
| ***G. elliotii*** | no | margins thickened, midrib and veins not thickened | glabrous, sometimes hairy |
| ***G. oppositiflorus*** | no | midrib and margins strongly thickened | hairy |
| ***G. sericeovillosus*** | yes | midrib, margins and secondary veins thickened | hairy or glabrous |

*Leaves form a fan.*

*Spikes bear up to 20 alternating flowers.*

*Inner bracts are fused around the ovary.*

*Dense speckling may present as a stripe.*

G. elliotii *in Golden Gate Highlands National Park, Free State.*

# Gladiolus ecklonii

**Named after Danish botanist and pharmacist Christian Friedrich Ecklon (1795–1868), who sent seeds of this species to the Hamburg Botanical Garden early in the 19th century. The species was described from plants grown from this seed.**

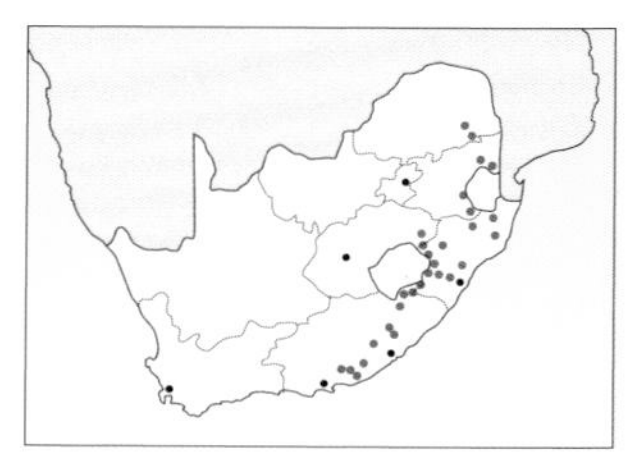

J F M A M J J A S O N D

*Gladiolus ecklonii* is widespread in summer-rainfall areas. Flowers are generally densely dotted with pink, red or purple, appearing from late December to early March. The species can be confused with *G. vinosomaculatus*.

*Flowers are usually densely dotted with cream-yellow nectar guides.*

**DESCRIPTION** **Plant** 15–35cm. **Corm** depressed-globose, with papery tunics fragmenting over time. **Cataphylls** ±3cm above ground, green. **Leaves** usually 7–9, with lower 4 in a two-ranked fan; margins thickened, midrib only slightly thickened. **Spike** inclined, unbranched, usually with 8–12 flowers. **Bracts** outer bract (±48mm) is longer than the perianth tube, green with strong reddish keels. **Flowers** white background covered in pink, red or purple spots; lower 3 tepals with cream to yellow nectar guides; heavy spotting sometimes produces the effect of a uniform colour. Perianth tube ±20mm. Flowers close at night. **Anthers** cream. **Pollen** white. **Capsules** obovoid with the top sunken. **Seeds** broadly winged. **Scent** unscented.

**DISTRIBUTION** Widespread, extending from the Amatola Mountains in the Eastern Cape, along the escarpment, through the Drakensberg to Haenertsburg in Limpopo.

**ECOLOGY & NOTES** Grows in well-watered, shortish grassland and sometimes on stony hillsides, in sun or partial shade. Despite their small size, the flowers are very showy. The most beautiful colour form is pink to red, from the Steenkampsberg in Mpumalanga, where they grow among rocks on a steep hillside together with the Critically Endangered *Protea roupelliae*.

**POLLINATORS** Long-tongued bees such as *Amegilla capensis* visit the flowers for nectar.

**SIMILAR SPECIES** Can be confused with *G. vinosomaculatus* from which it is distinguished by height, leaf shape and bract length.

| | Leaves | Bracts & flowers |
|---|---|---|
| ***G. ecklonii*** | lanceolate, 15–30mm wide, bright green, margins thickened, midrib slightly thickened, no pseudostem | 45–60mm, 2–3 internodes long, flowers speckled |
| ***G. vinosomaculatus*** | sword shaped, 4–12mm wide, whitish waxy bloom, margins & midribs thickened, with pseudostem | 40–80mm or longer, 4–5 internodes long, flowers speckled |

*Lanceolate leaves form a fan.*

*The inclined spike can have up to 12 flowers.*

*Deep purple spots make the dorsal tepal appear grey.*

*Dense speckling makes these flowers appear red.*

*Some flowers appear pink.*

*Note the long, pale green bract.*

*Flowering in grasslands, Steenkampsberg, Mpumalanga.*

# *Gladiolus vinosomaculatus*

***vinosomaculatus* = wine-spotted, referring to the tepals.**

*Gladiolus vinosomaculatus* is distributed in a swathe north of the Vaal River. Its unscented flowers are densely speckled with red, pink or purple, and appear from January to March. It is almost twice the height of *G. ecklonii* (with which it can be confused) and its leaves are a different shape.

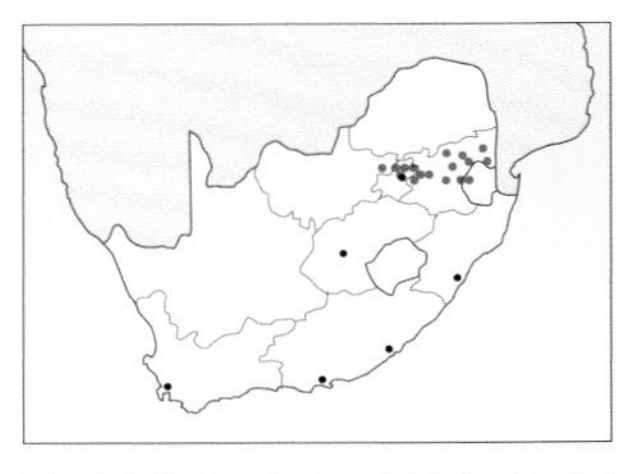

J F M A M J J A S O N D

**DESCRIPTION Plant** 65cm. **Corm** depressed-globose, with papery tunics fragmenting over time. **Cataphylls** dark above the ground, flushed with purple or brown, pale below. **Leaves** 5–8 forming a pseudostem, 6–12mm wide, distichous; margins and midrib moderately thickened. **Spike** strongly flexed at the base, may be branched, 9–14 flowers. **Bracts** 40–80mm long, grey-green, keeled. **Flowers** white with red to purple speckles; partly concealed by the bracts. Perianth 15–18mm. **Anthers** cream. **Pollen** yellow. **Capsules** oblong, 20–30mm long, concealed in the bracts. **Seeds** reddish brown, broadly winged. **Scent** unscented.

*Wine-coloured markings give* G. vinosomaculatus *its name.*

**DISTRIBUTION** Occurs north of the Vaal River, from west of Pretoria, across the Magaliesberg, to east of Barberton, Mpumalanga.

**ECOLOGY & NOTES** Grows in rocky sites in grassland, in exposed and often dry sites. We have found plants growing in grassland north of eMalahleni and in the hills close to Mbombela.

**POLLINATORS** Probably large anthophorid bees such as *Amegilla capensis*.

**SIMILAR SPECIES** *G. vinosomaculatus* can be confused with *G. ecklonii*.

| | Leaves | Flowers |
|---|---|---|
| ***G. vinosomaculatus*** | leaf sheaths form pseudostem, leaves 4–12mm wide, strongly thickened margins and midrib | concealed by bracts, bracts keeled, white flowers with red to purple speckles |
| ***G. ecklonii*** | no pseudostem, leaves 15–30mm wide, moderately thickened margins and midribs | bracts shorter than above and keeled, white flowers densely spotted, sometimes producing a uniform pink colour |

*Leaves form a pseudostem.*

*Bracts are long and keeled.*

*Flowers may be bee pollinated.*

*Long bracts partly conceal the flowers.*

*Tepals are lanceolate.*

G. vinosomaculatus, *here growing near Dennilton, Limpopo.*

# *Gladiolus rehmannii*

**Named after Anton Rehmann, a Polish botanist and explorer who first recorded the plant for science circa 1879.**

*Gladiolus rehmannii* grows in the northern parts of the Bushveld, from Mpumalanga into Botswana. Its large, white to lilac flowers are unscented. It flowers from January to March.

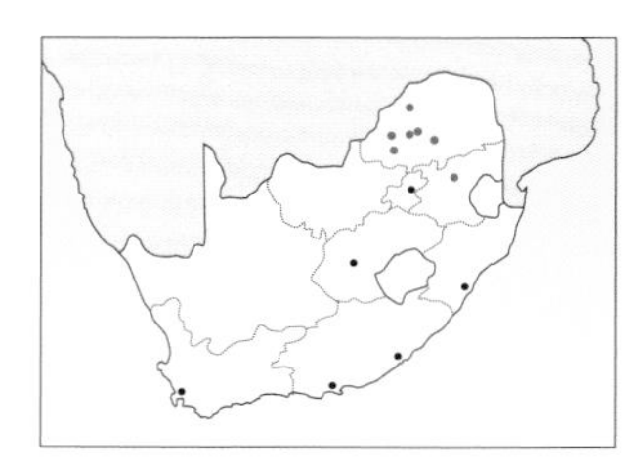

J F M A M J J A S O N D

**DESCRIPTION** **Plant** 30–50cm. **Corm** globose, tunics fragmenting with age. **Cataphylls** up to 18cm above ground, green or dull purple. **Leaves** ±8, 8–16mm wide in a lax, distichous fan; long sheaths form a pseudostem; midrib, not margins, moderately thickened. **Spike** inclined, occasionally branched, 6–10 flowers. **Bracts** 3–4 internodes long, green or purplish, without keels. **Flowers** white or pale lilac; with yellow nectar guides on lower tepals. Perianth tube 15–20mm, enclosed in the bracts. **Anthers** mauve. **Pollen** lilac to blue or yellow. **Capsules** oblong, up to 35mm. **Seeds** dark body, broadly winged with translucent edges. **Scent** unscented.

**DISTRIBUTION** Occurs from Groblersdal northwards through Limpopo and westwards through North West to eastern Botswana. The specimens that we saw were near Modimolle in the Waterberg.

**ECOLOGY & NOTES** Grows in sandy soils, often in red Kalahari sands. Plants are found in full sun to light shade. We have found specimens only twice; in both instances they were growing on the verge of the road. Both populations were associated with *Burkea africana*, *Dombeya rotundifolia*, *Faurea saligna*, *Sclerocarya birrea* subsp. *caffra* and *Senegalia caffra*. Most of the distribution range is grazed, either by cattle or game; we suspect that many plants are eaten either before they can flower or when in flower.

**POLLINATORS** Thought to be the same anthophorid bees that pollinate most summer-rainfall *Gladiolus* species.

**SIMILAR SPECIES** Unlikely to be mistaken for any other species as *Gladiolus* plants are not common in dry bushveld and this is the only species with large, white flowers in midsummer.

*Flowers have yellow nectar guides on lower tepals.*

*Leaves form a soft fan.*

*Bracts may be folded in the midline.*

*Perianth tubes are enclosed in bracts.*

G. rehmannii, *near Mookgophong, Limpopo.*

# Gladiolus antholyzoides

***antholyzoides* = resembling the genus *Antholyza*, especially the long perianth and wide open upper part of the flower.**

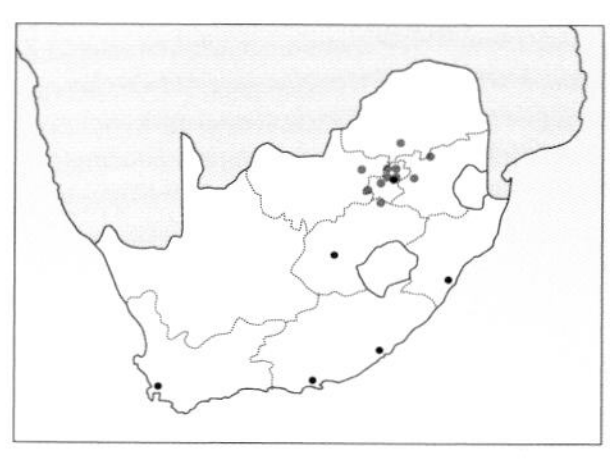

| J | F | M | A | M | J | J | A | S | O | N | D |
|---|---|---|---|---|---|---|---|---|---|---|---|

*Gladiolus antholyzoides* is a summer-rainfall species growing on the central Highveld. The yellow flowers appear from November to December, sometimes earlier. It is often confused with *G. dalenii* whose flowers are bigger with wider tepals, and with *G. aurantiacus* whose flowers are orange.

**DESCRIPTION** **Plant** 60–90cm. **Corm** depressed-globose with cormlets, tunics fragmenting with age. **Cataphylls** 2–4cm above ground, green. **Leaves** 5–8, 10–20mm wide, in a tight fan; margins and midribs lightly raised. Leafy shoots are produced from separate buds on the corm during flowering. **Spike** erect, unbranched, 8–12 secund flowers. **Bracts** ±30mm, pale green and shiny. **Flowers** yellow, sometimes streaked with orange or red, with orange ring around throat. Perianth tube long, up to 40mm. **Anthers** pale yellow. **Pollen** cream. **Capsules** ovoid-oblong. **Seeds** 8mm, ovate and broadly winged. **Scent** unscented.

*The flowers and anthers are pale yellow.*

**DISTRIBUTION** On the central Highveld, from the eastern Free State, through Gauteng, to Modimolle in Limpopo.

**ECOLOGY & NOTES** Flowering has been recorded in November and December before much rain has fallen; they are often found in areas that are damp all year round. Despite searching for several years, the only populations we found were at Sasolburg.

**POLLINATORS** Bird pollination is suggested based on the wide perianth tube and nectar production, although the colour is unusual for bird pollinators. While we were photographing the plants in Sasolburg, the flowers were visited by bees and flies.

**SIMILAR SPECIES** Vegetatively similar to its close relatives *G. aurantiacus* and *G. dalenii* whose distributions overlap. Flower structure and leaves are important in differentation.

| | Leaves | Flower size | Flower structure |
|---|---|---|---|
| ***G. antholyzoides*** | fan of leaves relatively short, reaching one-third up stem | flowers 55–70mm long, perianth tube 28–40mm, tepals lanceolate | upper 3 tepals larger than lower 3, dorsal tepal horizontal but not arching |
| ***G. aurantiacus*** | leaves on flowering stem, sheathing with short blades | perianth tube up to 70mm, slender for 30mm and then abruptly widening | dorsal tepal is nearly horizontal, lower tepals much shorter than upper |
| ***G. dalenii*** | fan of longer leaves reaching halfway up stem | flowers 65–100mm long, perianth tube 35–50mm, tepals broad | dorsal tepal broadly ovate, strongly arched and concealing anthers |

*Leaf midribs are raised.*

*Long perianths are cupped by green bracts.*

*The flower shape and nectar suggest a variety of pollinators.*

*A Black-banded bee (*Amegilla* species) visits the flower.*

*Erect spikes stand high above foliage.*

G. antholyzoides, *near Sasolberg, Free State.*

# *Gladiolus aurantiacus*

***aurantiacus*** **= orange-coloured.**

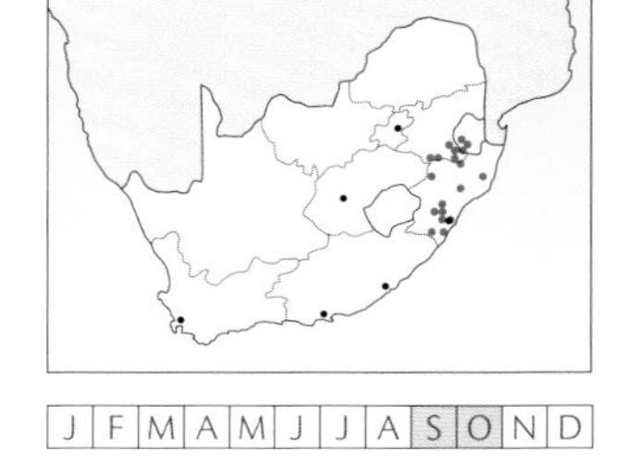


*Gladiolus aurantiacus* is centred in KwaZulu-Natal. Flowers are orange with a long perianth tube, appearing before the rainy season, in September and October. Often mistaken for *G. dalenii*, it can be differentiated by the shape of the perianth and tepals, and the foliage leaves.

**DESCRIPTION** **Plant** 45–75cm. **Corm** depressed-globose with cormlets, tunics fragmenting with age. **Cataphylls** up to 8cm above ground, firm textured. **Leaves** 4 or 5 on flowering stem; sheathing leaves overlap each other with short blades; midrib thickened and raised. **Spike** erect, unbranched, with 10–16 flowers. **Bracts** pale green, shiny. **Flowers** deep yellow, dotted and streaked with red or orange; lower tepals mostly not spotted, upper 3 curving forwards and much larger than lower 3. Perianth tube up to 70mm. **Anthers** yellow to orange. **Pollen** yellow. **Capsules** ovoid-oblong, ±20mm. **Seeds** broadly and evenly winged. **Scent** unscented.

**DISTRIBUTION** Concentrated in KwaZulu-Natal, moving northwards along the coast from Richmond into southern Eswatini and Mpumalanga. Plants used to be seen in large numbers in the areas around Vryheid and eMkhondo, but mining and forestry have decreased their presence; now they are rarely seen.

**ECOLOGY & NOTES** Found in grassland, often in low-lying damp areas, as well as on hillsides. Flowers early in summer, often before the rains. At the time of flowering, the plants have no basal leaves; these appear later in the season from the same corm but on a separate stem, and are fully developed only around March.

**POLLINATORS** Sunbirds are attracted to flowers of this colour. The flowers produce large quantities of nectar with a low sugar concentration.

**SIMILAR SPECIES** The closest relatives are *G. antholyzoides* and *G. dalenii*, whose distributions overlap. They are distinguished by flowering time, shape and leaf form.

| | Leaves | Flower size | Flower structure |
|---|---|---|---|
| ***G. aurantiacus*** | leaves on flowering stem, sheathing with short blades | perianth tube up to 70mm, slender for 30mm and then abruptly widening | dorsal tepal is nearly horizontal, lower tepals much shorter than upper |
| ***G. antholyzoides*** | fan of leaves relatively short, reaching one-third up stem | flowers 55–70mm long, perianth tube 28–40mm, tepals lanceolate | upper 3 tepals larger than lower 3, dorsal tepal horizontal but not arching |
| ***G. dalenii*** | fan of longer leaves reaching halfway up stem | flowers 65–100mm long, perianth tube 35–50mm, tepals broad | dorsal tepal broadly ovate, strongly arched and concealing anthers |

*Leaf midribs are raised.*

*Deep yellow flowers have long perianth tubes.*

*Erect spikes may bear up to 16 flowers.*

*Upper tepals are larger than the lower three.*

*The lower tepals of these flowers have strong yellow markings.*

G. aurantiacus, *in flower near Amsterdam, Mpumalanga.*

# *Gladiolus dalenii*

**Named after Dutch botanist Cornelius Dalen (1766–1852), who introduced the species to Europe.**

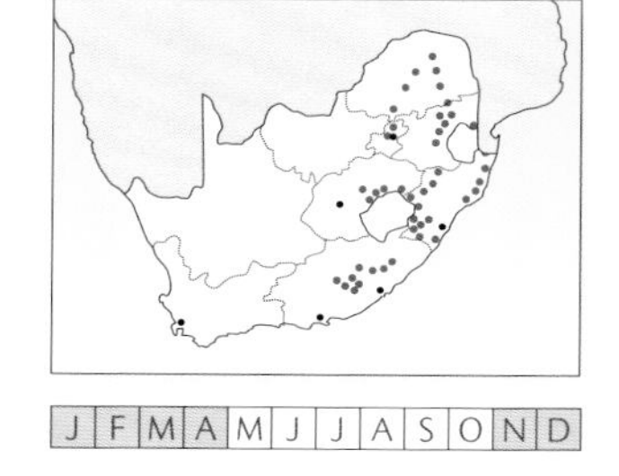

J F M A M J J A S O N D

The most common and widespread of gladioli, *Gladiolus dalenii* is absent only from winter-rainfall areas. Its flowers are red to orange with yellow markings and appear between November and April. Along the KwaZulu-Natal coast some flower between September and October, before the rains.

**DESCRIPTION** **Plant** 70–150cm. **Corm** globose with cormlets and leathery tunics; sometimes produces stolons with cormlets. **Cataphylls** up to 15cm, green or flushed purple. **Leaves** 4–6, firm, 10–30mm wide; margins and midrib moderately raised. **Spike** erect and unbranched, 3–8 flowers. **Bracts** 35–50mm on outside with shorter inner bracts, pale green to greyish purple, sometimes dry. **Flowers** red to orange; with lower 3 tepals marked either with yellow, or yellow or green with darker streaks; upper tepals broadly ovate, dorsal tepal strongly arched and almost spoon shaped. Perianth tube 35–50mm. **Anthers** yellow to brown. **Pollen** pale yellow. **Capsules** oblong, 30mm. **Seeds** light brown, unevenly winged. **Scent** unscented.

**DISTRIBUTION** From the Eastern Cape northwards through the eastern part of the Free State, Gauteng, Mpumalanga and eastern Limpopo. The species is widespread across sub-Saharan Africa.

**ECOLOGY & NOTES** Occurs in grassland, savanna and woodlands, usually in moderately moist areas. It is one of the most commonly seen *Gladiolus* species because it is widespread and tall, and has bright flowers. The colour is uniform within a population, but highly variable within the species. In the Eastern Cape we have only seen bright red flowers. However, in Mpumalanga we have seen red, orange and yellowish-green-flowered populations. In northern KwaZulu-Natal there is a large population with green flowers streaked with purplish black. We have seen a pure yellow form at Victoria Falls in Zimbabwe. Almost all the *Gladiolus* cultivars used as cut flowers today contain some *G. dalenii* genes. The corms are used medicinally in many places on the continent and as a food source in central Africa.

**POLLINATORS** Sunbirds forage for nectar; they seem to visit even if flowers are not red or orange.

**SIMILAR SPECIES** Differs from *G. antholyzoides* and *G. aurantiacus* in its distinctively shaped flowers.

| | Leaves | Flower size | Flower structure |
|---|---|---|---|
| ***G. dalenii*** | fan of longer leaves reaching halfway up stem | flowers 65–100mm long, perianth tube 35–50mm, tepals broad | dorsal tepal broadly ovate, strongly arched and concealing anthers |
| ***G. antholyzoides*** | fan of leaves relatively short, reaching one-third up stem | flowers 55–70mm long, perianth tube 28–40mm, tepals lanceolate | upper 3 tepals larger than lower 3, dorsal tepal horizontal but not arching |
| ***G. aurantiacus*** | leaves on flowering stem, sheathing with short blades | perianth tube up to 70mm, slender for 30mm and then abruptly widening | dorsal tepal is nearly horizontal, lower tepals much shorter than upper |

*Broad leaves are sword shaped.*

*Dorsal tepals are spoon shaped.*

*Spikes extend beyond leaves.*

*Loteni, KwaZulu-Natal.*

*Tonquani, Magaliesberg.*

*Near Ntendeka, KwaZulu-Natal.*

*Wolkberg, Limpopo.*

*Steenkampsberg, Mpumalanga.*

*Near Harrismith, Free State.*

# *Gladiolus flanaganii*

**Named after Henry Flanagan (1861–1919), a naturalist and collector from the Eastern Cape, who collected the type specimen from the northern Drakensberg in 1894.**

*Gladiolus flanaganii* is a species that grows on the high Drakensberg escarpment. Its deep red flowers appear in November and December.

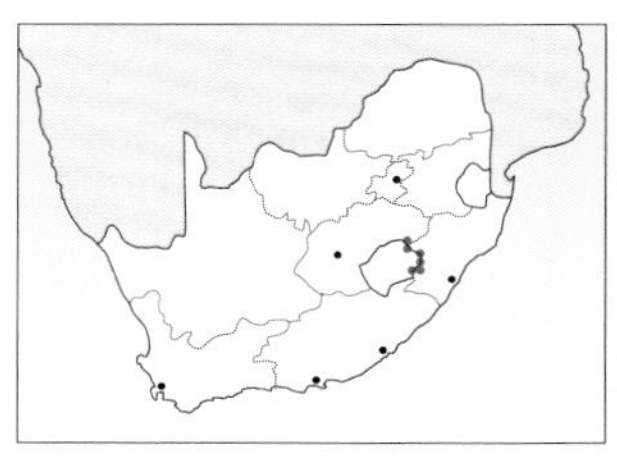

J F M A M J J A S O N D

**DESCRIPTION** **Plant** 35–60cm, erect or inclined. **Corm** globose, with cartilaginous layers fragmenting into vertical fibres. **Cataphylls** up to 4cm above ground, pinkish. **Leaves** ±5, 7–14mm wide; midrib slightly thickened and translucent, margins not thickened. **Spike** usually inclined and unbranched, with 4–7 flowers. **Bracts** 60–70mm, firm, green or purplish. **Flowers** bright red and urn shaped with a firm texture; lower tepals each with white streak in midline. Perianth tube 35–45mm. **Anthers** dark purple. **Pollen** pale lilac. **Capsules** oblong, 7mm. **Seeds** with pointed wings at each end. **Scent** unscented.

*Flowers are firm with white-streaked lower tepals.*

**DISTRIBUTION** Found only in the high Drakensberg from near Sani Pass in the south to Mont-aux-Sources in the north.

**ECOLOGY & NOTES** Most plants have been found hanging from damp basalt cliffs, with corms wedged into rock cracks. Sometimes the plants grow in semi-shade on east-facing cliffs. The common name 'Suicide gladiolus' is very apt.

**POLLINATORS** Flowers produce a large amount of nectar, suggesting that they are pollinated by sunbirds, particularly Malachite sunbirds. Birds have to perch on the stem above the flowers and reach down into them on account of the hanging growth habit of the plant.

**SIMILAR SPECIES** Once considered to be the same species as *G. cruentus*, the only commonality is their colour and *G. flanaganii* is unlikely to be confused with any other species.

| | Habitat | Leaves & flower shape | Pollinator |
|---|---|---|---|
| ***G. flanaganii*** | basalt cliffs at high altitude on the Drakensberg escarpment | firm leaves, cup-shaped flowers with narrow, longitudinal white streak | sunbirds |
| ***G. cruentus*** | sandstone cliffs, coastal area | soft leaves, open flowers with broad white transverse band on lower tepals | *Aeropetes tulbaghia* |

*Leaves have a thickened midrib, but margins are not thickened.*

*Stems are inclined, suggesting birds may perch to reach the nectar.*

*Plants grow high up on steep basalt cliffs, hence the common name 'Suicide gladiolus'.*

*Note the wide leaves and short, thick stems.*

*Corms wedge into cracks, securing the plants in rocky cliff faces.*

# Gladiolus saundersii

**Named after English horticulturalist William Wilson Saunders (1809–1879); the type specimen was collected in the Eastern Cape in 1861 by Thomas Cooper, a plant collector in his employ.**

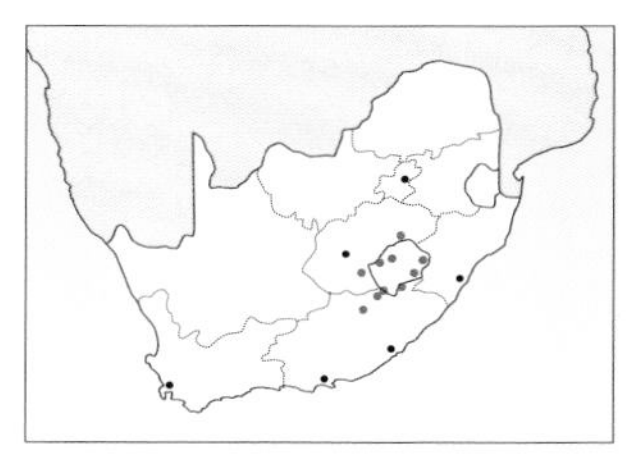

J F M A M J J A S O N D

*Gladiolus saundersii* grows in the Eastern Cape and central and southern regions of the Drakensberg in the Free State and Lesotho. Its bright red flowers appear between January and March.

**DESCRIPTION** **Plant** 40–60cm. **Corm** depressed-globose, with cormlets and papery tunics. **Cataphylls** 3–4cm above ground, orange to brown. **Leaves** 7–9, in basal fan, ±2cm wide; midrib and secondary veins are thickened. **Spike** erect, unbranched, with 4–9 flowers. **Bracts** 6cm, green, lightly folded in the midline. **Flowers** large, bright red, facing sidewards or drooping; lower 3 tepals are speckled red on a white background, with upper dorsal tepal horizontal and the end recurved. Perianth tube 33–37mm. **Anthers** purple. **Pollen** cream. **Capsules** ±30mm. **Seeds** broadly winged. **Scent** unscented.

*Flowers are bright red with distinctive spotty lower tepals.*

**DISTRIBUTION** Occurs from Joubert's Pass in the Eastern Cape, through Lesotho, to the northeastern Free State near Clarens. Seen frequently on Naude's Nek Pass and in the area around Rhodes.

*Natural hybrid with parent species* G. saundersii *and* G. oppositiflorus.

**ECOLOGY & NOTES** Usually found growing in dryish patches of seasonally wet sites, such as on the warmer western side of Naude's Nek; it is absent from the wetter, cooler eastern side. *G. saundersii* favours rocky places and is often on hillsides in exposed spots. This is one of only two species we have seen that form natural hybrids. On Naude's Nek we have seen two populations that are clearly hybrids between *G. saundersii* and *G. oppositiflorus*. Both parent species occur on this mountain pass, with *G. oppositiflorus* usually on the cooler, wetter, east-facing side of the pass and *G. saundersii* on the west-facing slopes. The hybrids have large flowers, similarly shaped to *G. saundersii*, but with the colour of *G. oppositiflorus*. Although capsules were found on these hybrids, the few seeds inside were found to be sterile. (See page 15 for more information about natural hybrids.)

**POLLINATORS** *Aeropetes tulbaghia*, the Mountain pride butterfly. This long-tongued butterfly is strongly attracted to the colour red and is the main pollinator for a number of late-flowering species including this one and *G. cardinalis*, and species from other genera such as *Brunsvigia marginata, B. radulosa, Crassula coccinea, Disa ferruginea, D. uniflora*, various *Kniphofia* species and *Watsonia schlechteri*. The butterfly occurs in high areas from Cape Town in the west, along mountain chains to the Eastern Cape, and then northwards along the Drakensberg. It pollinates summer- and winter-growing plants in a variety of montane habitats.

**SIMILAR SPECIES** The large red flowers with their distinctively spotty lower tepals and flower shape make *G. saundersii* unmistakable.

Leaves form a basal fan.

Filaments are long.

The large flowers have strongly recurved tepals.

Spikes of natural hybrid.

Natural hybrid, Naude's Nek, Eastern Cape.

Flowering on Naude's Nek Pass, Eastern Cape.

# Gladiolus cruentus

***cruentus* = blood-stained, referring to the vivid red colour.**

*Gladiolus cruentus* is restricted to damp cliffs in the coastal regions of central KwaZulu-Natal. Its blood-red flowers appear from mid-January to early March and are pollinated by butterflies.

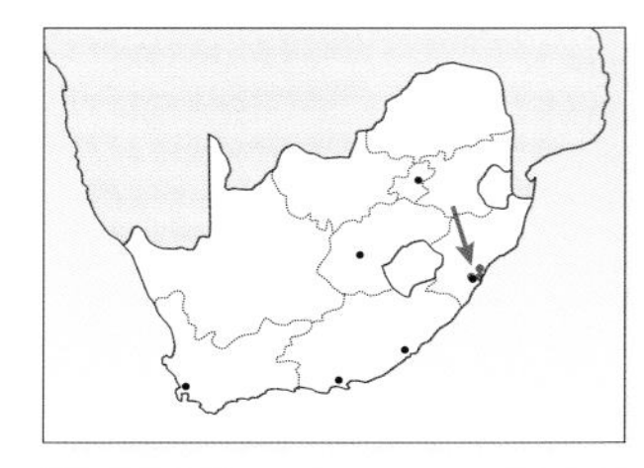

J F M A M J J A S O N D

**STATUS** Critically Endangered

**DESCRIPTION** **Plant** 50–70cm arching horizontally from cliffs. **Corm** globose with vertical fibres. **Cataphylls** up to 10cm above ground, becoming purple. **Leaves** 7 or 8, 15mm wide and drooping; with thickened midrib. **Spike** inclined and drooping, unbranched, 4–9 flowers. **Bracts** up to 60mm, green, sometimes keeled. **Flowers** scarlet; lower tepals marked with white. Perianth tube ±28mm; a very open flower. **Anthers** dark purple. **Pollen** whitish. **Capsules** oblong with a rounded apex. **Seeds** unevenly winged. **Scent** unscented.

*The species has open flowers.*

**DISTRIBUTION** Known only from coastal parts of central KwaZulu-Natal in the area around Durban.

**ECOLOGY & NOTES** Found on damp sandstone cliffs near waterfalls during the rainy summer season. Plants are extremely hard to reach in most of the areas where they occur, which makes photographing them very frustrating! Large populations grow on ledges in Krantzkloof Nature Reserve in Kloof, Durban. The seeds seem to fall from one ledge to the next and subsequently germinate in the rock cracks.

*Bracts may be keeled.*

**POLLINATORS** Thought to be *Aeropetes tulbaghia*, the Mountain pride butterfly, which is the most common pollinator of bright red open flowers with long slender tubes. See page 102 for more information on this butterfly.

**SIMILAR SPECIES** Has been confused with *G. flanaganii*; previously thought to be the same species. However, the only common feature is the colour of the flowers.

| | Habitat | Leaves & flower shape | Pollinator |
|---|---|---|---|
| ***G. cruentus*** | sandstone cliffs, coastal area | soft leaves, open flowers with broad white transverse band on lower tepals | *Aeropetes tulbaghia* |
| ***G. flanaganii*** | basalt cliffs at high altitude on the Drakensberg escarpment | firm leaves, cup-shaped flowers with narrow, longitudinal white streak | sunbirds |

*Drooping leaves have thickened midribs.*

*Inclined spikes bear up to nine flowers.*

*Seeds fall onto ledges and germinate in cracks.*

G. cruentus *flowers on damp sandstone cliffs in KwaZulu-Natal during the rainy season.*

*Rod Saunders records* G. aquamontanus, *Swartberg, Western Cape.*

# Gladiolus oreocharis

***oreocharis* = mountain grace, referring to its habitat and graceful appearance.**

*Gladiolus oreocharis* is restricted to the higher ranges of the winter-rainfall region. Its pink flowers with spear-shaped nectar guides appear mainly in December. It flowers prolifically after fire.

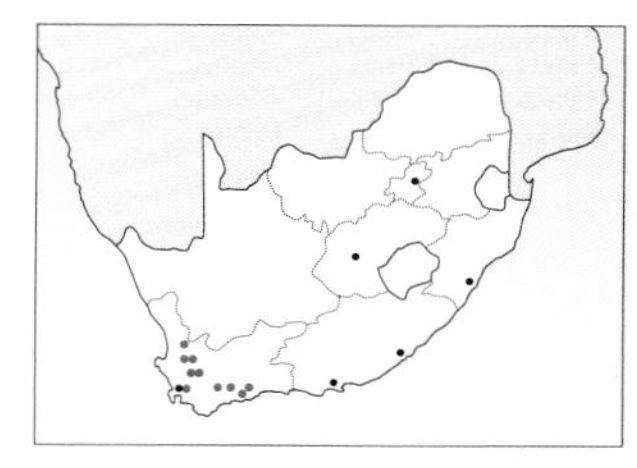

| J | F | M | A | M | J | J | A | S | O | N | D |
|---|---|---|---|---|---|---|---|---|---|---|---|

**STATUS** Rare

**DESCRIPTION** **Plant** 18–35cm. **Corm** globose, with fine, net-like tunics. **Cataphylls** uppermost just above ground, green to purple or dry, sometimes hairy. **Leaves** 3–5, linear, 2–6mm wide; margins and 1 or 2 main veins slightly thickened; sometimes leaves are so grooved as to appear pleated. **Bracts** green and firm textured, short and truncate, purple towards the apex. **Spike** inclined, with 2–7 flowers. **Flowers** pink to reddish purple to violet; lower 3 tepals with dark pink to purplish spear-shaped mark with white centre. Perianth tube 20mm. **Anthers** mauve. **Pollen** cream. **Capsules** obovoid. **Seeds** unknown. **Scent** unscented.

*Each stem bears between two and seven flowers.*

**DISTRIBUTION** Found in the mountainous regions of the southwestern Cape, extending from the Cederberg through the Hex River Mountains to the Langeberg and Klein Swartberg near Ladismith.

**ECOLOGY & NOTES** Not often seen as it flowers at the hottest time of year, mainly at altitudes of ±1,600m on the moist south-facing slopes. Almost all records show it flowering in the first year after a fire, although we found it on Matroosberg in old, unburned fynbos.

**POLLINATORS** Probably long-tongued bees, which are attracted to similarly shaped pink flowers.

**SIMILAR SPECIES** May be confused with *G. crispulatus* and *G. carneus*.

| | Flowers | Leaves |
|---|---|---|
| ***G. oreocharis*** | small, usually pink to purple with spear-shaped mark | 2–6mm wide, plane, venation visible on both sides of leaf |
| ***G. crispulatus*** | large, pink, tepals crisped along edges, triangular streak | 1–3mm, midribs thickened |
| ***G. carneus*** | large, white to pale or deep pink with spade-shaped marks, sometimes spotted or streaked | 6–14mm, usually forming a fan |

*Leaves may be closely pleated.*

*Truncated bracts are unusual in the genus.*

*The dorsal tepal is longer than the laterals.*

*The plant flowers well after fire.*

*Note the pale anthers.*

*A dark pink form without a white central marking.*

*Standard pink form.*

G. oreocharis *grows above 1,000m in the Western Cape mountains, seen here at Matroosberg.*

# Gladiolus crispulatus

***crispulatus* = tightly curled edges, referring to the crisped tepals.**

*Gladiolus crispulatus* occurs in the Langeberg Mountains of the Western Cape, a winter-rainfall area. It flowers well after fires. Deep pink, unscented flowers appear in November and December. It is sometimes mistaken for *G. oreocharis* or *G. carneus*.

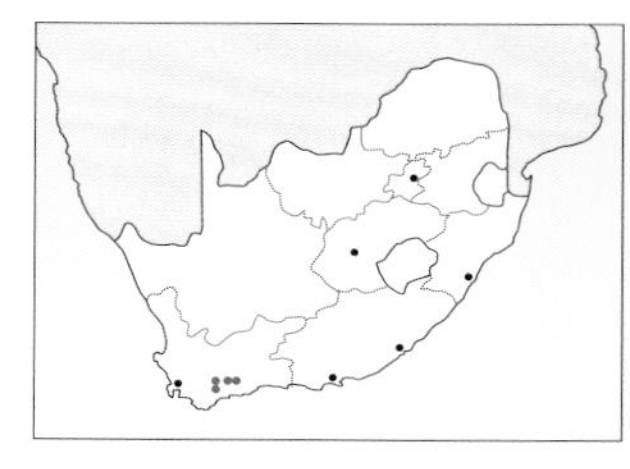

| J | F | M | A | M | J | J | A | S | O | N | D |
|---|---|---|---|---|---|---|---|---|---|---|---|

**STATUS** Rare

**DESCRIPTION** **Plant** 30–40cm. **Corm** globose, with fine, net-like tunics. **Cataphylls** up to 3cm above ground and dark green. **Leaves** 4 or 5 above one another, very narrow (up to 3mm wide); margin and midribs lightly thickened with a paired midrib on one leaf surface and single on the other. **Spike** slightly inclined, unbranched, with 2–4 flowers. **Bracts** up to 30mm long, pale green, soft textured. **Flowers** deep pink; lower 3 tepals marked with red streaks and speckles, tepal margins undulate to crisped. Perianth tube 26mm. **Anthers** lilac. **Pollen** white. **Capsules** truncate. **Seeds** evenly winged. **Scent** unscented.

G. crispulatus *flowers are short and brightly coloured.*

**DISTRIBUTION** Known only from the Langeberg, from Swellendam to Riversdale in the Western Cape.

**ECOLOGY & NOTES** Plants grow on south-facing slopes at altitudes of between 300 and 900m; they only seem to flower for one or two years after a fire. We found the plants flowering en masse early in December on the slopes above Swellendam in the year after a fire but did not find them again despite many visits to the same area.

*Midribs are different on either side of the leaf.*

**POLLINATORS** Unknown, assumed to be long-tongued insects such as bees and/or flies.

**SIMILAR SPECIES** May be mistaken for *G. oreocharis* or *G. carneus*. However, the leaves of all three species are different, and the flower size and markings differ.

| | Flowers | Leaves |
|---|---|---|
| ***G. crispulatus*** | large, pink, tepals crisped along edges, triangular streak | 1–3mm, superposed, midribs thickened |
| ***G. oreocharis*** | small, usually pink to purple with spear-shaped mark | 2–6mm wide, plane, venation visible on both sides of leaf |
| ***G. carneus*** | large, white to pale or deep pink with spade-shaped marks, sometimes spotted or streaked | 6–14mm, usually forming a fan |

*Bracts are long, soft and green.*

*Flexed and inclined stems give* G. crispulatus *a distinctive appearance.*

*Bright red triangular markings on lower tepals distinguish this species from* G. oreocharis.

*Seeds are evenly winged.*

G. crispulatus *grows at altitudes above 300m in the Langeberg, Western Cape.*

# Gladiolus phoenix

***phoenix* = rising from the ashes, referring to its habit of only flowering after a fire.**

*Gladiolus phoenix* is endemic to the Bainskloof Mountains of the Western Cape. It flowers only in the first year after fire. Its deep pink flowers with white spear-shaped marks on the lower tepals appear around November and December.

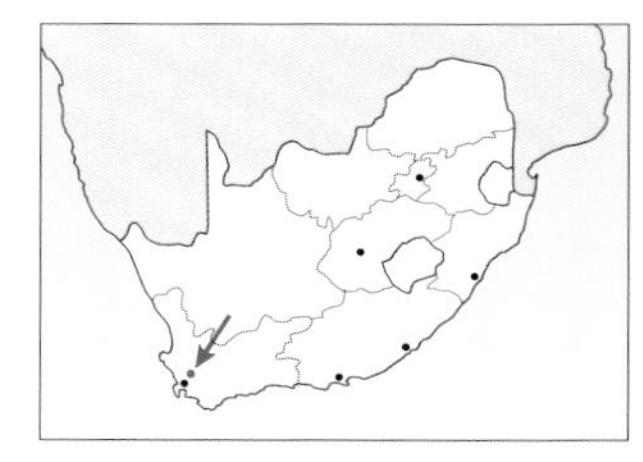

| J | F | M | A | M | J | J | A | S | O | N | D |
|---|---|---|---|---|---|---|---|---|---|---|---|

**STATUS** Critically Endangered

**DESCRIPTION** **Plant** 50–75cm. **Corm** bearing clusters of cormlets around the base on flattened stolons. **Cataphylls** up to 8cm above ground, green. **Leaves** 6–9, 3–5 basal, lanceolate, 6–12mm wide; midrib and secondary veins strongly raised, lowest leaves longest. **Spike** erect, up to 4 branches, with 9–12 flowers. **Bracts** ±15mm, dark green, lightly folded in the midline. **Flowers** deep pink; lower tepals each with a white spear-shaped mark outlined in deep pink. Perianth tube short, 16–20mm. **Anthers** dark purple. **Pollen** pale yellow. **Capsules** narrowly ovoid. **Seeds** broadly and evenly winged. **Scent** unscented.

**DISTRIBUTION** Known only from one small area in the Bainskloof Mountains above Wellington, Western Cape.

**ECOLOGY & NOTES** Grows on seasonally moist banks and slopes of the Bainskloof Mountains. We have seen it in three localities, all east facing. The plants were growing in light shade among the burned remains of woody vegetation in black, humus-rich, peaty soil. It has only ever been seen flowering in the year after a fire, and it appears not to flower or even produce leaves in the years between fires. This species is unusual in that it produces cormlets from the corm base, on short flattened stolons. This means that the plants form dense colonies, which are very showy when they flower.

**POLLINATORS** Flowers have relatively short perianth tubes, suggesting that they are pollinated by long-tongued bees such as *Amegilla spilostoma*.

**SIMILAR SPECIES** Owing to its extemely localised distribution on Bainskloof, it is not likely to be confused with any other species. The closest relatives are *G. oreocharis* and *G. crispulatus*. All have pink flowers with white markings on the lower tepals, shortish perianth tubes (±20mm), and all flower in late spring or summer in moist localities in the mountains. *G. phoenix* is a more robust plant than the other two, with broad, almost ribbed leaves and long flower spikes with up to 4 branches and many flowers per branch.

G. phoenix *is known only from the Bainskloof Mountains, where it flowers after fire.*

*Dense colonies of flowers are a result of vegetative reproduction.*

*Lanceolate to linear leaves form a basal fan.*

*Flower spikes are floriferous; note large dorsal tepal.*

*Plants grow in light shade in rich soils.*

*Bracts are dark green and slightly folded.*

*This photo was taken in the summer after a major fire in the Bainskloof area.*

# *Gladiolus gueinzii*

**Named after Wilhelm Gueinzius (1813–1874), a naturalist and apothecary, who collected this species near Durban in the 1840s.**

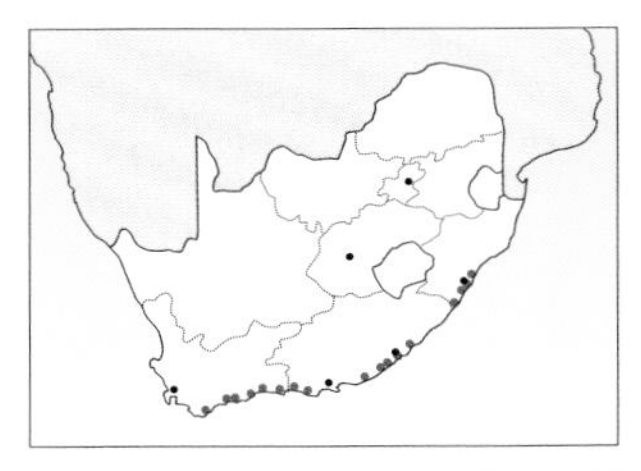

J F M A M J J A S O N D

*Gladiolus gueinzii* is a coastal dweller, adapted for saline conditions in both summer- and winter-rainfall regions. Its purple, self-pollinating flowers appear between October and January, depending on the rainfall region.

**DESCRIPTION** **Plant** 30–50cm. **Corms** globose and yellow, with cormlets and leathery tunics. **Cataphylls** up to 5cm above the soil, pale brown. **Leaves** 4–6, 4–6mm wide, grey-green, slightly succulent; only midrib raised. **Spike** upright, may be branched, with 4–8 flowers and clusters of cormlets at nodes. **Bracts** 18–27mm long, pale green or greyish pink. **Flowers** almost radially symmetrical, mauve to light purple; upper 3 tepals with red median streak, lower 3 with white streak surrounded with red, but sometimes all tepals have the same markings. Perianth tube 13–15mm. **Anthers** purple. **Pollen** grey-blue to cream. **Capsules** obovoid, 30mm. **Seeds** reddish brown, broadly and evenly winged. **Scent** unscented.

G. gueinzii *has almost radially symmetrical flowers, an unusual feature in* Gladiolus.

**DISTRIBUTION** Exclusively coastal, occurring from Arniston in the south to Durban in KwaZulu-Natal. Its range includes winter- and summer-rainfall areas.

**ECOLOGY & NOTES** Has a most unusual habitat and is found only on beaches and dunes right next to the sea. Plants are frequently immersed in sea water, which appears not to harm them at all. This habitat is extremely harsh as not only do the plants have to cope with saline conditions, but the sand is very well drained so that plants have very limited access to fresh water, even after rain. Although they produce good seed every year, they also reproduce vegetatively from two types of cormlet. Small cormlets are clustered at the base of the mature corm, and larger ones are produced when the mature corm is exposed as the sand in which they grow is moved by the tides. The larger cormlets can be up to 15mm long; they have a thick corky covering that is impermeable to water, and they float. It is thought that this is how the species spreads.

**POLLINATORS** Thought to be long-tongued bees, however *G. gueinzii* is self-compatible and self-pollinating, requiring no pollen transfer by insects. Presumably this adaptation is in response to the low number of insects found on beaches.

**SIMILAR SPECIES** This species is unmistakable for a variety of reasons. Very few *Gladiolus* species have almost radially symmetrical flowers; there are no other gladioli in South Africa that grow on the beach, and the leaves are slightly fleshy, which is unusual for the genus.

*The narrow, slightly succulent leaves are an adaptation to growing in beach soils.*

*Bracts are soft and long.*

G. gueinzii *is the only* Gladiolus *species found so close to the sea.*

*Although plants produce seed, reproduction appears to be largely vegetative.*

G. gueinzii *growing on the beach, Nature's Valley, Western Cape.*

# Gladiolus carneus

***carneus* = flesh-coloured or pink, for the flower colour.**

A winter-rainfall species, *Gladiolus carneus* has white or pink flowers marked with red nectar guides. It flowers from October to mid-November.

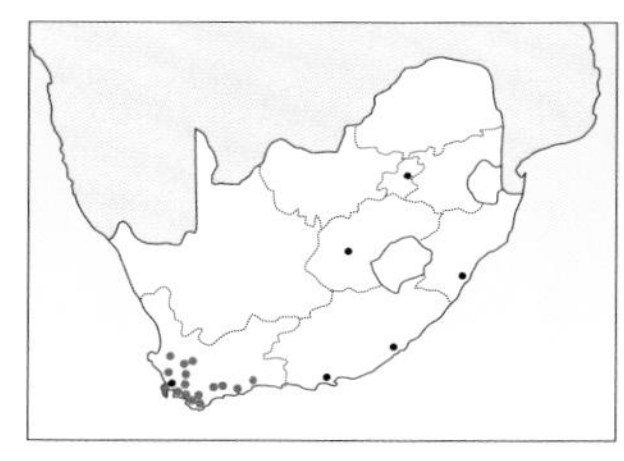

| J | F | M | A | M | J | J | A | S | O | N | D |
|---|---|---|---|---|---|---|---|---|---|---|---|

G. carneus *has variable markings; here, nectar guides are very pale.*

**DESCRIPTION** **Plant** 20–60cm. **Corm** globose, with papery tunics. **Cataphylls** to 8cm above ground, purple, mottled. **Leaves** 4 or 5, sometimes in a fan, to 14mm wide; midrib and 1 pair of veins lightly thickened, margins hyaline. **Spike** erect or inclined, unbranched or branched, 5–8 flowers in two rows. **Bracts** 35–45mm long, pale green or flushed with grey. **Flowers** large, pink or white; lower 3 tepals usually marked with dark pink to red 'spades' or 'diamonds', or with yellow blotches, sometimes unmarked. The tepals are undulate to crisped. Perianth tube up to 40mm. **Anthers** mauve. **Pollen** cream or purple. **Capsules** oblong. **Seeds** broadly and evenly winged. **Scent** unscented.

**DISTRIBUTION** Found from Bainskloof and Ceres southwards to the Cape Peninsula, and eastwards as far as the Outeniqua Mountains near George. It is common close to Cape Town and on the lower slopes of the Cape Peninsula mountains.

**ECOLOGY & NOTES** *G. carneus* grows in nutrient-poor shale and sandstone soils in fynbos, and flowers particularly well after fire.

**POLLINATORS** Long-proboscid flies *Philoliche rostrata* and *Prosoeca nitidula.*

**SIMILAR SPECIES** This species is extremely variable and in addition to its similarity to *G. oreocharis* and *G. crispulatus*, it may also be confused with the species listed in the table.

| | Leaves | Flowers | Location |
|---|---|---|---|
| ***G. carneus*** | 4 or 5, 14mm wide | white to pink, unmarked or with pink guides | widespread |
| ***G. pappei*** | 3 or 4, 3mm wide | dark pink with white guides | Cape Peninsula |
| ***G. undulatus*** | 4 or 5, 12mm wide | variable, with dark guides | north–south mountain axis of Western Cape |
| ***G. vigilans*** (sect. *Homoglossum*) | 4, to 1.5mm wide | pale pink with white guides | Cape Peninsula |
| ***G. ornatus*** (sect. *Homoglossum*) | 3, to 2mm wide | pink with white guides | Cape Peninsula and inland |

*Leaves may form a fan.*

*Tepals are usually lightly crisped.*

*Plants usually have a single stem which may be branched.*

*Pale pink form.*

*Tepal markings vary in shape.*

*Note the distinctive dark markings on lower tepals.*

*Sometimes tepal markings resemble stylised hearts.*

G. carneus *on Chapman's Peak, overlooking Hout Bay, Cape Town.*

# *Gladiolus pappei*

**After Ludwig Pappe (1803–1862), a Cape Colony government botanist who collected the type specimen in the 1850s.**

The pale to deep pink flowers of *Gladiolus pappei* appear from mid-October to mid-December. It is restricted to the Cape Peninsula and seldom flowers except after fire. The deep pink flowers distinguish it from *G. carneus*.

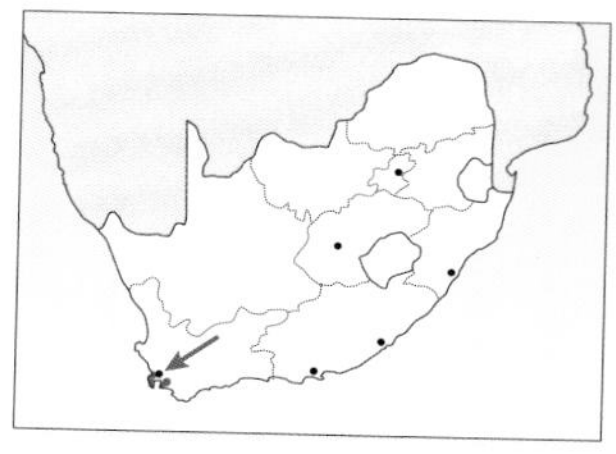

J F M A M J J A S O N D

**STATUS** Rare

*Dorsal tepals are lanceolate and may curve upwards.*

**DESCRIPTION** **Plant** up to 35cm. **Corm** globose, with membranous tunics. **Cataphylls** not very evident at flowering time. **Leaves** 3 or 4, very narrow (3mm); neither midrib nor margins thickened. **Spike** unbranched, with 2 or 3 flowers. **Bracts** grey-green and sometimes purple-grey on dorsal surfaces. **Flowers** pale to deep pink; the lower tepals with diamond- or heart-shaped markings, white in the centre and reddish on the edges. Perianth tube 30–35mm. **Anthers** pale grey-blue. **Pollen** pale mauve. **Capsules** unknown. **Seeds** broadly and evenly winged. **Scent** unscented.

G. pappei *growing on Table Mountain, Cape Town.*

**DISTRIBUTION** Originally known only from the Cape Peninsula, mainly Silvermine and Table Mountain, but now also recorded from Pringle Bay. We have also seen plants that resemble this species in the Jonkershoek Mountains near Stellenbosch but their identity has not been confirmed.

**ECOLOGY & NOTES** This species grows in marshy areas in sandy, peaty soils that are damp or wet most of the year. On Table Mountain the plants grow together with *Bulbinella nutans* subsp. *turfosicola*, also a peat-loving plant. Plants flower best after a fire, but our photographs on Table Mountain are of plants flowering in old vegetation. The photos from Silvermine were taken in the second year after a fire.

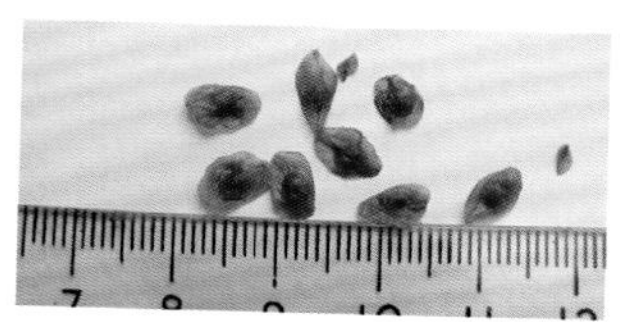

*Seeds have broad and even wings.*

**POLLINATORS** Probably long-tongued flies, which pollinate many of the long-tubed *Gladiolus* species.

**SIMILAR SPECIES** Similar to *G. carneus*, but *G. pappei* is localised in distribution.

| | Leaves | Flowers | Habitat |
|---|---|---|---|
| ***G. pappei*** | 3 or 4 very narrow leaves, one above the other on the stem, slender habit | deep pink flowers, lower tepals with white marks outlined in red | marshy and damp places |
| ***G. carneus*** | 4 or 5 leaves usually in a fan, 6–14mm wide, robust plants | white, cream or pale pink flowers with red markings or without markings | usually well-drained slopes or streambanks |

*Grey-green bracts have notched apices.*

*Anthers are pale grey-blue and pollen is mauve.*

G. pappei *at Silvermine in the second year after a fire, Cape Peninsula, Western Cape.*

# Gladiolus geardii

**Named after C Geard who owned the farm where this *Gladiolus* was first scientifically collected and recorded.**

*Gladiolus geardii* grows on moist slopes along the mountains of the Eastern Cape. Pinkish-purple flowers appear between November and mid-December. Its flowers are similar to those of *G. carneus*, but it is generally a larger plant with much larger flowers. The distributions of the two species do not quite overlap.

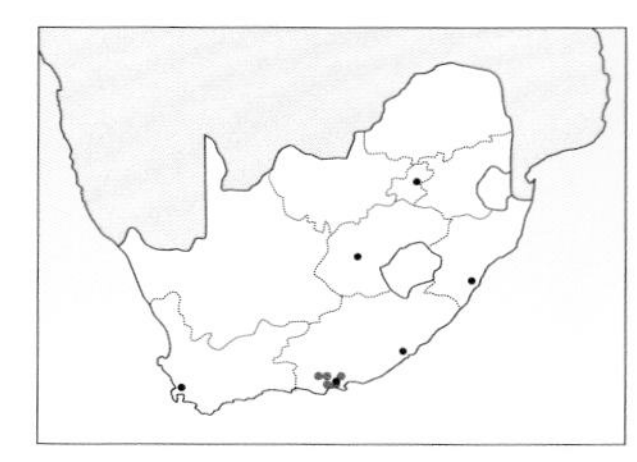

J | F | M | A | M | J | J | A | S | O | N | D

**STATUS** Near Threatened

**DESCRIPTION** **Plant** 80–150cm. **Corm** depressed-globose with membranous tunics. **Cataphylls** up to 15cm above ground, purple. **Leaves** 7–9, 14–28mm wide; midrib and margins lightly thickened. **Spike** usually branched, with 6–10 flowers. **Bracts** 30–50mm long, green. **Flowers** large, pale pinkish-purple; lower 3 tepals each with a darker spear-shaped mark, sometimes lighter in the centre. Perianth tube long, up to 40mm. **Anthers** purple. **Pollen** cream. **Capsules** ovoid-elliptic, 28mm. **Seeds** ovoid, broadly winged. **Scent** unscented.

**DISTRIBUTION** Known only from a few localities in the Eastern Cape, from the Baviaanskloof to the Kouga and Van Stadens River mountains west of Gqeberha (Port Elizabeth).

G. geardii *has large flowers with an arched dorsal tepal and extended lateral tepals.*

**ECOLOGY & NOTES** Plants grow in moist sites along streambanks, in seeps and in shady gullies. We photographed some very robust plants in December, growing along a stream in a pine plantation where they were in semi-shade. Some plants had up to 7 branches, and there were 14 flowers per spike. When we visited the site again in January, the plants were still green, and they appeared to be evergreen. They grow in an area that can receive rain at any time of year and they are found in moist localities, so perhaps they have a very short period of dormancy.

**POLLINATORS** Presumed to be pollinated by long-tongued flies.

*Spear-shaped markings on the lower tepals are characteristic.*

**SIMILAR SPECIES** The closest species is *G. carneus*, but one is unlikely to confuse them since *G. carneus* only occurs as far east as Knysna, whereas *G. geardii* is only found east of Knysna.

| | Height | Flower spike | Flowers |
|---|---|---|---|
| ***G. geardii*** | 80–150cm, 7–9 leaves | usually branched with 10 or more flowers | large flowers up to 10cm long |
| ***G. carneus*** | usually 50cm, 4 or 5 leaves | usually unbranched with 5–8 flowers | smaller flowers up to 8cm long |

*Plane leaves form a fan around the stem.*

*Branched spikes may bear as many as ten flowers.*

*Bracts are long and green.*

*The large dorsal tepal arches over the stamens.*

*Purple anthers bear cream pollen.*

G. geardii *is tall, reaching up to 150cm; here in a plantation near Gqeberha (Port Elizabeth), Eastern Cape.*

# *Gladiolus aquamontanus*

***aquamontanus*** **= of mountain waters, describing the waterfalls, cliffs and streams of the habitat.**

*Gladiolus aquamontanus* is restricted to the Swartberg Mountains in the winter-rainfall areas where it flowers between mid-November and December. Pale pink-purple, unscented flowers bloom on plants that grow in perennial streams or on wet cliffs.

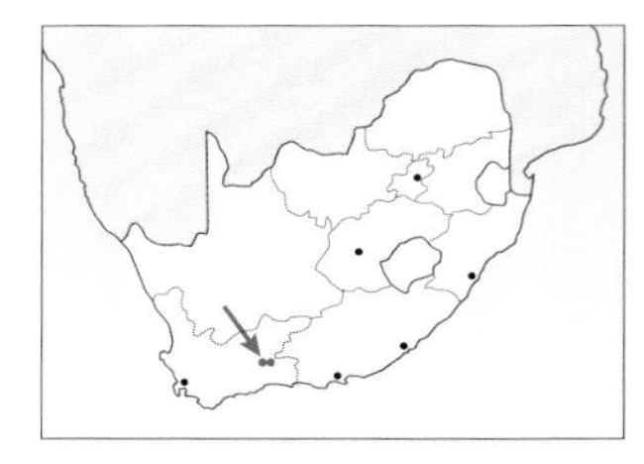

| J | F | M | A | M | J | J | A | S | O | N | D |
|---|---|---|---|---|---|---|---|---|---|---|---|

**STATUS** Vulnerable

**DESCRIPTION** **Plant** 40–100cm, inclined or drooping. **Corm** vestigial, similar to a short, swollen rhizome. **Cataphylls** up to 12cm, rapidly disintegrating. **Leaves** 5 or 6, narrowly lanceolate and up to 15mm wide, soft textured; with midrib lightly raised. **Spike** drooping, unbranched, with 4–8 flowers. **Bracts** green and soft textured. **Flowers** fairly large (70mm long), pale mauve-pink; lower 3 tepals each with a broad, dark purple streak. **Anthers** dark purple. **Pollen** cream. **Capsules** narrowly obovoid. **Seeds** narrowly oblong, unevenly winged. **Scent** unscented.

*The corms are rhizome-like and, unlike other species that grow on rocky outcrops, do not wedge into cracks.*

**DISTRIBUTION** Restricted to the Swartberg Mountains of the Western Cape; found over a range of ±70km.

**ECOLOGY & NOTES** The plants grow only in perennial streams and on wet cliffs, with the roots anchored in rock cracks. The rhizome-like corms are found on the surface of the rocks. As these plants grow in perennially wet places, they do not require a conventional corm to survive a dry period.

**POLLINATORS** Long-tongued flies.

**SIMILAR SPECIES** Owing to its unusual habitat and restricted distribution, this species is unlikely to be confused with any other. The flowers resemble those of *G. geardii* in colour, shape and markings, but the habitats and growth habits are markedly different. *G. geardii* grows in damp places, but the plants are upright and the inflorescences are branched, whereas *G. aquamontanus* is inclined with drooping leaves and flower spikes. *G. geardii* also has a conventional corm, whereas *G. aquamontanus* does not.

*Resembling a waterfall,* G. aquamontanus *spills alongside its namesake.*

*Soft-textured green bracts support narrow flowers.*

*This species has drooping spikes and leaves.*

*Mauve-pink flowers have dark purple anthers.*

*Note the dark purple median streaks on lower tepals.*

G. aquamontanus *is highly localised in the Swartberg Mountains of the Western Cape.*

# Gladiolus undulatus

***undulatus* = referring to the wavy margins of the tepals.**

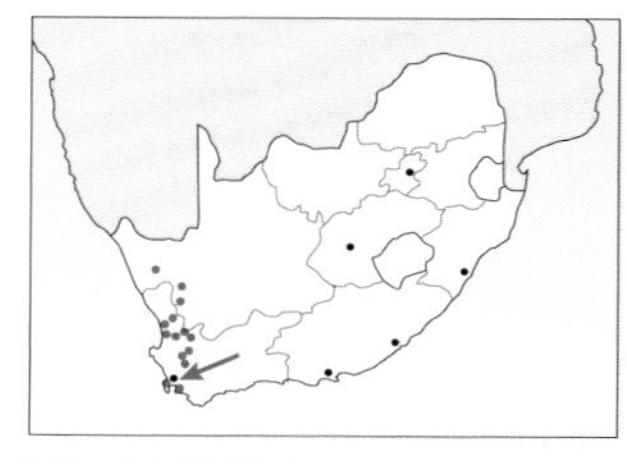

J F M A M J J A S O N D

The flowers of *Gladiolus undulatus*, which grows along mountains in the western regions of the Western and Northern Cape, are cream, pink, greenish or lilac. They appear between mid-November and December. *G. undulatus* resembles *G. carneus* and *G. angustus*.

**DESCRIPTION** **Plant** 45–150cm. **Corm** depressed-globose, with soft tunics. **Cataphylls** ±6cm above ground, firm, reddish purple speckled with white. **Leaves** 4 or 5, 5–12mm wide; midrib lightly raised, margins hyaline. **Spike** erect, unbranched or up to 2 branches, usually with 6–9 flowers, almost distichous. **Bracts** 30–40mm long, dull green or purple. **Flowers** pale lilac or pinkish cream to greenish; lower tepals with reddish to purple markings. Tepal margins strongly undulate. Perianth tube 50–70mm. **Anthers** cream above, purple below. **Pollen** purple. **Capsules** oblong-ellipsoid. **Seeds** ovate, broadly winged. **Scent** unscented.

*Two-ranked flower spikes bear flowers with undulating margins.*

**DISTRIBUTION** A widespread species found from the Cape Peninsula and Stellenbosch all the way north along the mountains to the Kamiesberg in the Northern Cape.

**ECOLOGY & NOTES** Plants grow in stony sandstone or granite soils (in the Kamiesberg), often along streams and in damp areas.

**POLLINATORS** This species has nectar with a low sugar concentration and is pollinated by long-tongued flies of the families Nemestrinidae and Tabanidae, including *Philoliche rostrata*.

*Plants often grow along streams.*

**SIMILAR SPECIES** *G. undulatus* resembles *G. carneus* but is distinguished by its flower colour and long perianth tube. It also resembles *G. angustus* but the latter has a much longer perianth.

| | Perianth tube | Tepals | Flower colour |
|---|---|---|---|
| ***G. undulatus*** | 50–60mm | tepals attenuate and strongly undulate | pale pink or white, with weakly developed nectar guides |
| ***G. carneus*** | 25–45mm | tepals lanceolate, sometimes undulating | pinkish cream to greenish, weakly developed nectar guides |
| ***G. angustus*** | 80–90mm | tepals lanceolate, lower tepals shorter than upper | greenish cream to ivory, distinctive dark red markings on lower tepals |

G. undulatus *has very long, flared perianth tubes.*

*The undulating tepal margins are clearly visible.*

*Colours and markings are variable.*

*Markings are almost invisible.*

*The dorsal tepal is large, extending over the stamens.*

*Cream flowers with red markings.*

G. undulatus *in the Gifberg, Western Cape.*

# Gladiolus angustus

***angustus* = narrow, for the long and narrow perianth tube.**

The creamy yellow flowers of *Gladiolus angustus* have very long perianth tubes. They appear from October to November in the winter-rainfall region. The species can be confused with *G. carneus* and *G. undulatus* but is distinguished by its longer perianth tube.

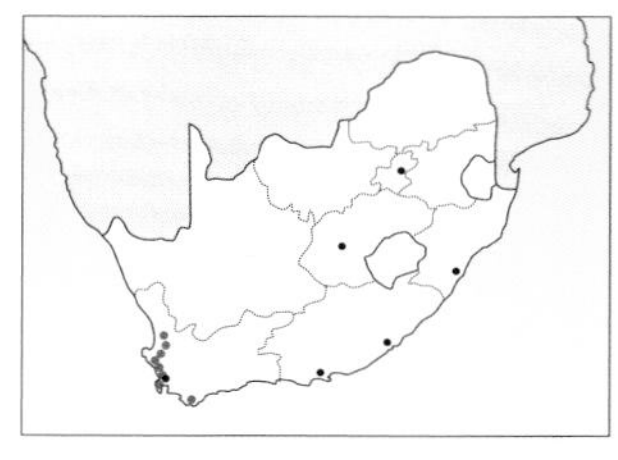

| J | F | M | A | M | J | J | A | S | O | N | D |
|---|---|---|---|---|---|---|---|---|---|---|---|

**DESCRIPTION** **Plant** 60–120cm. **Corm** globose, with papery tunics. **Cataphylls** purple above ground and mottled. **Leaves** 4 or 5, 5–10mm wide; midrib lightly thickened, margins hyaline. **Spike** branched or unbranched, with 5–9 flowers. **Bracts** 50–65mm long, green, sometimes red at apices. **Flowers** cream to pale yellow; with spade-shaped yellow markings on lower tepals outlined in dark red; tepals unequal with the upper 3 larger than the lower 3. Perianth tube 60–110mm long. **Anthers** whitish. **Pollen** cream. **Capsules** ellipsoid. **Seeds** ovate, translucent yellow-brown, broadly winged. **Scent** unscented.

**DISTRIBUTION** From the Cape Peninsula up the West Coast to Piketberg and the Cederberg Mountains of the Western Cape.

**ECOLOGY & NOTES** Usually grows in damp places such as marshes, seeps and streambanks. We have seen it in flower a few metres from the sea in damp, boggy soil near Cape Point, and along riverbanks in the Cederberg. It flowers particularly well after fire.

G. angustus *flowering after fire, Kouebokkeveld.*

**POLLINATORS** The very long perianth tube of this species requires an insect with mouthparts long enough to reach the nectar; Goldblatt & Manning (1998) identified *Moegistorhynchus longirostris* on the West Coast. The Cederberg plants have slightly shorter tubes and it is not clear whether the same insect is involved in their pollination.

**SIMILAR SPECIES** *G. angustus* is vegetatively identical to *G. carneus* and *G. undulatus*, but their flowers differ.

| | Perianth tube | Tepals | Flower colour |
|---|---|---|---|
| ***G. angustus*** | 80–90mm, longer than dorsal tepal | tepals lanceolate, lower tepals shorter than upper | greenish cream to ivory, distinctive dark red markings on lower tepals |
| ***G. carneus*** | 25–45mm, as long as dorsal tepal | tepals lanceolate, similar size | pinkish cream to greenish, weakly developed nectar guides |
| ***G. undulatus*** | 50–60mm, longer than dorsal tepal | tepals attenuate, strongly undulate, upper and lower tepals similar size | pale pink or white with weakly developed nectar guides |

*The two upper leaves sheath the stem; note thickened midribs.*

*The perianth tubes are long and slender.*

*Pale yellow form.*

*Tepals are unequal.*

*Cream form.*

*Note the spade-shaped markings.*

G. angustus, *here at Heuningvlei, Cederberg, Western Cape. Plants from this region have shorter perianth tubes than their counterparts elsewhere.*

# *Gladiolus buckerveldii*

**Named after MH Buckerveld who collected the plant and sent it to Kirstenbosch in 1927.**

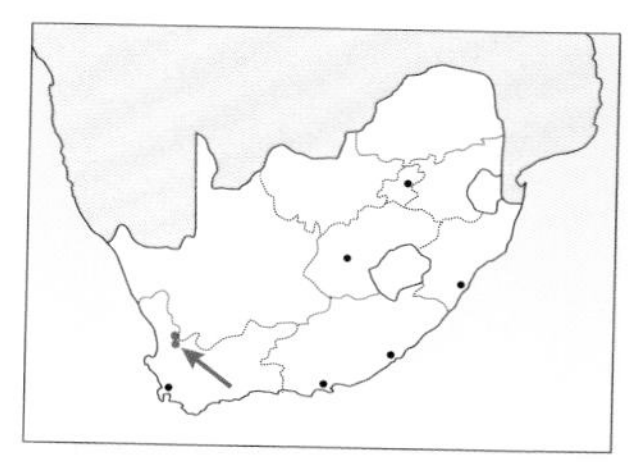

J F M A M J J A S O N D

**STATUS** Rare

*Gladiolus buckerveldii* occurs in the Cederberg of the Western Cape. Large ivory flowers with red nectar guides appear in January.

**DESCRIPTION** **Plant** 80–125mm. **Corm** globose, with papery tunics. **Cataphylls** light purple mottled with pale green above ground. **Leaves** 5 or 6, 25–30mm wide, trailing; margins not thickened, prominent midrib and veins. **Spike** inclined to drooping, unbranched, with 12–20 flowers all secund. **Bracts** 50–80mm long, green. **Flowers** ivory to greenish cream; lower tepals each with dark red mark in the centre, tepals unequal with top 3 larger than lower 3. Perianth tube 50mm long. **Anthers** cream becoming purple. **Pollen** cream. **Capsules** oblong. **Seeds** long and narrow, broadly winged at one end, weakly winged at sides. **Scent** unscented.

G. buckerveldii *has distinctively shaped flowers with the upper tepals longer than the lower.*

**DISTRIBUTION** Known only from the northern Cederberg in the area of Algeria Forest Station. When the plants were in flower in 2014, we visited several rivers and cliffs close to Algeria looking for additional sites, but found none.

**ECOLOGY & NOTES** Plants of this species grow in permanent water on vertical cliffs, with their corms wedged into the rocks. The leaves hang down and the flowering stems are horizontal. They grow among moss and in the same habitat as *Disa uniflora*, which flowers at a similar time.

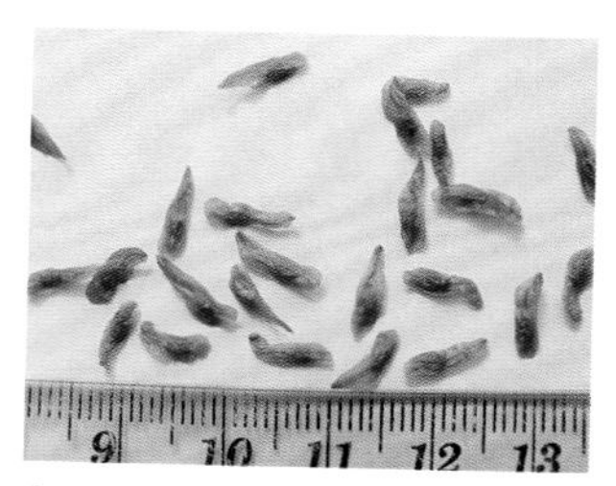


*Seeds are unusually shaped; long and slender and barely winged.*

**POLLINATORS** The shape of the perianth tube (fairly wide in the upper half) and the presence of a large quantity of nectar suggest that the pollinator may be a bird such as a sunbird. This is unusual in that most bird-pollinated flowers are red or orange. However, the pollinators could also be long-tongued flies similar to those that pollinate *G. angustus*. Whatever pollinates these flowers does a good job as when we returned to the plants a month after photographing them, almost every flower had set seed.

**SIMILAR SPECIES** *G. angustus* is probably the closest relative of *G. buckerveldii*, and the flowers of the two species are similar. The perianth tube of *G. angustus* is longer than that of *G. buckerveldii*, and the flowers of *G. buckerveldii* are strongly secund. However, the biggest difference between the two species is their choice of habitat, with *G. buckerveldii* found only on rock faces and in waterfalls.

*Bracts are long and soft, sometimes longer than the flower.*

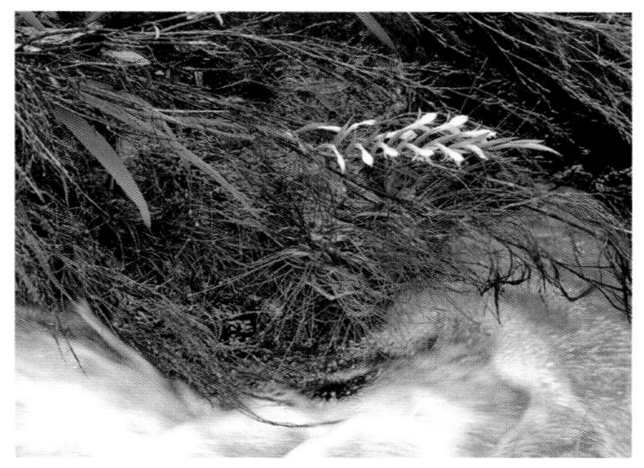

*Stems are inclined or horizontal.*

*The Algeria Waterfall in the Cederberg is one of only a few known locations of this species.*

# *Gladiolus bilineatus*

***bilineatus* = two-lined, for the paired dark streaks on the lower tepals.**

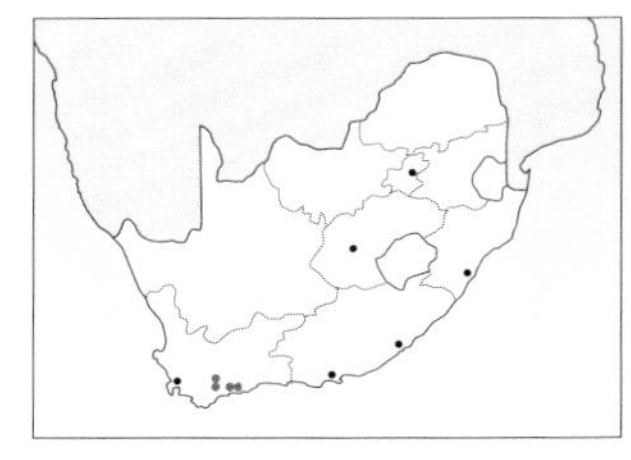

| J | F | M | A | M | J | J | A | S | O | N | D |
|---|---|---|---|---|---|---|---|---|---|---|---|

**STATUS** Vulnerable

Flowering in March and April, *Gladiolus bilineatus* is endemic to the Langeberg of the Western Cape. Its flowers are creamy pink to salmon. Its flowering time at the end of summer is unusual for a winter-rainfall plant, and it is unlikely to be mistaken for any other species.

**DESCRIPTION** **Plant** 20–35cm. **Corm** globose, with fine, net-like tunics. **Cataphylls** green, sometimes speckled above ground. **Leaves** 3, 6–8mm wide; midrib very lightly thickened, margins hyaline. Flowering stems have cauline leaves; non-flowering plants have 1 long foliage leaf. **Spike** slightly inclined, unbranched, with 2–5 flowers. **Bracts** green or dull purple. **Flowers** creamy pink to salmon; lower tepals each with light streak outlined in dark pink, tepals unequal. Perianth tube long, 50–70mm. **Anthers** purple. **Pollen** lilac. **Capsules** ovoid. **Seeds** ovate, reddish brown, with broad wing. **Scent** unscented.

**DISTRIBUTION** Found only on the lower southern slopes of the Langeberg from Tradouw Pass in the west to Albertinia in the east.

**ECOLOGY & NOTES** This species grows in a variety of soils and habitats. At the western end of its range, it grows among Proteaceae in fynbos on nutrient-poor sandstone-derived soils. In most other areas, plants grow in renosterveld on richer soils at the interface between sandstone and shale.

**POLLINATORS** *Prosoeca longipennis* – a long-tongued fly of the family Nemestrinidae – is the sole pollinator.

*The eponymous streaks of* G. bilineatus *are clearly visible.*

**SIMILAR SPECIES** Owing to its autumn flowering time and its restricted distribution on the southern slopes of the Langeberg, this species can be mistaken for only one other species, *G. engysiphon* (section *Homoglossum*). They grow in the same locality and flower in autumn.

| | Leaves of non-flowering plants | Outer bracts | Flowers |
|---|---|---|---|
| ***G. bilineatus*** | flat leaf, 6–8mm wide | 40–50mm long | cream to pink, red streaks on lower tepals |
| ***G. engysiphon*** (sect. *Homoglossum*) | terete with 4 longitudinal grooves | 15–20mm long | white to cream, red streaks on lower tepals |

*Leaves sheath the stem for about half their length.*

*The dorsal tepal is longer than the laterals; the lower tepals are shorter than the upper ones.*

*A pale cream-pink form.*

*Salmon-coloured flowers; note the long, soft bracts.*

G. bilineatus *growing near Swellendam with the Langeberg Mountains in the background.*

# Gladiolus dolichosiphon

***dolichosiphon* = long tube, referring to the perianth tube.**

The short *Gladiolus dolichosiphon* bears pale cream flowers with red markings from January to mid-February. It grows in the drier regions of the winter-rainfall area, in the Little Karoo. It is distinguished from *G. bilineatus* by growing at higher altitudes, flowering earlier and having more leaves.

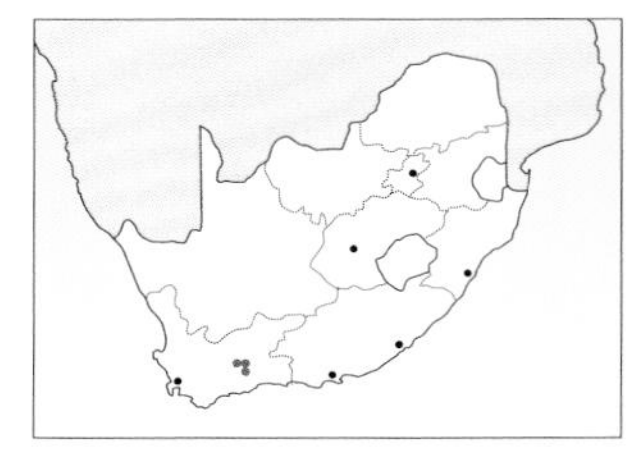

| J | F | M | A | M | J | J | A | S | O | N | D |
|---|---|---|---|---|---|---|---|---|---|---|---|

**STATUS** Rare

**DESCRIPTION** **Plant** up to 40cm. **Corm** subglobose, with outer tunics becoming soft fibres. **Cataphylls** flushed purple above ground and slightly mottled. **Leaves** 6 or 7, linear to sword shaped; with midrib and secondary veins thickened, margins hyaline; upper cauline leaves with no blades. **Spike** inclined, sometimes branched, with 1–7 flowers. **Bracts** up to 3.5cm, green flushed with purple. **Flowers** pale cream to salmon pink; lower 3 tepals with pale marks outlined in dark pink or red. Perianth tube long, up to 55mm. **Anthers** purple. **Pollen** cream. **Capsules** distinctly tri-lobed, concealed in bracts. **Seeds** oval, evenly winged. **Scent** unscented.

**DISTRIBUTION** Known only from a few localities in the Klein Swartberg and Rooiberg mountains of the Little Karoo.

**ECOLOGY & NOTES** The plants grow at higher altitudes; on the Klein Swartberg they are found at 1,900m above sea level. The plants we photographed were on south- and east-facing slopes. They grew in black peaty soil at the base of rocks where they benefitted from increased run-off after rain. They appear to flower best in the year after a fire, although we found many flowering in the second year after fire as well. At this altitude, recovery of vegetation after fire is slow, and the *Gladiolus* plants were not threatened by competition from other plants.

**POLLINATORS** Likely to be long-proboscid flies, which pollinate several other members of the Iridaceae family with long-tubed pink flowers.

**SIMILAR SPECIES** Distinctive and unlikely to be confused with any other species. Not many other *Gladiolus* species flower in the Little Karoo in midsummer, and none of them has the same distinctive leaves or flowers as *G. dolichosiphon*.

G. dolichosiphon *was only recently described in the botanical literature, in 2009.*

*Flowers may be cream to salmon-pink with purple anthers.*

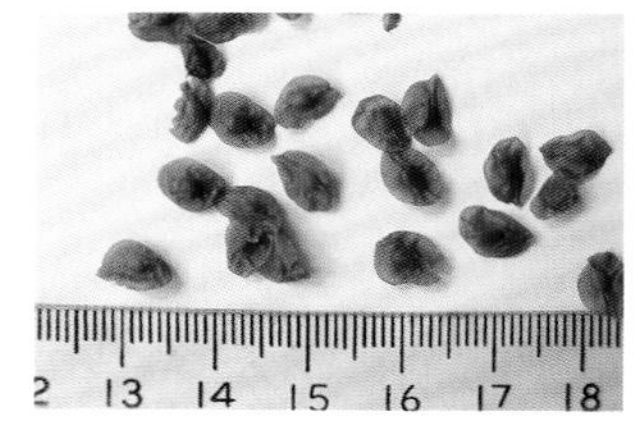


*Oval seeds are evenly winged.*

*Leaves are narrow with the cauline leaves lacking blades.*

*The perianth tube is longer than the dorsal tepal.*

*Capsules ripening.*

*Plants grow at higher altitudes, seen here in the Swartberg.*

*Long, lanceolate dorsal tepals arch over stamens.*

*Although the flowers are striking, they may easily be overlooked when in bud because of the dull-coloured bracts and stems.*

# Gladiolus insolens

***insolens* = radiant, shining like the sun, for the bright red flower colour.**

The radially symmetrical scarlet flowers of *Gladiolus insolens* are unmistakable. Flowering between December and mid-January, it is endemic to the mountains near Piketberg in the Western Cape.

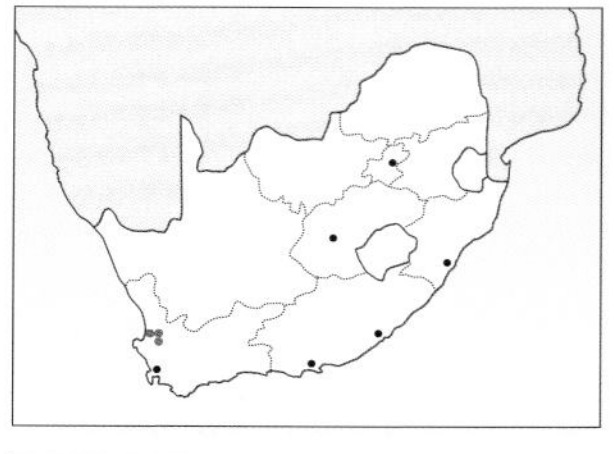

| J | F | M | A | M | J | J | A | S | O | N | D |
|---|---|---|---|---|---|---|---|---|---|---|---|

**STATUS** Vulnerable

**DESCRIPTION** **Plant** 30–50cm. **Corm** globose, with papery tunics becoming fibrous with age. **Cataphylls** uppermost green or dry and brown. **Leaves** 5 or 6, grey-green, lower 3 basal, 4–6mm wide; midrib thickened and lightly raised, margins hyaline. **Spike** inclined, usually unbranched, with 1 or 2 flowers. **Bracts** outer 30–40mm long, light green. **Flowers** bright scarlet; with radially symmetrically arranged tepals forming a wide bowl. Perianth tube 38mm long. **Anthers** scarlet. **Pollen** pale yellow. **Capsules** unknown. **Seeds** unknown. **Scent** unscented.

*It is hard to capture the vivid red accurately in a photograph.*

**DISTRIBUTION** Known only from the northeastern end of the Piketberg Mountain in the southwestern Cape. This mountain behind the small town of Piketberg is isolated from all surrounding mountains and reaches a height of just over 1,450m. Like the rest of the southwestern Cape, it receives winter rain and has hot, dry summers.

**ECOLOGY & NOTES** Plants are found in rocky as well as non-rocky areas, in seeps or on the edges of streams. We found plants in flower in thick fynbos on the edge of a permanent river, which remains moist until midsummer. They were growing among typical riverine plants such as *Brachylaena neriifolia*, *Metrosideros angustifolia* and *Psoralea aphylla*, at an altitude of ±700m. Apparently other localities include moist cliffs and rocky areas, but despite searching for several hours, we found no flowering plants elsewhere.

**POLLINATORS** The pollinator has not been observed, but it is presumed to be the butterfly *Aeropetes tulbaghia* (Mountain pride butterfly). This butterfly is attracted to summer-flowering plants with bright red flowers, including species of *Crassula*, *Cyrtanthus*, *Kniphofia*, *Nerine* and *Watsonia*.

**SIMILAR SPECIES** The only species that resembles this one is *G. stokoei* (section *Linearifolii*) from the Riviersonderend Mountains. However, as the localities are so far apart and the flowering times of the two species differ, there is no chance of mistaking one for the other.

| | Locality | Leaves | Flowers |
|---|---|---|---|
| ***G. insolens*** | Piketberg | 6 long grey-green leaves at the same time as the flowers | 1 or 2 bright scarlet, slightly cupped flowers on inclined stem |
| ***G. stokoei*** (sect. *Linearifolii*) | Riviersonderend | 1 hairy foliage leaf after flowering, flowering stem has reduced sheathing leaves | erect spike of 3–5 bright scarlet bowl-shaped flowers, stamens exserted |

*Leaves have raised midribs and hyaline margins.*

*Note the radial symmetry and scarlet anthers.*

*Spikes are usually inclined, bearing few flowers.*

*Tepals form a wide bowl. Bracts are long and green.*

G. insolens *is known only from the Piketberg in the Western Cape.*

# Gladiolus cardinalis

***cardinalis* = named for the bright red flower colour.**

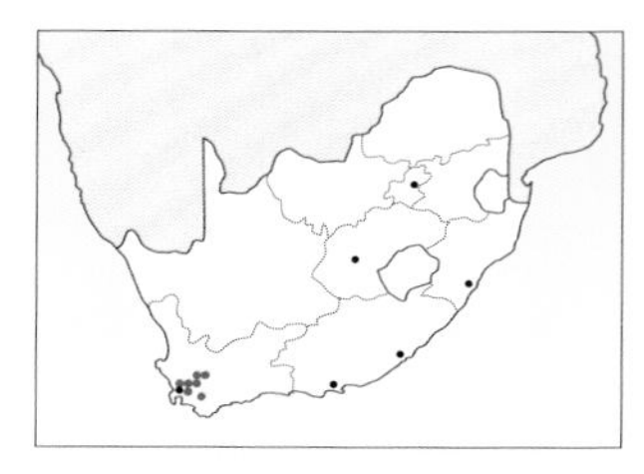

J F M A M J J A S O N D

Known as the 'New year lily' or 'Waterfall lily', the bright red flowers of *Gladiolus cardinalis* have white, spear-shaped marks. This species grows in the southwestern part of the Western Cape and flowers from mid-December to mid-January.

**DESCRIPTION** **Plant** 55–90cm. **Corm** poorly developed, attached to rocks. **Cataphylls** green. **Leaves** 5–7, drooping, 11–20mm wide; midrib and 1 pair of veins thickened and prominent, margins not thickened. **Spike** drooping, unbranched, with 8–11 flowers borne on the upper side. **Bracts** bright green, diverging sharply from the stem. **Flowers** bright red; lower tepals each with a broad white spear-shaped mark. Perianth tube 30–40mm long. **Anthers** upper surface red, lower surface white. **Pollen** cream. **Capsules** obovoid. **Seeds** yellow-brown. **Scent** unscented.

*The bright red flower with bold white markings is pollinated by the Mountain pride butterfly.*

**DISTRIBUTION** Found only in the mountains of the southwestern Cape, in the areas around Paarl, Wellington and Worcester, where it is restricted to wet cliffs close to waterfalls.

**ECOLOGY & NOTES** The plants flower in midsummer, the driest time of the year. The corms are wedged into cracks in the rocks where they are protected. Corms and roots must be constantly wet, and have been found in very fast-flowing rivers. We have sometimes seen the most amazing floral displays in the Hex River Mountains, and at Du Toitskloof and Bainskloof. Often found flowering together with *Disa uniflora*, an orchid of a similar colour; the two species share the same pollinator. The perianth may be twisted to best position the flower for pollination.

*Corms are wedged into crevices.*

**POLLINATORS** The pollinator of this red-flowered species is *Aeropetes tulbaghia* (Mountain pride butterfly), which pollinates many other plants with red flowers in mid- to late summer. *G. cardinalis* has very long filaments (up to 40mm as opposed to less than 20mm for most other species) that extend up to 30mm from the tube, ensuring that the butterflies brush against the anthers or style branches as they look for nectar. See page 102 for more information on this butterfly.

**SIMILAR SPECIES** In flower, this species is unmistakable and cannot be confused with any other *Gladiolus*. The plants have bright red flowers on inclined spikes in midsummer and grow in waterfalls. In leaf, it could be mistaken for some of the *G. carneus* relatives, but none of these usually hangs in waterfalls.

*Leaves and stems are inclined.*

*Bright green bracts hold flowers erect from the stem.*

*Floriferous, drooping stems bear erect flowers.*

*This species presents a showy display when flowering en masse.*

*Flowering in midsummer,* G. cardinalis *grows on wet cliffs.*

G. cardinalis *flowering high in the Matroosberg above Ceres.*

# Gladiolus sempervirens

**sempervirens = always green, alluding to the evergreen growth habit.**

*Gladiolus sempervirens* grows along the Outeniqua–Tsitsikamma mountains in the Western and Eastern Cape provinces. Bright red flowers appear from March to May. It can be mistaken for *G. cardinalis* or *G. stefaniae*.

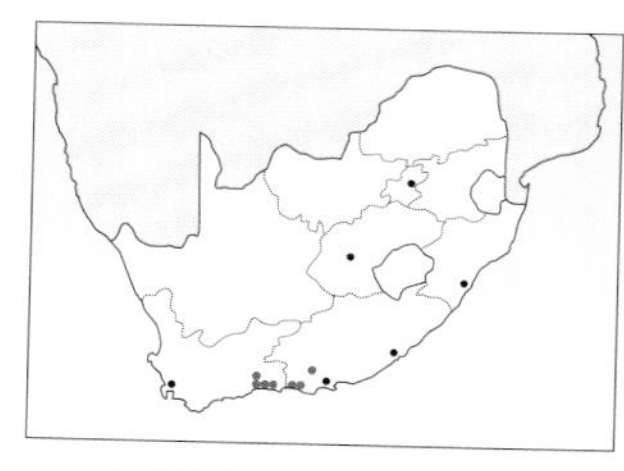

| J | F | M | A | M | J | J | A | S | O | N | D |
|---|---|---|---|---|---|---|---|---|---|---|---|

**STATUS** Rare

**DESCRIPTION** **Plant** 35–100cm tall. **Corm** shallow and vestigial, with rhizome-like stolons from the base and leathery tunics. **Cataphylls** green. **Leaves** 7–12, lanceolate, 6–18mm wide; midrib and 1 pair of veins lightly thickened, margins not thickened. **Spike** inclined, sometimes branched, with 4–8 flowers on the upper side. **Bracts** up to 55mm, green. **Flowers** large (up to 10cm across), deep red; each of the lower 3 tepals has a narrow white mark on the midline. Perianth tube 25–40mm long. **Anthers** purple. **Pollen** yellow. **Capules** oblong. **Seeds** oval, golden brown, wing well developed. **Scent** unscented.

*Carmine flowers are pollinated by the Mountain pride butterfly.*

**DISTRIBUTION** Known from only a few sites on the southern slopes of the Outeniqua–Tsitsikamma mountain chain. Plants have been found from George in the west to Kareedouw in the east, at altitudes of ±400m.

**ECOLOGY & NOTES** The plants grow in seeps and in damp on the banks of rivers, in sandstone soils in areas where they are often exposed to the southeasterly wind, which blows in summer. At higher altitudes, this wind brings mist, which presumably allows this species to remain evergreen through the dry summer. The vestigial corm is more like a rhizome; the stolons produced from the corm mean that the species reproduces vegetatively. We found plants flowering in damp places in old fynbos verging on forest, where it occurred with tall plants of *Cunonia capensis*, *Leucadendron eucalyptifolium* and *Passerina* species, together with *Blechnum tabulare*.

**POLLINATORS** The bright red, long-tubed flowers are pollinated by *Aeropetes tulbaghia* (Mountain pride butterfly), which is attracted to red flowers of many different genera and families. See page 102 for more information on this butterfly.

**SIMILAR SPECIES** Although the flowers of *G. sempervirens*, *G. cardinalis* and *G. stefaniae* are similar, their distribution ranges are different and they are unlikely to be mistaken for one another.

| | Locality | Flowering | Growth |
|---|---|---|---|
| ***G. sempervirens*** | George to Kareedouw, moist sites | autumn | evergreen |
| ***G. cardinalis*** | southwestern part of the Western Cape, waterfalls | midsummer | does not go fully dormant |
| ***G. stefaniae*** | Potberg and near Montagu in the Langeberg, dry sites | autumn | dormant in summer |

*This species is evergreen, unlike one of its nearest relatives,* G. stefaniae.

*Flower spikes are inclined, bearing flowers on the upper side.*

*Anthers are purple.*

*Typical diamond-shaped markings on lower tepals.*

G. sempervirens, *here in the Outeniqua Mountains, Eastern Cape.*

# Gladiolus stefaniae

**Named after Stefanie Pienaar, whose father Tom Pienaar assisted in collecting plants for the type specimens.**

*Gladiolus stefaniae* grows in the Langeberg and Potberg mountains of the Western Cape. It has bright red flowers with white streaks, which are pollinated by butterflies. It can be mistaken for *G. cardinalis* and *G. sempervirens*.

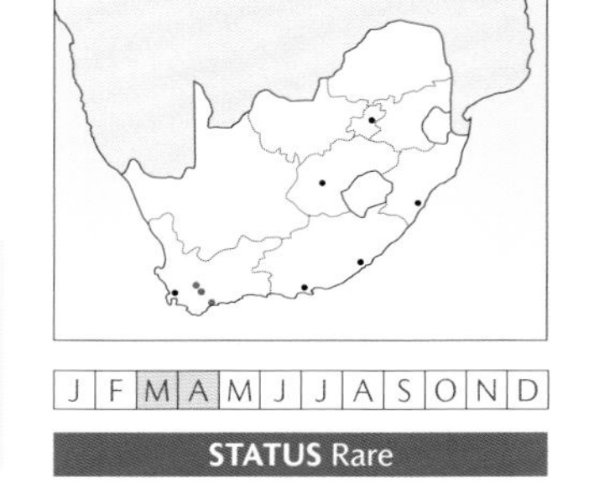

| J | F | M | A | M | J | J | A | S | O | N | D |
|---|---|---|---|---|---|---|---|---|---|---|---|

**STATUS** Rare

**DESCRIPTION** **Plant** 40–65cm. **Corm** globose, with papery tunics becoming fibrous with age. **Cataphylls** up to 8cm above ground, purple sometimes mottled with cream. **Leaves** 3–5 sheathing leaves on flowering stems, or 3 or 4 foliage leaves on non-flowering stems, ±5mm wide; with margins lightly raised and midrib thickened. **Spike** erect, unbranched, with 2–4 flowers. **Bracts** ±60mm, pale green or grey, slightly twisted. **Flowers** bright scarlet or carmine red; lower 3 tepals with a median white streak, tepals in outer whorl larger than those in inner. Perianth tube up to 45mm. **Anthers** yellow with purple lines, or dark purple. **Pollen** cream. **Capsules** up to 25mm, oblong. **Seeds** ovate, broadly and evenly winged. **Scent** unscented.

**DISTRIBUTION** Rare with a strange, disjunct distribution. Populations are known from the Potberg on the coast east of Bredasdorp, and from the Langeberg near Montagu.

**ECOLOGY & NOTES** At Potberg, a mountain of only 600m, plants grow on the lower west- and south-facing slopes, and have carmine flowers. In the Langeberg populations, plants are found at much higher elevations (between 700 and 1,400m), on rocky slopes in slightly wetter gullies, and have bright scarlet flowers. The plants have an interesting growth strategy that enables them to flower in autumn after a hot, dry summer. Large corms with sufficient stored nutrients sprout after the first rains in autumn and produce a flowering spike with only bract-like sheathing leaves. However, this same corm will probably not flower in the next season, as the sheathing leaves do not photosynthesise sufficient food to replace the large corm. Smaller corms produce a leafy shoot that manufactures carbohydrates and enables the corm to grow. Once the corm is sufficiently big and can sustain the production of a flowering stem and fruit, it will flower the following season.

**POLLINATORS** The butterfly *Aeropetes tulbaghia* (Mountain pride butterfly). See page 102 for more information on this butterfly.

**SIMILAR SPECIES** *G. stefaniae* is most likely to be confused with *G. cardinalis* and *G. sempervirens* as they are all red flowered. *G. carmineus* is sometimes mistaken for *G. stefaniae* because of similarities in form and habitat, but flowers are mauve-pink not red.

| | Locality | Flowering | Growth |
|---|---|---|---|
| ***G. stefaniae*** | Potberg and near Montagu in the Langeberg, dry sites | autumn | dormant in summer |
| ***G. cardinalis*** | southwestern part of the Western Cape, waterfalls | midsummer | does not go fully dormant |
| ***G. sempervirens*** | George to Kareedouw, moist sites | autumn | evergreen |

*The scarlet or carmine flowers of* G. stefaniae *attract Mountain pride butterflies.*

*Seed capsules dry and split, dispersing the winged seeds.*

*Note the purple lines on the anthers.*

*Note the purplish cataphylls and strongly raised leaf midrib.*

*Plants from Potberg, such as this one, grow at lower altitudes than those of the Langeberg.*

# Gladiolus carmineus

***carmineus* = referring to the dark red (carmine) flowers.**

Although endemic to the southwestern coast in the winter-rainfall area, *Gladiolus carmineus* flowers from February to March. It can be confused with *G. carneus* and *G. stefaniae*, but the latter has much larger flowers which are not radially symmetrical.

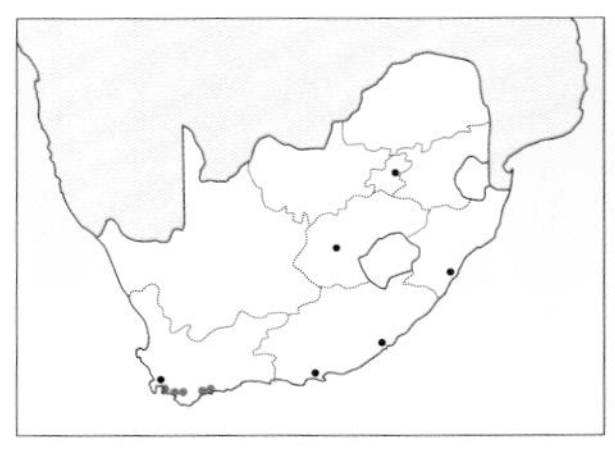

J F M A M J J A S O N D

**STATUS** Vulnerable

**DESCRIPTION** **Plant** 30–50cm. **Corm** globose, with membranous tuncis becoming fibrous and forming a neck around the stem. **Cataphylls** up to 8cm above ground, purple mottled with cream or green. **Leaves** 3–5 sheathing leaves on flowering spikes, non-flowering plants have 2 foliage leaves up to 10mm wide; with midrib lightly thickened. **Spike** erect, unbranched, with 2–6 flowers. **Bracts** up to 40mm, pale green to grey. **Flowers** pale to deep pink; lower 3 tepals with a median white streak surrounded by a mauve halo. Perianth tube up to 35mm. **Anthers** yellow. **Pollen** cream. **Capsules** obovoid, up to20mm. **Seeds** light brown, ovate, evenly winged. **Scent** unscented.

**DISTRIBUTION** Known only from the coast between Pringle Bay in the west to Cape Infanta in the east.

**ECOLOGY & NOTES** Plants are never out of sight of the sea, and grow with their corms wedged among rocks, often on cliffs. This species has a similar strategy to *G. stefaniae*: the flowering stems have reduced leaves, whereas non-flowering plants have foliage leaves. These leaves are totally dry by the time the flowers appear (on flowering stems from different corms).

**POLLINATORS** Thought to be *Aeropetes tulbaghia* (Mountain pride butterfly.) See page 102 for more information on this butterfly.

G. carmineus *is a summer-flowering plant that grows in the winter-rainfall area.*

**SIMILAR SPECIES** *G. carmineus* resembles *G. carneus*, but the latter flowers in spring. It is closely related to *G. stefaniae*, with which it could be confused.

| | Leaves | Flowering time | Flower |
|---|---|---|---|
| ***G. carmineus*** | sheathing leaves on flowering stems, foliage leaves on non-flowering plants | late summer to autumn | open pink flower, tube 30mm long, dorsal tepals up to 45mm and spreading |
| ***G. carneus*** | foliage leaves on flowering stems | early to midsummer | perianth tube 38mm, dorsal tepal erect or arched, up to 40mm |
| ***G. stefaniae*** | sheathing leaves on flowering stems, foliage leaves on non-flowering plants | autumn | large flowers with tube up to 45mm long, dorsal tepal up to 58mm |

Lanceolate leaves have thickened midribs.

*The median white streak has a mauve halo effect.*

*Flowers are almost radially symmetrical.*

G. carmineus *has a restricted range and always grows within sight of the sea, here at Hermanus.*

# Gladiolus rudis

***rudis*** **= raw, or 'scoop' or 'ladle', presumably referring to the scoop-shaped nectar guides**

*Gladiolus rudis* is endemic to the area between Caledon and Bredasdorp in the Western Cape. Its pale flowers appear around September and October. It has remarkable speckled leaf sheaths. It can be confused with *G. grandiflorus*, but the latter has a longer perianth and favours clay soils.

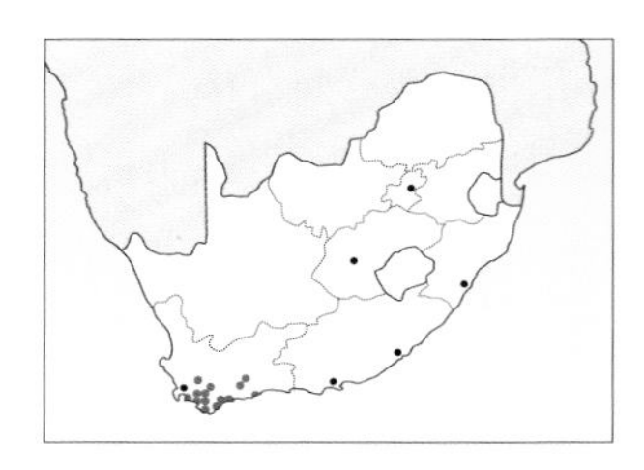

| J | F | M | A | M | J | J | A | S | O | N | D |
|---|---|---|---|---|---|---|---|---|---|---|---|

*Spade-shaped marks are outlined in a darker colour.*

**DESCRIPTION** **Plant** 15–55mm. **Corm** globose, with papery layers becoming fibrous. **Cataphylls** up to 10cm above ground, purple mottled with green and white. **Leaves** 4–7, sword shaped, up to 15mm wide; margins, midrib and 1 or 2 other veins thickened. **Spike** inclined to horizontal, unbranched, with 2–5 flowers. **Bracts** up to 50mm, green to flushed with brown or purple. **Flowers** cream to pale pink; lower tepals with yellowish spade-shaped mark outlined in a darker colour. Perianth tube up to 20mm. **Anthers** white to blue. **Pollen** white. **Capsules** unknown. **Seeds** up to 15mm long, dull brown, evenly winged. **Scent** unscented.

**DISTRIBUTION** Known only from the Stanford, Caledon and Bredasdorp areas of the Western Cape.

**ECOLOGY & NOTES** Plants grow in rocky areas in nutrient-poor sandstone soils on the lower and middle slopes of the Riviersonderend and Bredasdorp mountains, and on ridges close to Elim. We found plants on the hills close to Stanford, flowering among old fynbos, suggesting that they were not dependent on fire to flower.

*Cataphylls are mottled and rough, rising up the sword-shaped leaves.*

**POLLINATORS** Thought to be bees.

**SIMILAR SPECIES** *G. rudis* could be confused with *G. grandiflorus*, a closely related species.

| | Flower size | Flower markings | Habitat |
|---|---|---|---|
| ***G. rudis*** | perianth tube 20mm, filaments 12–14mm | lower tepals with spade-shaped mark outlined in darker colour | stony sandstone soil in fynbos |
| ***G. grandiflorus*** | perianth tube 27–35mm or longer, filaments 14–22mm | upper and lower tepals with dark linear stripe | heavy clay soil in renosterveld |

*Dull seeds are evenly winged.*

*Long green bracts extend the length of the tepals, which may be undulate.*

*A cream form with pale blue anthers.*

*A pale pink form.*

*Tepals may be darker on the reverse. Recorded at Stanford, Western Cape.*

G. rudis *growing along the Van der Stel Pass, between Bot River and Villiersdorp, Western Cape.*

# *Gladiolus grandiflorus*

***grandiflorus*** **= large-flowered.**

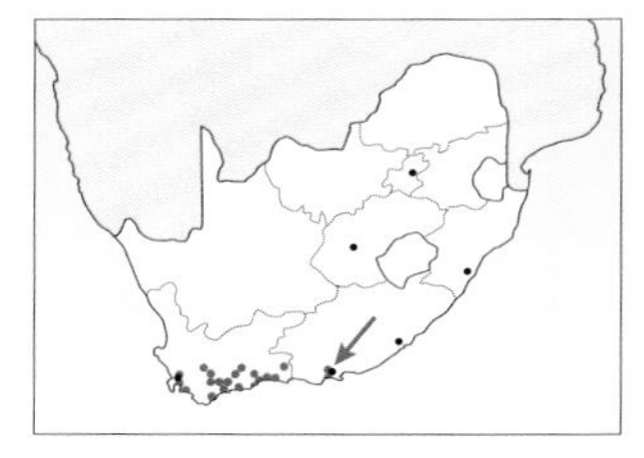

J F M A M J J A S O N D

Widespread in the winter-rainfall region, *Gladiolus grandiflorus* favours the clay soils of the renosterveld. Its cream or pale pink flowers have red median streaks. It can be confused with *G. rudis* or *G. floribundus*.

**DESCRIPTION** **Plant** 35–50cm. **Corm** globose, with papery tunics decaying into fibres. **Cataphylls** uppermost up to 10cm above ground, purple mottled green or white, in time forming a neck around the stem base. **Leaves** 4–7, sword shaped, 6–15mm wide; with margins, midrib and 1 or 2 other veins lightly thickened. **Spike** inclined, unbranched, with 3–6 flowers. **Bracts** green or brownish or purple. **Flowers** whitish, cream or pale pink; upper and lower tepals with a pinkish median streak. Upper tepals larger than the lower 3. Perianth tube 27–35mm. **Anthers** white, cream or purple. **Pollen** white or mauve. **Capsules** obovoid. **Seeds** light yellow-brown, large, broadly and evenly winged. **Scent** sometimes lightly scented.

*The species has large flowers.*

**DISTRIBUTION** From Bot River in the Overberg eastwards to Gqeberha (Port Elizabeth), Eastern Cape.

**ECOLOGY & NOTES** Plants are usually found in renosterveld in heavy clay soils. A distinctive form from the coast between Mossel Bay and Cape Infanta grows in coastal fynbos in calcareous sand.

**POLLINATORS** Sweet nectar attracts long-tongued bees, mostly belonging to the family Anthophoridae, including *Anthophora diversipes*.

*Dark tepals lighten as they open.*

**SIMILAR SPECIES** *G. rudis* and *G. floribundus* are very similar.

| | Habitat | Spike | Flower colour | Flower size |
|---|---|---|---|---|
| ***G. grandiflorus*** | clay, renosterveld | inclined | pink or white with median red stripes | perianth tube 27–35mm |
| ***G. rudis*** | stony sandstone soil in fynbos | inclined, unbranched | lower tepals with spade-shaped mark | perianth tube 20mm |
| ***G. floribundus*** | rocky sandstone in dry fynbos | strongly inclined to horizontal | tepals cream to greenish with median red stripes | perianth tube 40–70mm |

*Bracts are soft and green-purple. They may be folded in the midline.*

*Spikes are inclined or sometimes horizontal, as seen here.*

*Tepal margins may be undulate.*

G. rudis *has been confused with this cream form of* G. grandiflorus.

*Upper tepals are longer than the lower ones. Note the pale purple anthers.*

*Pale pink form at Stilbaai, Western Cape.*

# *Gladiolus floribundus*

***floribundus* = profusely flowering, up to 3 or 4 branches each with up to 8 flowers.**

The pale green to cream flowers of *Gladiolus floribundus* grow on strongly inclined spikes. It has a wide range across the Western and Eastern Cape and is usually found in drier areas.

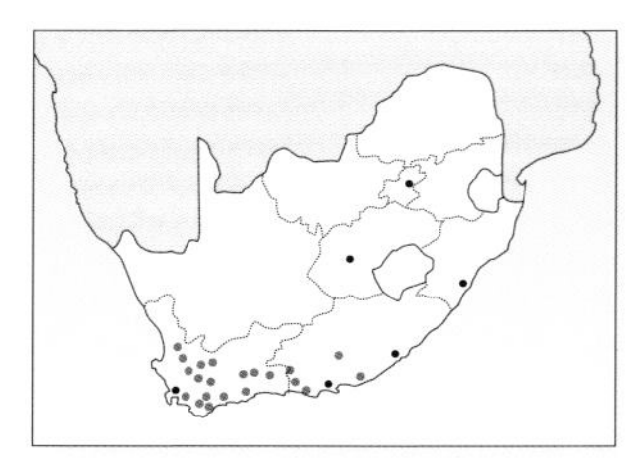

J F M A M J J A S O N D

**DESCRIPTION** **Plant** 25–45cm. **Corm** globose, with tunics coarsely textured. **Cataphylls** upper up to 8cm above ground, purplish and usually mottled with green or white, accumulating around leaf base to form a neck. **Leaves** 6–8, sword shaped, sometimes flaccid, 12–20mm wide. **Spike** inclined to nearly horizontal, often branched, with 3–8 flowers. **Bracts** outer 40–50mm long, dull green to grey-purple. **Flowers** cream or ivory to greenish; all tepals with a broad dark streak along the midlines. Tepals unequal, with the top 3 larger than the lower 3; tepal margins may be undulate. Perianth tube 40–60mm, usually longer than the bracts. **Anthers** purple. **Pollen** cream to purple. **Capsules** oblong, up to 40mm. **Seeds** yellowish brown, oval, wing broadly and evenly developed. **Scent** unscented.

*The tepals are unequal, with the lower tepals shorter than the laterals and dorsal.*

**DISTRIBUTION** Widespread from the Cederberg in the west to the Hlanganani district in the east. It is most commonly found between Worcester and Riversdale.

**ECOLOGY & NOTES** Usually found on sandstone-derived soil in stony places, but also known from heavier clay soils (in the Langkloof near Uniondale and at Bot River) and limestone (along the southern Cape coast).

**POLLINATORS** The long perianth tube and dark median stripes on the tepals suggest that the species is pollinated by long-tongued flies; two flies, *Philoliche gulosa* and *P. rostrata*, have been observed visiting the flowers.

**SIMILAR SPECIES** *G. floribundus* and *G. grandiflorus* are difficult to tell apart. Populations of *G. floribundus* with typical horizontal spikes and long perianth tubes are found on clay slopes near Bot River and in the Langkloof on the road to Uniondale. Similarly, we have photographed plants of *G. grandiflorus* with slightly inclined spikes on sandy soils close to Gqueberha (Port Elizabeth). The flower colour of the two species is similar as are tepal markings. Typical specimens can be separated and are distinctive, but there are many populations found on atypical 'wrong' substrates or whose perianth tubes or spikes differ from the type specimen.

| | Habitat | Spike | Flower colour | Flower size |
|---|---|---|---|---|
| ***G. floribundus*** | rocky sandstone in dry fynbos | strongly inclined spike, almost horizontal | tepals deep cream to greenish with median red stripes | lower tepals smaller than upper, perianth tube 40–70mm |
| ***G. grandiflorus*** | clay, renosterveld | lightly to strongly inclined spike | tepals pink or white with or without median red stripes | tepals more or less equal, perianth tube 27–35mm, about as long as bracts |

*Stems are flexed and usually branched, with strongly inclined spikes.*

*Old leaves and cataphylls form a neck around the base of the stem.*

*Tepals may be undulate.*

*Flower colour is variable; here an ivory form.*

G. floribundus *flowering en masse along the edge of the road between Bot River and Villiersdorp, Western Cape.*

# Gladiolus miniatus

***miniatus* = orange-coloured, named for the flower colour.**

*Gladiolus miniatus* has cream to salmon flowers with copious nectar, an adaptation for pollination by sunbirds. This species is endemic to limestone outcrops along the south coast of the Western Cape. Although the vegetative form of *G. miniatus* is identical to *G. floribundus* and *G. grandiflorus*, its flowers are unique.

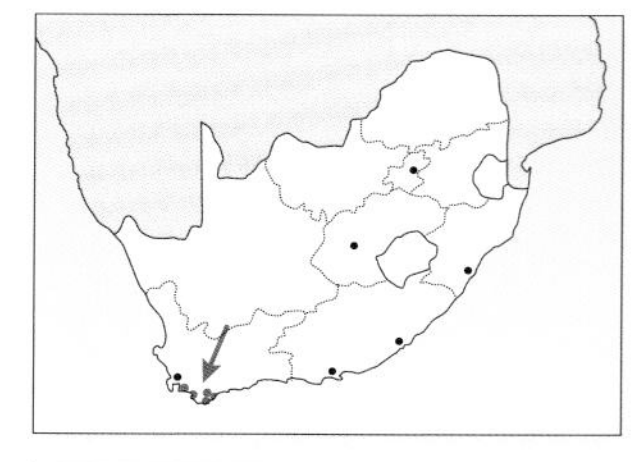

| J | F | M | A | M | J | J | A | S | O | N | D |
|---|---|---|---|---|---|---|---|---|---|---|---|

**STATUS** Vulnerable

**DESCRIPTION** **Plant** 15–40cm. **Corm** globose, with medium- to coarse-fibred tunics. **Cataphylls** upper up to 6cm above ground, purple, mottled with green and white. Old leaves and cataphylls form a thick neck around stem. **Leaves** 6, lanceolate, 7–18mm wide; margins and midrib thickened. **Spike** strongly inclined to horizontal, unbranched, 3–7 flowers. **Bracts** 40–55mm, dull greyish green can be flushed with purple. **Flowers** cream to light salmon at first, becoming deep salmon-pink to apricot; with dark median stripes on all tepals. Perianth tube 50–65mm. **Anthers** purple. **Pollen** purple. **Capsules** ovoid, up to 30mm. **Seeds** brown with darker body, ovate, broadly winged. **Scent** unscented.

*The cream to pale salmon flowers deepen to darker shades with age.*

**DISTRIBUTION** Occurs on limestone outcrops and hills from the mouth of the Bot River in the west to Riversdale in the east.

**ECOLOGY & NOTES** Populations are known from limestone outcrops along the coast close to the sea, and also from the limestone hills near Gansbaai. Plants grow with their corms in cracks between the rocks. We often found them flowering together with *Pelargonium betulinum*, *P. elegans* and *Satyrium carneum*. Threatened by urban development and invasive aliens.

*Spikes are horizontal with erect flowers on the upper side.*

**POLLINATORS** It is thought that these salmon- to orange-coloured flowers are pollinated by sunbirds. The flowers have large quantities of nectar (10 microliters per flower) with a low sugar concentration, which is usually associated with bird pollinators. This means that the four related species are adapted for different pollinators: *G. miniatus* by sunbirds, *G. rudis* by bees, *G. grandiflorus* by long-tongued bees and *G. floribundus* by long-tongued flies.

**SIMILAR SPECIES** There are many similarities between *G. miniatus*, *G. rudis*, *G. grandiflorus* and *G. floribundus*. They all flower in the same months. They all have a dense mass of old corm tunics, leaf bases and cataphylls around the corm and base of the stem; fans of short narrow leaves; and similarly shaped flowers on more or less inclined spikes. However, *G. miniatus* has only been found on limestone. It differs from the others in flower colour and perianth shape.

*Lanceolate leaves form a fan.*

*Decaying cataphylls and leaf bases are characteristic features.*

*Bracts reach less than half the length of the very long perianths.*

*Tepals may be lightly undulate.*

*Anthers and pollen are dark purple and form a striking contrast with the flowers.*

G. miniatus *grows on limestone outcrops, here at Arniston, Western Cape.*

*Rod Saunders stands amid* G. stokoei, *Riviersonderend, Western Cape.*

# Gladiolus woodii

**Named after the first curator of the Natal Botanic Gardens, John Medley Wood (1827–1915).**

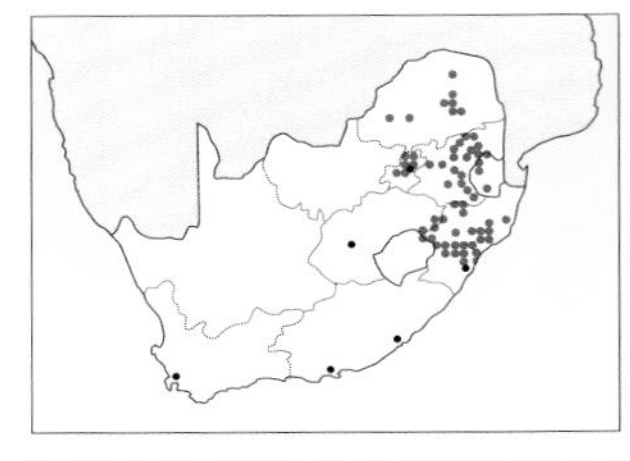

| J | F | M | A | M | J | J | A | S | O | N | D |
|---|---|---|---|---|---|---|---|---|---|---|---|

A common summer-rainfall plant, the flowers of *Gladiolus woodii* vary from maroon to reddish, yellow, lilac or grey.

**DESCRIPTION** **Plant** 20–45cm. **Corm** globose, with fine to coarse vertically fibred tunics. **Cataphylls** upper 3–5cm above ground, green to purple, hairy. **Leaves** 2 or 3 sheathing leaves on flowering plants, hairy. Non-flowering plants have 1 narrow, lanceolate leaf, densely hairy; margins and midrib thickened and raised. **Spike** inclined, unbranched, 6–11 flowers arranged one above the other. **Bracts** outer 12–16mm, green to purple-brown. **Flowers** pale yellow, pale lilac, oyster coloured to maroon or reddish brown, sometimes bicoloured; usually with median markings on the lower 2 lateral tepals, but very variable. Perianth tube ±8mm. **Anthers** blackish. **Pollen** cream to pale yellow. **Capsules** ovoid, 9–12mm long. **Seeds** reddish brown and opaque, with wing evenly developed. **Scent** variable: different colour flowers are differently scented; may be unscented.

*Delicate flowers alternate on an unbranched spike.*

**DISTRIBUTION** A widespread species occurring from the KwaZulu-Natal Midlands and the Drakensberg in the south, through Eswatini and Mpumalanga, to the Soutpansberg in Limpopo.

**ECOLOGY & NOTES** Grows in well-watered areas both at sea level and at higher elevations. Usually found in grassland in stony areas and on a variety of soils. Plants flower in early spring after the first rains have fallen. Saving resources, plants that flower produce only sheathing leaves that remain green for the growing season. Plants that do not flower produce a single long leaf. If plants flower in September, their seeds will ripen in early November, so they probably germinate in the same growing season.

*Pale lilac form, growing in the Wolkberg, Limpopo.*

**POLLINATORS** Thought to be long-tongued anthophorid bees.

**SIMILAR SPECIES** Mauve forms could be confused with *G. malvinus* and *G. oatesii* (section *Heterocolon*).

| | Size | Flower colour & markings | Leaves |
|---|---|---|---|
| ***G. woodii*** | 20–45cm | variable; 2 lower tepals with markings | pubescent |
| ***G. malvinus*** | 55–70cm | mauve-purple; all lower tepals with markings | sparsely pubescent or glabrous |
| ***G. oatesii*** (sect. *Heterocolon*) | 40–50cm | white-purple; lower tepals marked | no hair on sheaths and cataphylls |

*Leaves are hairy on both flowering and non-flowering plants.*

*The drooping spike straightens as the flowers open.*

*Bracts become dry as the flowers fade.*

*Lilac form, near eMkhondo, Mpumalanga.*

*Striking orange-yellow form, near Barberton, Mpumalanga.*

*Pink form, in the Wolkberg, Limpopo.*

*This pale form was recorded in the Wolkberg, Limpopo.*

*Despite being common in the summer-rainfall area,* G. woodii *was formally recorded only in 1880.*

# Gladiolus malvinus

***malvinus* = mauve, for the flower colour.**

The mauve and purple flowers of *Gladiolus malvinus* appear in October and early November. The species is restricted to the mountains of Mpumalanga. It can be confused with *G. woodii* in its lilac form but has larger flowers.

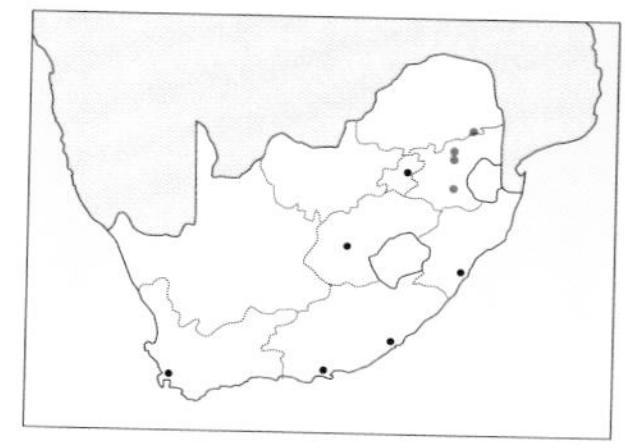

| J | F | M | A | M | J | J | A | S | O | N | D |
|---|---|---|---|---|---|---|---|---|---|---|---|

**STATUS** Vulnerable

**DESCRIPTION** **Plant** 55–70cm. **Corm** globose, with coarse-fibred tunic. **Cataphylls** upper reaching up to 8cm above ground, green and lightly hairy, accumulating with corm tunics to form a thick, fibrous neck at the base of the stem. **Leaves** 3 on flowering stems, the lowest 2 sheathing the stem and sparsely hairy or glabrous; non-flowering plants have 1 narrow, hairy leaf; raised margins and veins. **Spike** unbranched, 9–15 flowers ±20mm apart. **Bracts** 20mm long, grey-green flushed with purple. **Flowers** pale mauve; all 3 lower tepals marked with a dark purple streak and shaded with purple in the lower half; dorsal tepals larger than the lower, lower tepals clawed. Perianth tube ±10mm. **Anthers** blackish. **Pollen** cream. **Capsules** obovoid, 14mm. **Seeds** reddish brown, ovate, wing evenly developed. **Scent** unscented.

*The plant was described by botanists only in 1994.*

**DISTRIBUTION** Known only from one small area between Dullstroom and eMakhazeni in Mpumalanga. The species is classified as Vulnerable, owing to plantations and displacement by invasive alien plants.

**ECOLOGY & NOTES** Occurs in dolerite outcrops in clay in the hills between Dullstroom and eMakhazeni; grows in grassland. We saw plants flowering particularly well after a fire, together with species of *Berkheya*, *Helichrysum*, *Hypoxis* and *Peucedanum*, all also responding to the burn. *G. malvinus* flowers early in spring while the grass is still short and the relatively tall flower spikes with their distinctive-coloured flowers are easy to see.

*Lower tepals are lanceolate and have bold markings.*

**POLLINATORS** Probably the same long-tongued anthophorid bees that pollinate other short-tubed, small-flowered species of gladioli.

**SIMILAR SPECIES** Can be confused with *G. woodii*; although *G. oatesii* (section *Heterocolon*) looks similar, their ranges differ.

| | Size | Flower colour & markings | Leaves |
|---|---|---|---|
| ***G. malvinus*** | 55–70cm | mauve-purple; all lower tepals with markings | sparsely pubescent or glabrous |
| ***G. woodii*** | 20–45cm | variable including mauve; 2 lower tepals with markings | pubescent |

*Leaves may be glabrous or sparsely villous, as seen here.*

*Grey-green bracts are flushed with purple.*

*The flower is unusual in having black anthers.*

G. malvinus, *in flower near Belfast, Mpumalanga.*

# Gladiolus pardalinus

***pardalinus*** **= leopard-like, for the yellow and red flower markings.**

*Gladiolus pardalinus* is found in the Bushveld of Mpumalanga and Limpopo. It bears pale yellow, speckled flowers on floriferous spikes between mid-October and November. It is larger and has more leaves than closely related *G. woodii*.

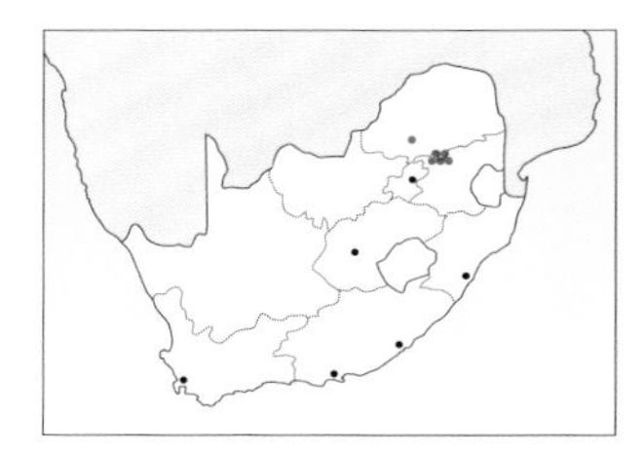

| J | F | M | A | M | J | J | A | S | O | N | D |
|---|---|---|---|---|---|---|---|---|---|---|---|

**STATUS** Rare

**DESCRIPTION** **Plant** 35–60cm. **Corm** ovoid to conic, with cartilaginous tunics decaying into vertical fibres. **Cataphylls** upper up to 5cm above ground, pale green, lightly hairy, decaying to form a fibrous neck at the base of the stem. **Leaves** 3 or 4 sheathing the stem, glabrous; non-flowering plants have 3 leaves; ±6mm wide and hairy; raised margins and midrib. **Spike** unbranched, 10–15 flowers. **Bracts** ±20mm, pale grey-green, often flushed with purple. **Flowers** pale yellow, with dark red speckles and streaks on all tepals except the uppermost; upper 3 tepals very widely separated from one another with 'windows' between them; all tepals have crisped margins. Perianth tube ±8mm. **Anthers** dark red-purple. **Pollen** cream. **Capsules** obovoid, up to 20mm. **Seeds** rust-brown, wing evenly developed. **Scent** faint acrid scent.

**DISTRIBUTION** Known only from a small area in western Mpumalanga and southern Limpopo. It occurs from Roossenekal in the east to Modimolle in the west.

**ECOLOGY & NOTES** A bushveld species growing among dolerite rocks together with grass and trees. Trees in the area include *Senegalia caffra* and *Cussonia transvaalensis*; there were several other bulbous plants including *Hypoxis* species, *Merwilla plumbea*, *Zantedeschia pentlandii* and *Gladiolus dalenii*. We found flowering plants in both burned and unburned veld, suggesting that flowering is not fire dependent. Plants were particularly abundant in rock cracks and with their corms firmly wedged under large boulders where they are safe from predation.

**POLLINATORS** Thought to be pollinated by long-tongued anthophorid bees.

**SIMILAR SPECIES** Although closely related to *G. woodii*, *G. malvinus* and *G. pubiger*, *G. pardalinus* is unlikely to be confused with them when they are in flower owing to its striking flower colour and speckles and streaks. Non-flowering plants are slightly more difficult to identify and are differentiated by leaf number and form.

*Dorsal and lateral tepals are widely separated with 'windows'.*

*Unbranched spikes bear as many as 15 flowers.*

*Leaves of non-flowering plants are hairy while those of flowering plants are smooth.*

*Spikes droop while in bud and straighten as flowers open.*

*Note the unbranched spike, speckled tepals and dark anthers.*

*The leopard-spotted tepals give the species its name.*

G. pardalinus *growing among dolerite boulders at Stofberg, Limpopo.*

# Gladiolus pubiger

***pubiger*** **= referring to the hairs on the cataphylls and leaves.**

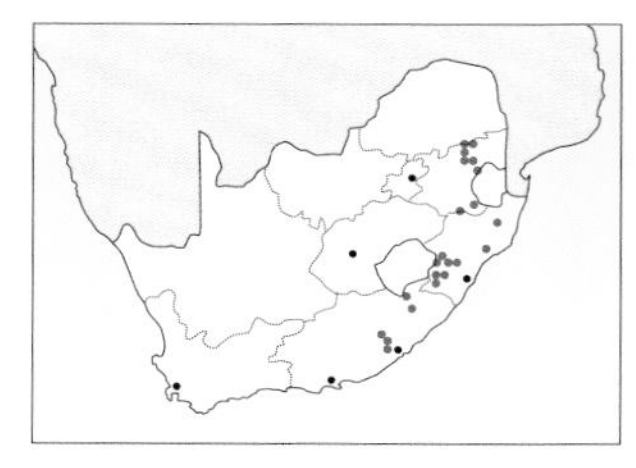

J F M A M J J A S O N D

*Gladiolus pubiger* is widespread in the eastern parts of South Africa. Its pale yellow flowers usually appear between September and November but can appear later at higher altitudes. When in flower it can be confused with *G. woodii*.

**DESCRIPTION** **Plant** 30–50cm. **Corm** globose, leathery tunics decaying into vertical fibres. **Cataphylls** upper reaching to 6cm above ground, densely hairy. **Leaves** 2 or 3, the lowermost sheathing the stem and very hairy, with short 4mm-wide blade; margins thickened and hyaline, midrib slightly raised; non-flowering plants have 1 hairy leaf. **Spike** inclined, unbranched, 4–9 flowers. **Bracts** up to 18mm, pale grey-green to pink or purplish. **Flowers** usually pale yellow tinged with green, occasionally blueish mauve; tepals almost equal, facing forward, usually unmarked. Perianth tube 8mm. **Anthers** dull yellow or purple. **Pollen** cream. **Capsules** up to 10mm. **Seeds** rusty brown, small, wing evenly developed. **Scent** may have strong gardenia scent in the morning.

*Flowers are usually pale yellow (as seen above) or blue-mauve, and sweetly scented.*

**DISTRIBUTION** A disjunct distribution, possibly due to poor collecting. Plants have been recorded in the Eastern Cape in the Stutterheim area, in KwaZulu-Natal inland of Durban, and in Eswatini and Mpumalanga as far north as Pilgrim's Rest and Mashishing. The species has not been recorded in the large areas between these sites.

**ECOLOGY & NOTES** Grows in rocky grassland; seems to flower only in the season after a fire. We saw plants close to Barberton and south of Mashishing; both populations were hill slopes. Flowering time seems to vary considerably, perhaps relative to the first spring rains in different areas. It is very difficult to see the flowering plants as they are not tall and the flowers 'disappear' in the green grass as they are a similar colour.

*Globose capsules release evenly winged seeds on drying.*

**POLLINATORS** Thought to be long-tongued bees.

**SIMILAR SPECIES** The closest relative is thought to be *G. parvulus*; the distributions of these two species overlap in KwaZulu-Natal.

| | Flower colour | Cataphylls | Flowering leaves |
|---|---|---|---|
| ***G. pubiger*** | yellow, all tepals equal | hairy, appearing above ground | 2 or 3 hairy, 1 sheathing, others arising below the flowers |
| ***G. parvulus*** | pink, all tepals equal | hairless, not visible above ground | 2 hairy, 1 sheathing, 1 minute leaf just below flowers |
| ***G. woodii*** | variable, tepals unequal | hairy, appearing above ground | 2 or 3 sheathing, hairy |

*Leaves and stem are hairy.*

*Spikes bear several flowers; note the pale green-and-pink bracts.*

G. pubiger *is not always easy to see, which may account for its disjunct distribution record.*

# Gladiolus parvulus

***parvulus*** **= small, referring to plant and flower size.**

The small, pale pink, nodding flowers of *Gladiolus parvulus* are borne on fine spikes from October to early December. This species is restricted to southern KwaZulu-Natal and Lesotho, where it grows on sandstone soils.

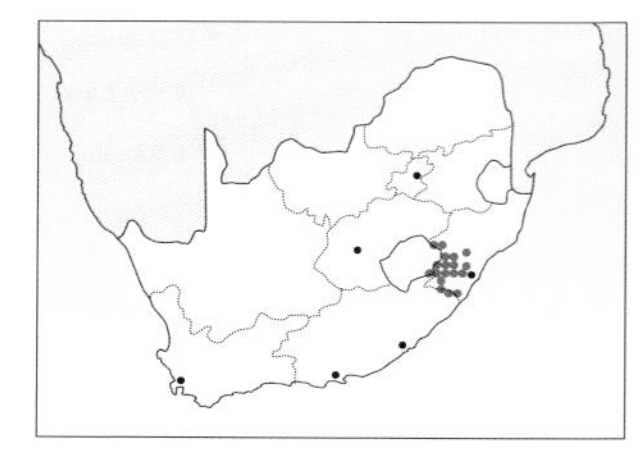

| J | F | M | A | M | J | J | A | S | O | N | D |
|---|---|---|---|---|---|---|---|---|---|---|---|

**DESCRIPTION** **Plant** 15–35cm. **Corm** globose, with leathery to woody tunics. **Cataphylls** do not reach above ground. **Leaves** 2, lower almost entirely sheathing, ribbed and slightly hairy, upper leaf minute and barely visible. Non-flowering plants have 1 slightly hairy leaf, ±2mm wide; with moderately thickened margins and midrib. **Spike** slightly inclined, unbranched, slender and wiry, 1–5 flowers. **Bracts** ±10mm, grey-purple. **Flowers** pale pink; tepals equal in size and flowers radially symmetrical; lower tepals may have dark markings on the midline, but some are unmarked. Perianth tube short, up to 5mm. **Anthers** pale pink. **Pollen** creamy yellow. **Capsules** globose to oblong, up to 7mm. **Seeds** light brown, ovate, evenly winged. **Scent** unscented.

*The pale pink flowers are radially symmetrical.*

**DISTRIBUTION** Found only in southern KwaZulu-Natal and eastern Lesotho. The distribution range is from the Zuurberg and Sehlabathebe in the south, to Howick and Cathedral Peak in the north.

**ECOLOGY & NOTES** Plants grow among rocks, often on rock pavements, but also in short grassland. Associated plants include *Watsonia lepida*, *Rhodohypoxis* and *Ledebouria* species, as well as low-growing *Helichrysum* and *Gerbera* species. At first sight, these small-flowered plants do not resemble a gladiolus; they look very ixia-like, and it is only when one realises that ixias do not occur in the same region as *G. parvulus* that this plant can be correctly identified. Many of the plants that we saw and photographed were only ±10cm in height, but because of their habitat on rock pavements, they did not need to be taller.

**POLLINATORS** Thought to be small, short-tongued bees.

*Spikes bear up to five flowers.*

**SIMILAR SPECIES** *G. parvulus* cannot be mistaken for any other species when it is in flower. When not in flower, it could be confused with *G. pubiger*.

| | Flower colour | Cataphylls | Flowering leaves |
|---|---|---|---|
| ***G. parvulus*** | pink, all tepals equal | not visible above ground | 2 hairy, 1 sheathing, 1 minute leaf just below flowers |
| ***G. pubiger*** | yellow, all tepals equal | hairy, appearing above ground | 2 or 3 hairy, 1 sheathing, others arising below the flowers |

*Note the short perianth tube.*

*Flowers appear to float on wiry stems and spikes.*

*Some flowers have no markings.*

*A delicate form with fine lines in the throat.*

G. parvulus, *encountered near Boston, KwaZulu-Natal.*

# Gladiolus hirsutus

***hirsutus*** **= hairy, referring to the leaf sheaths and blades.**

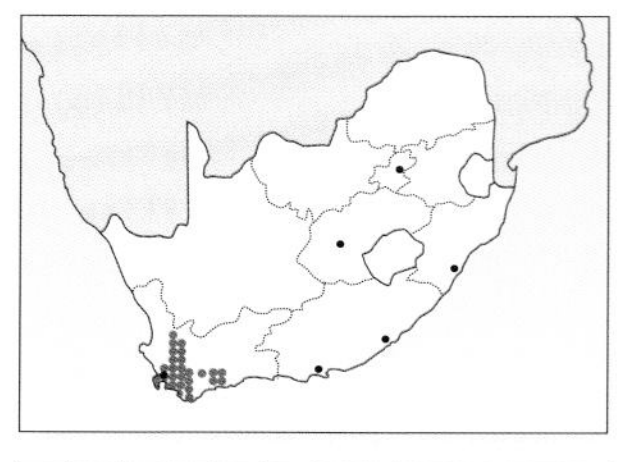

J F M A M J J A S O N D

*Gladiolus hirsutus* is widespread in the winter-rainfall area. Its flowers range from pink to mauve and occasionally white, and it has very hairy leaves. The flowering period is mid-July to September. It can be confused with *G. caryophyllaceus*, but the latter has larger, scented flowers.

**DESCRIPTION** **Plant** 35–50cm. **Corm** globose, with coarsely fibred tunics. **Cataphylls** uppermost dark green or purple, usually hairy. **Leaves** 3 or 4, usually hairy with relatively short blades, ±8mm wide; margins and midrib lightly thickened. **Spike** slightly inclined, unbranched, 3–6 flowers. **Bracts** up to 26mm, green, may be flushed grey or purple. **Flowers** pale to deep pink to mauve, occasionally white; tepals are unequal with the lower 3 having nectar guides of dark streaks or spots on a yellow or cream zone. Perianth tube up to 26mm. **Anthers** pink. **Pollen** white. **Capsules** ovate, up to 20mm. **Seeds** ovate, wing usually well developed, seed body large. **Scent** unscented.

*A Shaggy monkey beetle visits; monkey beetles are opportunistic pollinators.*

**DISTRIBUTION** Occurs in the southwestern Cape from the Kouebokkeveld and the Cape Peninsula in the west to the Langeberg near Robinson Pass in the east.

**ECOLOGY & NOTES** Found in a variety of habitats ranging from sandy flats near Bot River to lower mountain slopes in the Bainskloof and Du Toitskloof mountains, to rocky outcrops on the higher-altitude slopes of the Kouebokkeveld. The soils are usually derived from sandstone or granite, and plants are generally in rocky areas where the corms are protected from predators. Flowering seems to vary from year to year with the same population of plants flowering at different times, perhaps depending on the rain. Plants flower particularly well in the first few years after a fire, but we have seen them flowering in fairly old fynbos at Bainskloof.

**POLLINATORS** Appear to be long-tongued bees including honeybees. Later in the spring, flower visitors include nectar-seeking solitary bees in genus *Anthophora*, and monkey beetles, which visit opportunistically.

**SIMILAR SPECIES** *G. hirsutus* is sometimes confused with *G. caryophyllaceus*, which has much larger, highly scented flowers, and short, broad leaves with no hairs. On Franschhoek Pass we found plants of *G. hirsutus* flowering together with the unrelated species *G. blommesteinii* (section *Homoglossum*); the flowers of the two are very similar in appearance.

| | Leaves | Perianth tube | Markings on lower tepals |
|---|---|---|---|
| ***G. hirsutus*** | 3 or 4, up to 8mm wide, hairy | 15–26mm | irregularly spotted or streaked lines of dark pink |
| ***G. blommesteinii*** (sect. *Homoglossum*) | 4, 1.5mm wide, not hairy | 13–24mm | streaks, lines and spots of red, blue or purple |

*The somewhat hairy leaves and stem give the species its name.*

*Spikes are unbranched and inclined, with three to six flowers.*

*White form, seen in the Kouebokkeveld.*

*Pale pink form; note nectar guides on lower tepals.*

*Deep pink form, found near Franschhoek.*

*Mauve form, Stanford.*

G. hirsutus *on the mountain slopes near Bainskloof, Western Cape.*

# Gladiolus caryophyllaceus

***caryophyllaceus*** **= from *Dianthus caryophyllaceus*, a pink carnation with the same colour flowers and a similar scent.**

*Gladiolus caryophyllaceus* occurs mostly in the western parts of the Western Cape, up to Namaqualand. Its deep pink or sometimes mauve flowers are strongly scented and appear from mid-August to end September.

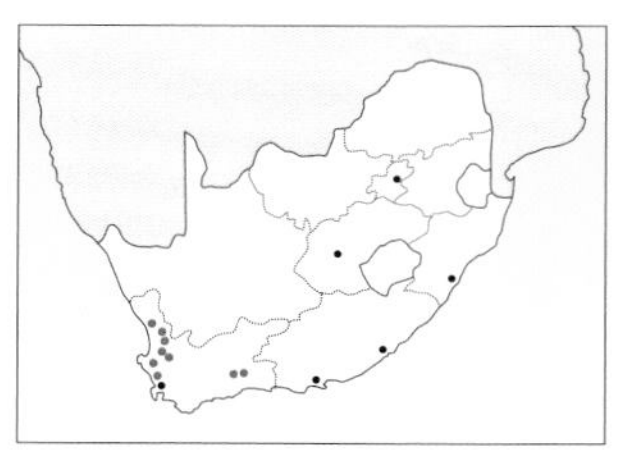

| J | F | M | A | M | J | J | A | S | O | N | D |
|---|---|---|---|---|---|---|---|---|---|---|---|

*This was one of the earliest species to be recorded, in 1685; it was scientifically described in 1768.*

**DESCRIPTION** **Plant** 60–110cm. **Corm** globose, with coarsely fibred tunics. **Cataphylls** upper grey-green to purple, usually hairy. **Leaves** 4 or 5, sword shaped, superposed or forming a fan, 10–27mm wide, hairy on the leaf sheaths; margins, midrib and secondary veins thickened, upper 2 leaves cauline with long blades. **Spike** erect, unbranched, 4–8 flowers. **Bracts** pale green to greyish. **Flowers** pink or mauve, occasionally cream; intensely pigmented along midline of tepals, lower tepals marked with streaks and spots. Perianth tube 30–40mm. **Anthers** cream. **Pollen** yellow. **Capsules** unknown. **Seeds** translucent to red-brown, ovate, wing evenly developed. **Scent** strong carnation-like scent.

**DISTRIBUTION** Extends from Mamre and Hopefield, northwards through the Cederberg and Nieuwoudtville areas, to the granite ridges of southern Namaqualand. It has also been recorded in the Hex River Mountains and the Swartberg.

**ECOLOGY & NOTES** Usually found in dryish sandy or stony areas in sandstone-derived soils. Often associated with clumps of restios, which may offer protection from predators or provide cooler soil conditions. Although the distribution range is quite extensive and it has been classified as Least Concern because mountainous populations are relatively inaccessible, we rarely see these plants except on road verges. Rooibos tea and potatoes grow on the same dry, sandy soils as this species and large areas have been ploughed for their cultivation, to the detriment of many bulbous and cormous plants. Over the last 40 years this species has become increasingly uncommon and it is probably threatened in South Africa. However, it has naturalised in parts of Western Australia where it is regarded as a weed.

**POLLINATORS** Thought to be long-tongued bees.

**SIMILAR SPECIES** May be confused with *G. hirsutus* (smaller flowers and little to no scent) and *G. guthriei*.

| | Perianth tube | Flowering time | Leaves |
|---|---|---|---|
| ***G. caryophyllaceus*** | 30–40mm | Aug–Sep | 10–27mm wide, hairy |
| ***G. hirsutus*** | 15–20mm | Jun–Sep | 3–8mm wide, hairy, long, leathery, trailing in summer |
| ***G. guthriei*** | 20–27mm | Apr–Jun | 3–20mm wide, hairy, margins not thickened |

*Striking flowers make this species a popular feature at flower shows.*

*Long bracts support elongate perianth tubes.*

*The erect spikes carry up to eight flowers.*

*Note the hooded dorsal tepal, cream anthers and yellow pollen.*

*This pale form shows off deep colouring along the midlines.*

*A pink form thrives in sandy soil at Nieuwoudtville.*

*Cream anthers with yellow pollen are seen on this bright mauve form at Nieuwoudtville.*

G. caryophyllaceus, *growing on the side of Moedverloor Road, near Nieuwoudtville.*

# Gladiolus rhodanthus

***rhodanthus* = rosy-flowered, referring to the flower colour.**

*Gladiolus rhodanthus* is known from only one locality in a winter-rainfall area, high in the mountains near Villiersdorp. Its deep pink flowers marked with white appear around December and January.

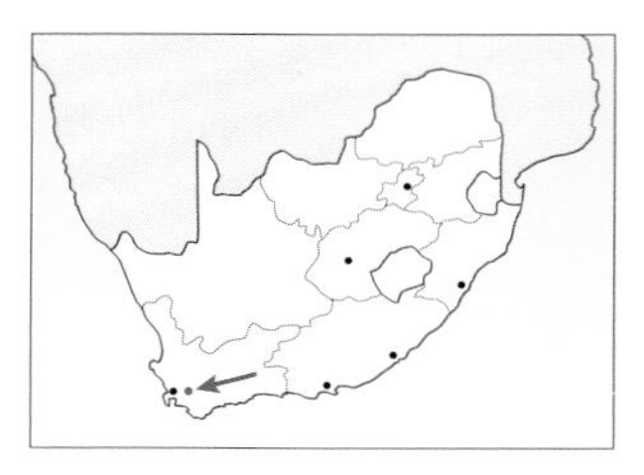

J F M A M J J A S O N D

**STATUS** Rare

**DESCRIPTION** **Plant** 30–50cm. **Corm** globose, with medium to finely textured tunics forming a neck around the stem. **Cataphylls** uppermost green or purple above ground. **Leaves** 3, all hairy, 2 lower ones sheathing, blades 3–6mm wide; midrib and second pair of veins thickened, margins not thickened. **Spike** unbranched, 2–5 flowers. **Bracts** 25–40mm, green. **Flowers** pink; tepals unequal, lower 3 tepals with a spear-shaped whitish streak edged in dark pink. Perianth tube 25–36mm. **Anthers** light mauve. **Pollen** cream. **Capsules** ellipsoid. **Seeds** golden brown, evenly winged. **Scent** unscented.

*The delicate flowers are streaked on the lower tepals.*

**DISTRIBUTION** Known only from the Stettynsberg mountain near Villiersdorp, on north-facing slopes, at an altitude of 1,700–1,800m.

**ECOLOGY & NOTES** The plants we found had their corms firmly wedged into rock cracks and many were on small ledges on cliffs. The plants presumably manage to grow and flower in midsummer because of the frequent mist driven into the high mountains by the southeast wind. It was extremely misty and damp when we photographed the plants.

**POLLINATORS** Pollinated by a long-proboscid fly of the genus *Moegistorhynchus* (family Nemestrinidae). This fly has a 20mm-long proboscis, enabling it to reach the nectar held in the lower third of the tube. It appears that different species of plants have adapted to make use of a common pollinator; other pink-flowering plants found in these mountains are pollinated by the same insect and form a pollination guild. On the day that we photographed this species, we also saw large numbers of pink *Watsonia paucifolia* and an erica species that we tentatively identified as *Erica savilea*. Both of these species were flowering well, probably owing to a fire two years previously.

*Specialist long-proboscid flies pollinate this species and other pink-flowered plants in the area.*

**SIMILAR SPECIES** Most similar to *G. hirsutus* and *G. caryophyllaceus*, but flowering times differ.

| | Perianth tube | Flowering time | Habitat |
|---|---|---|---|
| ***G. rhodanthus*** | up to 36mm | Dec–Jan | rocky crevices, high altitude |
| ***G. hirsutus*** | up to 26mm | Jul–Sep | among rocks, low to medium altitudes |
| ***G. caryophyllaceus*** | 30–40mm | Aug–Sep | dry habitats, open slopes or deep sands |

*Stems are inclined, sometimes almost horizontal; note the thickened midrib and second pair of veins.*

*Frequent mists provide sufficient moisture for summer flowering.*

*This rare species has a unique habitat.*

*Corms wedge into rock cracks at high altitudes, such as on Stettynsberg, Villiersdorp.*

G. rhodanthus *was scientifically described in 1999, here growing at Stettynsberg, Villiersdorp.*

# Gladiolus guthriei

**Named after Francis Guthrie (1831–1899), who made many plant collections in the southern Cape in the mid-19th century.**

*Gladiolus guthriei* is found from the Bokkeveld to the Potberg mountains of the Western Cape. Sweetly scented, pink to brown flowers appear between April and June, and are pollinated by night-flying moths.

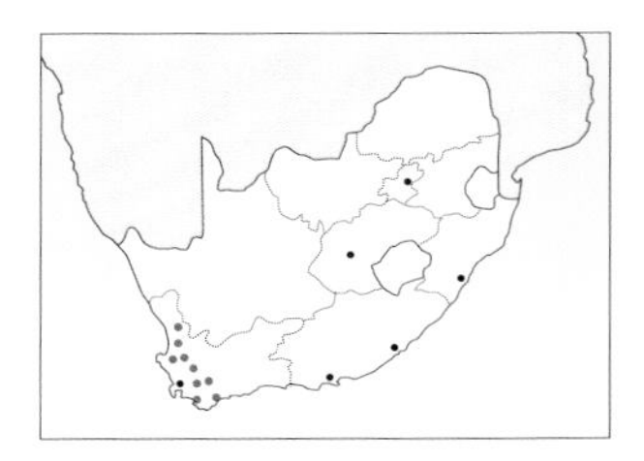

J F M A M J J A S O N D

**DESCRIPTION** **Plant** 40–70cm. **Corm** globose, with coarsely fibred tunics. **Cataphylls** of flowering shoots, up to 3cm above ground, brownish purple and may be hairy. **Leaves** usually 3, mostly sheathing the stem, smooth or hairy; margins, midrib and sometimes a second pair of veins thickened on the short blades; non-flowering plants have 1 leaf, 12–20mm wide; seedlings always have hairy leaves. **Spike** unbranched, 3–9 flowers. **Bracts** 20–40mm, dull green to greyish. **Flowers** very variable but distinctive, mainly dull pink to purple; tepals unequal, with dull yellow lower tepals often speckled and streaked with brown. Perianth tube up to 27mm. **Anthers** brownish. **Pollen** yellow. **Capsules** ellipsoid. **Seeds** ovate with a broad wing. **Scent** strong clove-like scent.

*Flowering in autumn and winter,* G. guthriei *was scientifically poorly known and was described only in 1917.*

**DISTRIBUTION** Found from the Bokkeveld escarpment in the north, southwards towards the Cape Peninsula (although it has not been recorded on the peninsula itself), and eastwards to the western end of the Langeberg and Potberg.

**ECOLOGY & NOTES** Usually found in rocky situations in sandstone – either on rocky pavements (as on the Gifberg) or on rocky slopes. *G. guthriei* flowers particularly well after fire. Cryptically coloured, it is often not seen, but can be detected by its strong, distinctive scent. Within a couple of years after fire, most of the plants stop flowering and only the broad leaves are visible in rock cracks.

*The clove-scented flowers are pollinated by moths.*

**POLLINATORS** Small night-flying moths attracted to the clove-like scent.

**SIMILAR SPECIES** Although variable, the flowers of *G. guthriei* have a distinctive shape and scent. *G. caryophyllaceus* is sometimes mistaken for *G. guthriei*, but they flower at different times.

*Leaves on flowering plants are short and sheath the stem.*

*Note how the lower tepals curve deeply downwards.*

*The lower and lateral tepals may have pronounced median streaks.*

*Erect stems bear three to nine flowers, sometimes more.*

*Unusual markings, flowering time and scent make this species unlikely to be mistaken for any other.*

G. guthriei *stands amid rocks at Bainskloof, Western Cape.*

# *Gladiolus emiliae*

**Named after the plant collector Emily Ferguson, who was active in the Riversdale and Swellendam areas in the 1920s and '30s.**

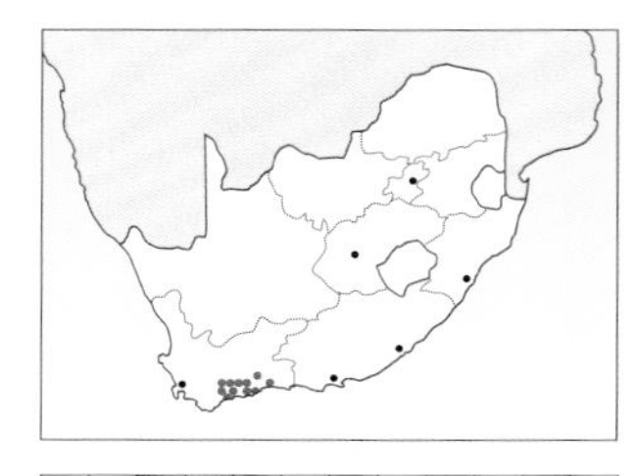

| J | F | M | A | M | J | J | A | S | O | N | D |
|---|---|---|---|---|---|---|---|---|---|---|---|

**STATUS** Near Threatened

The pale yellow to brownish flowers of *Gladiolus emiliae* are borne on inclined spikes between March and April, an unusual time for plants to flower in the winter-rainfall area. This uncommon species is found along the Langeberg and Riviersonderend mountains; it is unlikely to be confused with any other.

**DESCRIPTION** **Plant** 45–60cm. **Corm** globose, with papery tunics. **Cataphylls** upper up to 9cm above ground, green and sparsely hairy. **Leaves** 2, entirely sheathing, 1 arising near the ground, and the other on the upper third of the stem; non-flowering plants and seedlings have 1 hairy leaf with a hairy purple cataphyll; leaf is ±7mm wide and 60–75mm long; margins and midrib lightly thickened. **Spike** inclined, unbranched, 3–8 flowers. **Bracts** ±20mm long, green. **Flowers** brownish to dull or pale yellow, dotted or streaked with reddish brown to maroon; lower 3 tepals each have a dark red streak in the midline. Perianth tube 35–45mm, sharply curved at the upper end. **Anthers** brownish on the upper side, dark yellow below. **Pollen** cream. **Capsules** obovoid. **Seeds** yellow-brown, wing well developed. **Scent** spicy to sweetly scented.

*Flowering after fire in the Langeberg.*

**DISTRIBUTION** This uncommon species occurs in the southern Cape coastal plain, and has been recorded from the foothills of the Riviersonderend Mountains in the west, 250km eastwards along the Langeberg to George.

**ECOLOGY & NOTES** Grows in rocky areas on a variety of soil types (clay to sand) and at the sandstone–shale interface; found in either fynbos or mixed fynbos and renosterveld. We first saw this species in the foothills of the Langeberg in the first year after a fire. As it was in flower in autumn, in an area where not much else had grown after a recent fire, it was easily seen (and smelled). However, we have also seen this species in flower in older vegetation, and in these areas it is more difficult to notice on account of its cryptically coloured flowers.

**POLLINATORS** Apparently moths, which are attracted to the rich, sweet scent and the yellowish-brown speckled flowers. A number of night-flying moths such as *Cucullia extricata* and *C. inaequalis*, with long tongues of up to 30mm, have been recorded feeding on this species.

**SIMILAR SPECIES** In flower, *G. emiliae* cannot be mistaken for any other species. It flowers in autumn, has unusual maroon or brown to yellow densely speckled flowers with a long tube, and is strongly scented by day and night. It also has entirely sheathing leaves on the flowering stems.

*Narrow leaves sheath the stem; cataphylls may reach as high as 9cm above ground.*

*The perianth is long with a sharp curve at the upper end.*

*Dense maroon speckles make these flowers appear deep red.*

*It is believed that the flowers are pollinated by moths attracted to the scent and speckled flowers.*

*This form was encountered in older vegetation near Riviersonderend.*

G. emiliae *flowers most prolifically after fire; here seen in the foothills of the Langeberg.*

# *Gladiolus overbergensis*

**From the Overberg, the southwestern part of the Western Cape.**

*Gladiolus overbergensis* is restricted to a small area in the Western Cape. Its red and orange flowers bloom from late July to mid-September and are assumed to be pollinated by sunbirds.

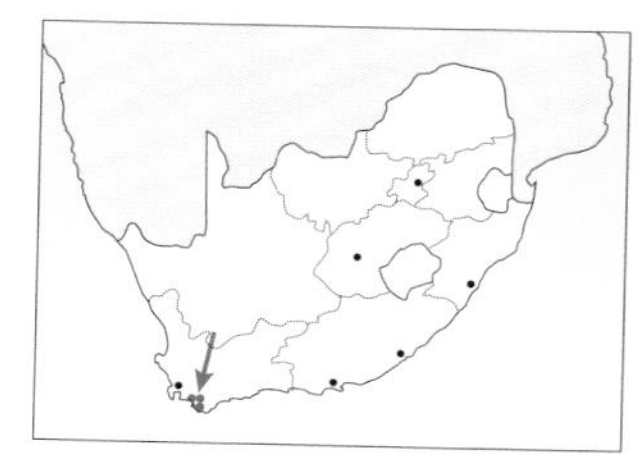

| J | F | M | A | M | J | J | A | S | O | N | D |
|---|---|---|---|---|---|---|---|---|---|---|---|

**STATUS** Vulnerable

**DESCRIPTION** **Plant** 35–50cm. **Corm** globose, with leathery to woody tunics. **Cataphylls** uppermost not usually above ground. **Leaves** 4, the upper one sheathing, 3 basal, up to 4mm wide; midrib and 1 pair of veins lightly raised. Leaves appear hairless but have very small, recurved, prickly scabrid hairs on the veins, which can be felt and seen using a hand lens. **Spike** occasionally branched, 2–5 flowers. **Bracts** up to 40mm long, pale green or flushed grey-purple. **Flowers** scarlet; with lower 3 tepals orange or yellow, tepals spotted and streaked with darker red; upper tepals larger than lower. Perianth tube long, up to 55mm. **Anthers** yellow. **Pollen** pale yellow. **Capsules** unknown. **Seeds** unknown. **Scent** unscented.

**DISTRIBUTION** Known only from the Hermanus, Napier and Elim areas of the southwestern Cape. Although it has been reported to be in the Bredasdorp Mountains and on the Soetanysberg, no specimens have been collected from those areas.

**ECOLOGY & NOTES** The plants that we saw were growing on sandy clay soil, possibly on the contact zone between granite and Table Mountain sandstone. It is thought that the plants flower best after fire. The specimens that we photographed had not been burned, but were in an area that had been cleared and the vegetation cut as a firebreak. They were flowering among fynbos, including species of *Elegia* and *Berzelia*. This striking plant is rarely seen, probably owing to habitat destruction as a result of farming. It occurs in an area that is highly cultivated with vines, fruit trees, cut flower orchards and wheat, and its habitat is severely threatened.

**POLLINATORS** It is assumed that this species with its long-tubed, bright red flowers is adapted for pollination by sunbirds.

**SIMILAR SPECIES** Owing to the striking flower colour, and its limited distribution, it is unlikely that one would mistake this species for any other. While it resembles *G. merianellus*, the distribution ranges of the two species do not overlap.

*The unusual flowers are distinctive in shape although similar in colour to those of* G. merianellus.

*The bright red flowers are assumed to be adapted for pollination by sunbirds.*

*The upper leaf sheaths the stem; leaf veins are prickly and scabrid.*

*Perianth tubes are long and sharply curved; note the elongate grey-purple bracts.*

G. overbergensis *flowering near Napier, with vineyards in the background.*

# *Gladiolus merianellus*

(= *Gladiolus bonaspei*)

**Named after Maria Sibylla Merian (1647–1717), a German naturalist and illustrator.**

Known only from the Cape Peninsula, *Gladiolus merianellus* flowers from July to September. Its bright orange flowers are thought to be pollinated by sunbirds.

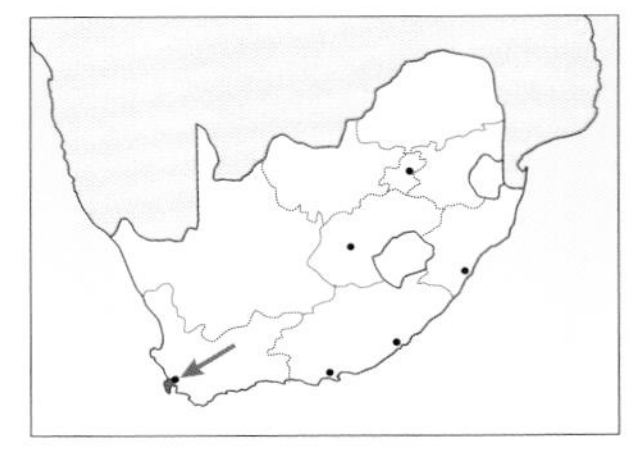

| J | F | M | A | M | J | J | A | S | O | N | D |
|---|---|---|---|---|---|---|---|---|---|---|---|

**STATUS** Near Threatened

**DESCRIPTION** **Plant** 35–50cm. **Corm** globose, with leathery to woody tunics. **Cataphylls** not usually seen above ground. **Leaves** 3, hairy, ±5mm wide; margins and midribs lightly thickened; upper leaf short and bract-like. **Spike** erect, unbranched, 2–7 flowers. **Bracts** up to 20mm, green or flushed grey-purple. **Flowers** bright orange; sometimes lower tepals yellow, usually speckled with scarlet; tepals almost round and sub-equal, with the upper 3 slightly larger than the lower 3. Perianth tube ±40mm. **Anthers** purple. **Pollen** pale yellow. **Capsules** ovoid. **Seeds** translucent golden yellow, broadly winged. **Scent** unscented.

**DISTRIBUTION** Found only on the Cape Peninsula at Cape Point, around Simonstown and in Silvermine.

**ECOLOGY & NOTES** Always found in peaty sands both near the coast and at slightly higher elevations. Flowers best in the year or two after a fire, but also seen flowering in quite old fynbos.

**POLLINATORS** Presumed to be sunbirds, the common pollinators of red- or orange-flowered species with long tubes.

**SIMILAR SPECIES** *G. merianellus* is distinctive, with its orange flowers and very limited distribution range. There is no other species that it could be confused with. It was formerly named *G. bonaspei*, a reference to its distribution on the Cape of Good Hope.

G. merianellus *flowering after fire in peaty sand. The flowers are more inflated than those of* G. overbergensis.

*The brilliant orange hues give some flowers a glowing quality.*

*The dorsal tepal is almost parallel with the purple anthers.*

*Perianth tubes curve abruptly before flaring.*

*Orange flowers are adapted for sunbird pollination.*

*An uncommon and cheerful sight,* G. merianellus *flowering at Silvermine, Cape Town.*

# Gladiolus aureus

***aureus* = golden, named for the flower colour.**

*Gladiolus aureus* is one of the rarest gladioli. It is known only from the Cape Peninsula. Small, bright yellow flowers distinguish it from the otherwise vegetatively similar *G. merianellus*.

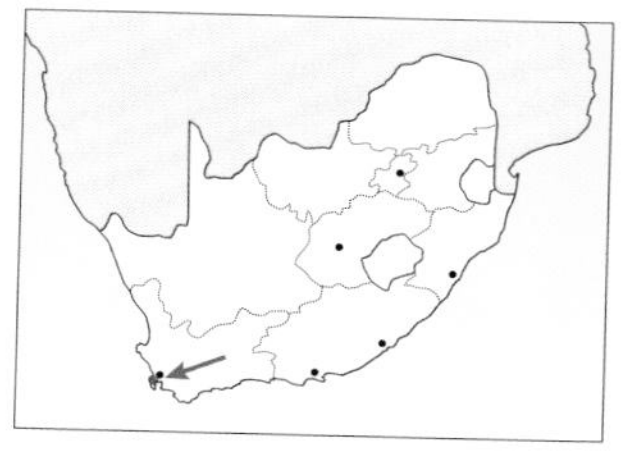

| J | F | M | A | M | J | J | A | S | O | N | D |
|---|---|---|---|---|---|---|---|---|---|---|---|

**STATUS** Critically Endangered

**DESCRIPTION** **Plant** 35–50cm.**Corm** globose, with leathery to woody tunics. **Cataphylls** upper up to 6cm above ground, green, hairy. **Leaves** 3, the lower 2 basal, 3–5mm wide; midrib, veins and margins raised and prominent, hairy. **Spike** erect, unbranched, 3–8 flowers. **Bracts** ±20mm, green or flushed with red or brown. **Flowers** bright yellow; the tube and base of tepals flushing red with age; tepals slightly unequal with tapering apex. Perianth tube short and narrow, 20mm. **Anthers** cream. **Pollen** white. **Capsules** oblong, 18mm. **Seeds** oval, evenly winged. **Scent** unscented.

**DISTRIBUTION** Currently known from only one site on the Cape Peninsula near Ocean View. Historically it has been recorded near Kommetjie and Simonstown, but it has not been seen at Simonstown for many years.

*The species has actinomorphic (radially symmetrical) flowers.*

**ECOLOGY & NOTES** The plants grow in peaty sand in areas that remain damp in late spring and early summer. The remaining population is in a very vulnerable position close to houses and roads. The plants that we photographed were among Australian *Acacia saligna* infestations. Luckily plants cultivated from seed are being grown by several people. It responds well to cultivation and sets seed abundantly if the flowers are hand pollinated. We hope the species will not become extinct.

**POLLINATORS** The only insect seen on *G. aureus* flowers is the honeybee *Apis mellifera*. It is assumed that this bee is attracted to the pollen as the flowers have no nectar.

**SIMILAR SPECIES** Owing to its bright yellow colour and very localised distribution, *G. aureus* is unlikely to be mistaken for any other species.

*Stems and spikes are erect and unbranched, bearing up to eight flowers.*

*Hairy leaves and stems are visible with natural backlighting.*

*The species is named for its golden flower colour.*

G. aureus, *seen here near Ocean View, Cape Peninsula, Western Cape.*

# *Gladiolus brevifolius*

***brevifolius* = with short leaves, referring to the leaves on the flowering stems.**

The pale to deep pink, cream or grey flowers of *Gladiolus brevifolius* appear in March and April in the western and southwestern parts of the Western Cape. In the east of its range, it is scented but elsewhere not.

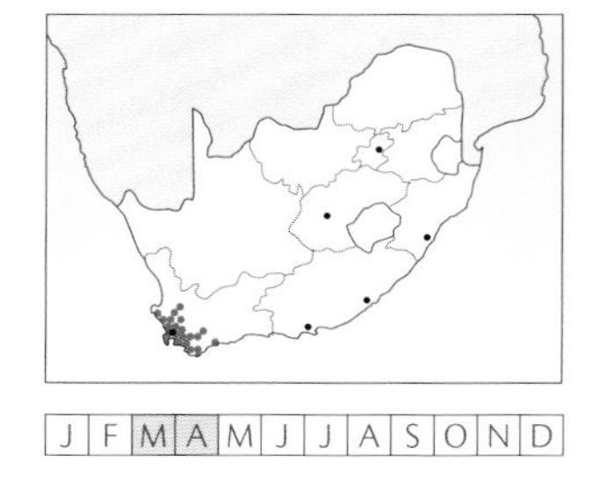

| J | F | M | A | M | J | J | A | S | O | N | D |
|---|---|---|---|---|---|---|---|---|---|---|---|

**DESCRIPTION** **Plant** normally 25–55cm but can be shorter or taller. **Corm** tunics of leathery texture, decaying into coarse vertical fibres. **Cataphylls** upper 1 just above ground, usually dry, may be hairy. **Leaves** 1–3 sheathing leaves with short blades on the flowering stem; 1 foliage leaf produced after flowering, narrow and hairy; margins and veins lightly thickened. **Spike** slightly inclined, unbranched, 8–12 flowers, sometimes more. **Bracts** up to 20mm, grey-green flushed purple above. **Flowers** small and usually pink, but some populations are cream, mauve or greenish grey; lower tepals with a yellow band sometimes with a darker edge. Perianth tube short, ±12mm. **Anthers** cream. **Pollen** pale yellow. **Capsules** ellipsoid, up to 20mm. **Seeds** oval, broadly and evenly winged, light yellow-brown. **Scent** variable: usually unscented, but may be rose scented.

**DISTRIBUTION** Common in the southwestern Cape, *G. brevifolius* is found from Piketberg in the west, south to the Cape Peninsula, east to Agulhas, and inland to Montagu.

**ECOLOGY & NOTES** Found on a variety of soil types: clay, stony sandstone and granitic soils; it grows in full sun. The species flowers at the end of summer, both in mature vegetation and in burned areas. When the plant flowers it uses reserves from the corm to produce a flower spike, which only has sheathing leaves. After flowering, a foliage leaf emerges from a separate shoot, replenishing the reserves and allowing for the production of a new corm. The species is very variable. There is a small, very brightly coloured form from Napier and Bredasdorp, a robust large form from Saldanha and Tulbagh, a cream-coloured form from Malmesbury, and a pale, almost unmarked form from the slopes of Table Mountain.

**POLLINATORS** Nectar-seeking long-tongued bees such as *Amegilla fallax* and flies such as *Psilodera valida*, both with mouthparts that are ±12mm long.

**SIMILAR SPECIES** Vegetatively identical to *G. monticola* but when in flower it is most likely to be confused with *G. martleyi* (section *Homoglossum*) which has similar flowers and flowering leaves. However, the foliage leaves and the corms of the two species differ.

| | Corm | Leaves | Scent |
|---|---|---|---|
| ***G. brevifolius*** | hard, leathery | 1 foliage leaf after flowering, sword shaped, plane, hairy | seldom scented |
| ***G. martleyi*** (sect. *Homoglossum*) | fleshy | 1 or 2 foliage leaves after flowering, needle-like with 4 fine grooves | usually scented |

*Narrow, short leaves sheath the spike; a single foliage leaf is produced after flowering.*

G. brevifolius *usually has up to 12 flowers on a spike.*

*Grey-green bracts surround the perianth tube.*

*Flower colour may be variable. Note the darker edge to the yellow band.*

*Mauve form, seen near Napier, Western Cape.*

*Cream form, flowering near Tulbagh, Western Cape.*

*Bright pink form, Honingklip.*

G. brevifolius *growing in stony ground, Napier, Western Cape.*

# Gladiolus monticola

***monticola*** **= growing on mountains.**

*Gladiolus monticola* is restricted to the Cape Peninsula, where it flowers from mid-December to February. Pale apricot to cream flowers are pollinated by long-tongued flies. It can be confused with *G. brevifolius* but the latter's flower colour is more variable.

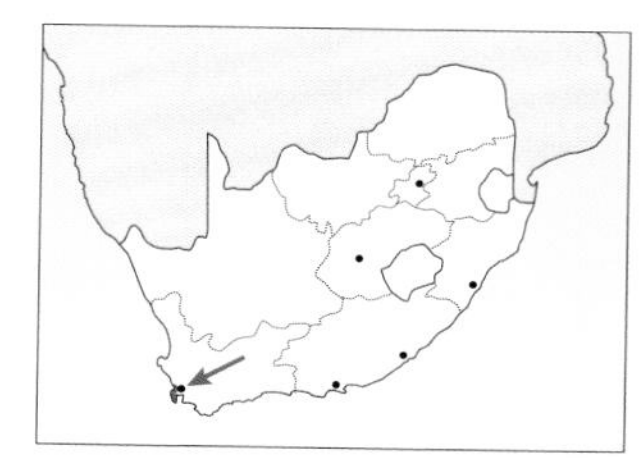

J F M A M J J A S O N D

**STATUS** Rare

**DESCRIPTION** **Plant** 30–45cm. **Corm** globose, with tunics of fine fibre. **Cataphylls** uppermost up to 15cm above ground; green or reddish brown. **Leaves** 2 on flowering stem, sheathing, smooth or slightly hairy; foliage leaf produced after flowering, narrowly sword shaped, hairy. **Spike** slightly inclined, unbranched, 3–9 flowers. **Bracts** up to 24mm, grey-green, flushed purple on upper surface. **Flowers** pale apricot to cream, often flushed pale pink; tepals unequal, lower tepals each with a yellow median stripe, sometimes outlined in dark pink. Perianth tube up to 30mm. **Anthers** cream. **Pollen** pale yellow. **Capsules** ellipsoid, 13mm. **Seeds** oblong, golden brown, well-developed wing. **Scent** unscented.

*Spikes are gently inclined, bearing three to nine flowers.*

**DISTRIBUTION** Known only from the Cape Peninsula. It is most commonly seen on Table Mountain although it also occurs further south through Silvermine to Simonstown.

**ECOLOGY & NOTES** Grows in rocky sandstone soils among fynbos. Flowers best in the two or three years after fire, but plants are also seen in older vegetation in disturbed places such as along paths.

**POLLINATORS** Long-tongued flies such as the horse fly (*Philoliche rostrata*) and *Prosoeca nitidula*, both of which have tongues up to 30mm, allowing them to reach the nectar in the base of the perianth tube.

*The dorsal tepal is largest, with lower tepals slightly longer than the laterals.*

**SIMILAR SPECIES** *G. monticola* can be confused with *G. brevifolius* when not in flower, as vegetatively they are very similar. However, they flower at different times.

| | Leaves | Flowers | Flowering time |
|---|---|---|---|
| ***G. monticola*** | sheathing leaves on flowering stem; one hairy foliage leaf from separate shoot in winter and spring | perianth tube up to 30mm with abrupt bend just below tepals; salmon to cream with yellow stripe on lower tepals | Dec–Feb |
| ***G. brevifolius*** | sheathing leaves on flowering stem; one hairy foliage leaf from separate shoot in winter and spring | perianth tube ±12mm, no abrupt curve at tepals; cream, pink or grey with nectar guides | Mar–May |

*An apricot form has a soft yellow stripe on the pale lower tepals.*

*Bracts are grey-green and surround the skinny perianth tube.*

G. monticola *growing in sandstone soils on Table Mountain, Cape Town, Western Cape.*

# *Gladiolus nerineoides*

***nerineoides* = like a nerine, in particular *Nerine sarniensis*, which is similar in colour and appearance.**

*Gladiolus nerineoides* is known from the mountains above Stellenbosch, which is a winter-rainfall area. It flowers in summer, between December and February. Its bright scarlet flowers are pollinated by the Mountain pride butterfly.

**DESCRIPTION** **Plant** 35–60cm. **Cataphylls** uppermost just above ground, green and hairy. **Leaves** 2 on flowering stem, entirely sheathing and hairy; foliage leaf 3–6mm wide, slightly hairy, produced after flowering; midrib and veins thickened. **Spike** flexed outwards and inclined to nearly horizontal, unbranched, 5–10 flowers crowded on the upper side. **Bracts** pale green. **Flowers** scarlet, sometimes orange; tepals sub-equal, upper tepal often bent backwards. Perianth tube ±30mm **Anthers** yellow. **Pollen** cream. **Capsules** obovoid, ±12mm. **Seeds** unknown. **Scent** unscented.

**DISTRIBUTION** Known only from the mountains around Somerset West and Stellenbosch. It has been recorded from the Jonkershoek area, Helderberg, Somerset Sneeukop and Simonsberg.

**ECOLOGY & NOTES** The localities of this species are all above 1,000m in rocky areas that face south or southwest. This is the cool side of mountain ranges in the region. Here the plants grow either among rocks or on cliff faces and are partly protected from the harsh summer sun. The slopes are hot and dry at flowering time, except when mist is driven in by the southeast wind. The plants have adapted by growing in winter and spring, and flowering in midsummer when the leaves have already gone dormant. The flowering stems only have sheathing leaves with reduced blades, thus reducing water loss. It is currently not considered threatened.

**POLLINATORS** Pollinated by the Mountain pride butterfly, *Aeropetes tulbaghia*. This butterfly is the sole pollinator of a number of completely unrelated plants, all flowering in midsummer and all with bright red flowers. If one walks in the mountains in summer, wearing a bright red shirt or hat, one is almost assured of having a close encounter with this butterfly! See page 102 for more information on this butterfly.

**SIMILAR SPECIES** Bright red flowers resemble those of *G. stokoei* and *G. insolens* (section *Blandi*), but distribution ranges do not overlap.

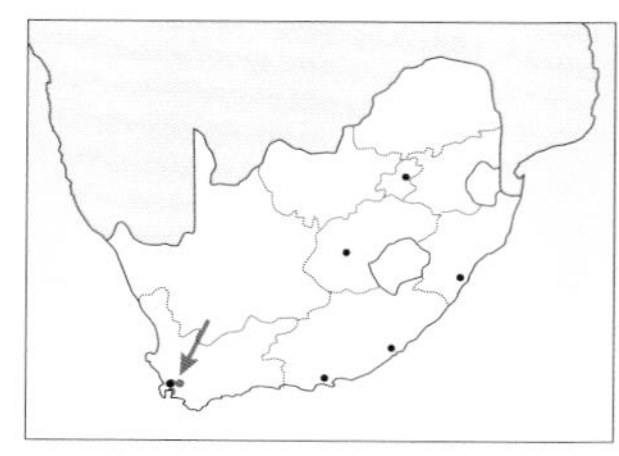

J F M A M J J A S O N D

**STATUS** Rare

G. nerineoides *has strongly inclined spikes bearing many erect flowers.*

*Flowering plants have hairy, sheathing leaves and grow on south- or southwest-facing slopes.*

*Pale green bracts are about half as long as the perianth tubes.*

*Bright red flowers with yellow anthers are pollinated by the Mountain pride butterfly.*

*Plants grow above 1,000m, here pictured in the Helderberg, overlooking Somerset West, Western Cape.*

# Gladiolus stokoei

**Named after TP Stokoe (1868–1959), a plant collector who walked in the mountains of the southwestern Cape and collected about 150 species new to science, including this one.**

*Gladiolus stokoei* is restricted to the high peaks of the Riviersonderend Mountains in the Western Cape. Scarlet flowers appear between March and April and are pollinated by butterflies.

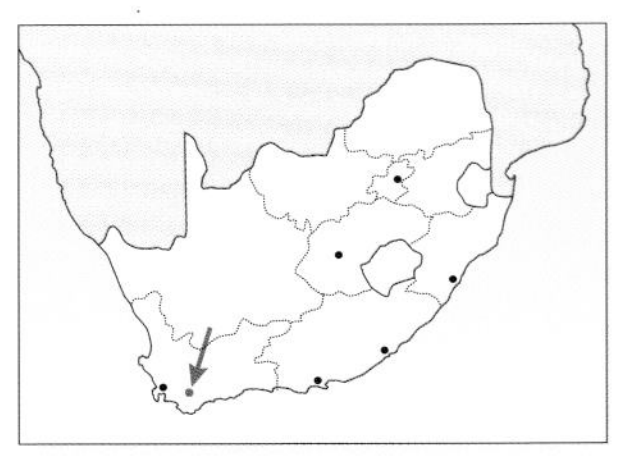

| J | F | M | A | M | J | J | A | S | O | N | D |
|---|---|---|---|---|---|---|---|---|---|---|---|

**STATUS** Critically Endangered

**DESCRIPTION** **Plant** 30–45cm. **Corm** globose, with finely fibred tunics. **Cataphylls** upper 1 just above ground, brownish purple. **Leaves** usually 2, largely sheathing on the flowering stem, 1 foliage leaf produced after flowering, 2–3mm wide; margins, midrib and veins lightly raised, hairy. **Spike** erect, unbranched, 1–3 flowers. **Bracts** up to 30mm, green. **Flowers** bright scarlet; tepals almost equal and symmetrical, forming a bowl, lower 3 tepals with dark red streak, stamens exserted. Perianth tube up to 35mm. **Anthers** yellow. **Pollen** yellow. **Capsules** ovoid. **Seeds** ovoid-ellipsoid, up to 8mm. **Scent** unscented.

**DISTRIBUTION** Known only from a small area (concentrated around Pilaarkop) in the Riviersonderend Mountains of the Western Cape.

**ECOLOGY & NOTES** Found on the southern slopes of mountains at moderate altitudes of ±700m above sea level. These cooler, damper slopes receive mist and some rain. Plants appear to flower in response to fire, and we saw several hundred flowering a full 12 months after a fire in the summer of 2014. Its conservation status is a result of unmanaged alien invasive pines and hakea.

*The flowers make a bright splash on a dull day at Riviersonderend, Western Cape.*

**POLLINATORS** Thought to be *Aeropetes tulbaghia* (Mountain pride butterfly). See page 102 for more information about this pollinator.

**SIMILAR SPECIES** *G. stokoei* has bright scarlet flowers like *G. nerineoides* and *G. insolens* (section *Blandi*), but their distributions do not overlap.

| | Locality | Leaves | Flowers |
|---|---|---|---|
| ***G. stokoei*** | Riviersonderend | 1 hairy foliage leaf after flowering, flowering stem has reduced sheathing leaves | erect spike of 3–5 bright scarlet bowl-shaped flowers, stamens exserted |
| ***G. nerineoides*** | Stellenbosch and Somerset West | 1 hairy foliage leaf after flowering, flowering stem leaves as above | 3–10 flowers on inclined spike, stamens included |
| ***G. insolens*** (sect. *Blandi*) | Piketberg | 6 long grey-green leaves at the same time as the flowers | 1 or 2 bright scarlet, slightly cupped flowers on inclined stem |

*Note how the bracts are almost as long as the perianth tube.*

*Erect spikes support bowl-shaped flowers.*

*Seeds awaiting dispersal are stacked in the capsule.*

*The protruding anthers are bright yellow, as is the pollen.*

G. stokoei *in the Riviersonderend Mountains.*

G. filiformis *blooming near Zeerust, North West Province.*

# *Gladiolus oatesii*

**Named after British explorer Frank Oates (1840–1875), who first collected and scientifically recorded this species in Zimbabwe in 1874.**

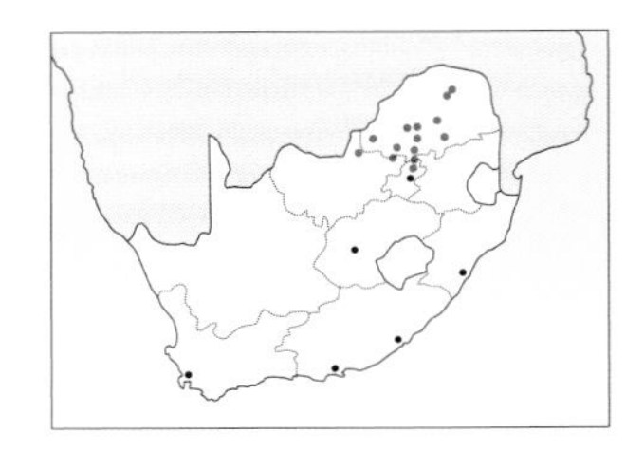

J F M A M J J A S O N D

*Gladiolus oatesii* is widespread in the northern parts of South Africa, extending into Zimbabwe and Botswana. Its whitish, unscented flowers appear between mid-October and early December.

*This species is widespread in the northern parts of South Africa.*

**DESCRIPTION** **Plant** ±40–50cm. **Corm** globose to obconic, with tunics of matted fibres extending up the stem. **Cataphylls** up to 3cm above ground, hairless with purple veins. **Leaves** 2 or 3, 3mm wide, hairless, sheathing the flowering stem; non-flowering plants have a single, glabrous, grey-green leaf with lightly thickened margins and midrib. **Spike** erect, 5–12 flowers. **Bracts** pale green, sometimes flushed reddish purple. **Flowers** white flushed with purple, darker towards the tips of the tepals; with unequal tepals, lower tepals with white or yellow band outlined in purple. Perianth tube 10mm. **Anthers** greenish or mauve with dark lines on the sides. **Pollen** yellow. **Capsules** ellipsoid. **Seeds** light brown with broad wing. **Scent** unscented.

**DISTRIBUTION** From Gauteng to Limpopo and North West, extending into eastern Botswana and northwards into Zimbabwe.

**ECOLOGY & NOTES** Plants are found in light woodland or in open grassland, always among rocks. Corms that have reached flowering size can produce flowers early in the season while the grass is still short. Their seeds ripen and may even germinate early, in the same growing season. Flowering plants do not produce long-bladed leaves at the base of the stem, but only short, sheathing stem leaves, whereas non-flowering plants produce a solitary, long, narrow foliage leaf. Once flowering is complete and the seeds have ripened and the new corm has been produced, the plants become dormant. Similarly, *G. woodii* and *G. malvinus* do not produce foliage leaves in the same season that they flower.

*Non-flowering plants produce a single lanceolate, glabrous leaf.*

**POLLINATORS** Probably long-tongued bees.

**SIMILAR SPECIES** *G. woodii* and *G. malvinus* (section *Linearifolii*) are also early flowering species, similar in colour and flower shape, with hairy sheathing leaves and cataphylls; leaves of *G. oatesii* are not hairy. *G. pretoriensis* and *G. filiformis* have similar flowers.

*Reddish-purple bracts surround the short perianth tube.*

*Note the greenish anthers, recurved dorsal tepal and clawed lower tepals.*

G. oatesii *is typically found amid rocks, here in grasslands near Zeerust, North West Province.*

# Gladiolus pretoriensis

**Referring to the type locality in Pretoria.**

*Gladiolus pretoriensis* is restricted to areas north of the Vaal River. The pale lilac to pinkish flowers appear between mid-January and February.

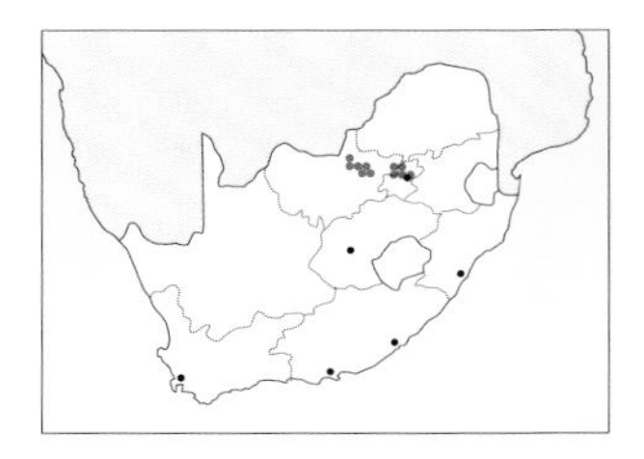



**DESCRIPTION** **Plant** 40–70cm. **Corm** medium sized, with fibrous tunic. **Cataphylls** up to 5cm above ground, green or flushed with purple. **Leaves** 3–5, with the basal leaf being the longest; margins and midrib are thickened and raised so much that the surface of the leaf is concealed in grooves running the length of the leaf. **Spike** may be branched, 9–13 flowers. **Bracts** green or purplish, becoming dry at flowering. **Flowers** pale lilac to pinkish with unequal tepals; lower tepals have a white to yellow band across, outlined with purple. Perianth tube ±14mm. **Anthers** yellow. **Pollen** yellow. **Capsules** 10mm long, showing outline of seeds. **Seeds** angular, wingless. **Scent** unscented.

*Flowers have unequal tepals.*

**DISTRIBUTION** From Krugersdorp and Pretoria in Gauteng, to Rustenburg and Zeerust in the North West.

**ECOLOGY & NOTES** Plants grow on stony hill slopes in grassland or among trees. Flowering specimens have been identified near Zeerust, where it grows in clay derived from shale. The plants we found were among trees and aloes in light shade, and were in full flower in mid-December. The area had just received a week of good rain, which may have stimulated the early flowering. The interesting leaf shape occurs in a number of unrelated species, most of which come from dry habitats or from nutrient-poor or 'toxic' soils with high levels of minerals such as cobalt, copper or nickel.

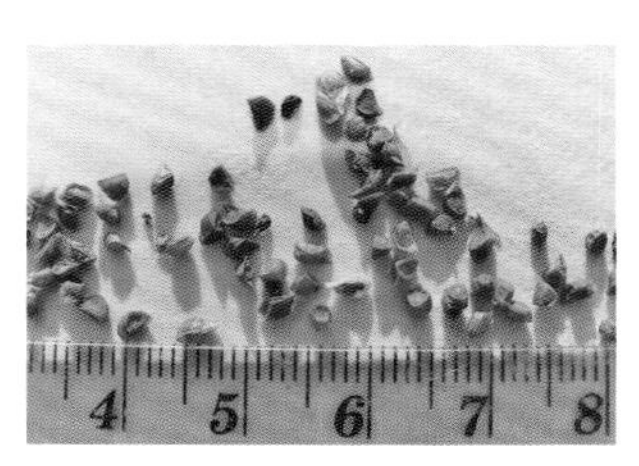



*Seeds are angular and wingless.*

**POLLINATORS** Probably pollinated by bees, together with bee flies from family Bombyliidae.

**SIMILAR SPECIES** *G. filiformis* and *G. oatesii* are similar.

| | Leaves | Flowers | Seeds |
|---|---|---|---|
| ***G. pretoriensis*** | terete, appearing grooved | pink-lilac | small and angular, wingless |
| ***G. filiformis*** | terete, appearing grooved | blue-mauve, with reddish markings | irregular, winged at one end only |
| ***G. oatesii*** | plane, lanceolate, with thickened midrib and margins | white flushed with purple, markings on lower tepals | ovate, broadly and evenly winged. |

*The flowers are small with smooth bracts.*

*Leaves have thickened and raised margins and midribs.*

*Flowers are pale lilac to pinkish.*

*This pink form has distinct white bands on the lower tepals.*

*A Golden bee fly brushes against pollen-covered anthers in search of nectar.*

G. pretoriensis *in grasslands near Zeerust, North West Province.*

# *Gladiolus filiformis*

***filiformis*** **= thin and thread-like, referring to its very slender perianth tube.**

*Gladiolus filiformis* has only been found near Zeerust in North West Province. Blue-mauve flowers are borne on erect spikes and appear in December, depending on rains.

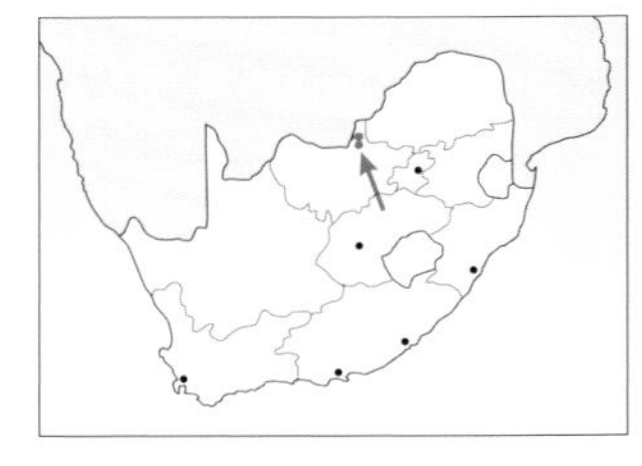

J F M A M J J A S O N D

**STATUS** Near Threatened

**DESCRIPTION** **Plant** 35–50cm. **Corm** obconic, with tunics of netted fibre. **Cataphylls** up to 2cm above ground, dry and dark brown. **Leaves** 4 terete blades; heavily thickened margins and midrib creating 4 grooves. **Spike** erect, ±5 flowers. **Bracts** dry and brown at flowering. **Flowers** blueish to light mauve; lower tepals with various white and reddish markings. Perianth tube slim and curved, up to 35mm. **Anthers** yellow. **Pollen** yellow. **Capsules** ellipsoid, 10mm. **Seeds** small, irregularly shaped, most with a small wing at one end only. **Scent** unscented.

**DISTRIBUTION** Known only from a small area between Zeerust and the Botswana border.

*Note the mauve and white coloration on the lower tepals.*

**ECOLOGY & NOTES** Plants grow mainly in open grassland on soils derived from sedimentary rock of the Transvaal Supergroup, in bands of ironstone, conglomerate, chert and dolomite, with the type locality being dolerite. Flowering specimens have been found in mid-December after good rains. Associated plants include species of *Trachyandra* and *Commelina* and some very handsome *Clematis villosa* (formerly *Clematopsis scabiosifolia*). Plants flower by the thousands in years with good rain; very few flower in years of poor rain, but the flowering period is extended when they do, starting in early October and continuing through to February.

*Reddish streaks are clearly visible on these lower tepals. Earlier descriptions lacked information about the markings.*

**POLLINATORS** Probably butterflies or bee flies with long probosces, which they use to feed from the long, slender perianth tube.

**SIMILAR SPECIES** *G. pretoriensis* and *G. oatesii*, from the same area.

| | Leaves | Flowers | Seeds |
|---|---|---|---|
| ***G. filiformis*** | terete, appearing grooved | blue-mauve, with reddish markings | irregular, winged at one end only |
| ***G. pretoriensis*** | terete, appearing grooved | pink-lilac | small and angular, wingless |
| ***G. oatesii*** | plane, lanceolate, with thickened midrib and margins | white flushed with purple, markings on lower tepals | ovate, broadly and evenly winged. |

*Strongly grooved leaves sheath the lower third of the stem.*

*Slender perianth tubes suggest pollination by butterflies or long-proboscid bee flies.*

*This plant's pale mauve flower has minimal markings.*

*Seeds are winged at one end only. Earlier studies did not include records of the seeds.*

*When rains are good, flowers can be seen dotted around the grasslands near Zeerust, North West Province.*

# *Gladiolus rufomarginatus*

***rufomarginatus* = rust-coloured margins, referring to the margins of the floral bracts.**

*Gladiolus rufomarginatus* occurs in a small part of Mpumalanga. It has distinctive leaves, rusty-coloured bracts and unmistakable, unusually coloured flowers – cream with dark red markings. It flowers very late in the season, around March and April, when most other species are already in seed.

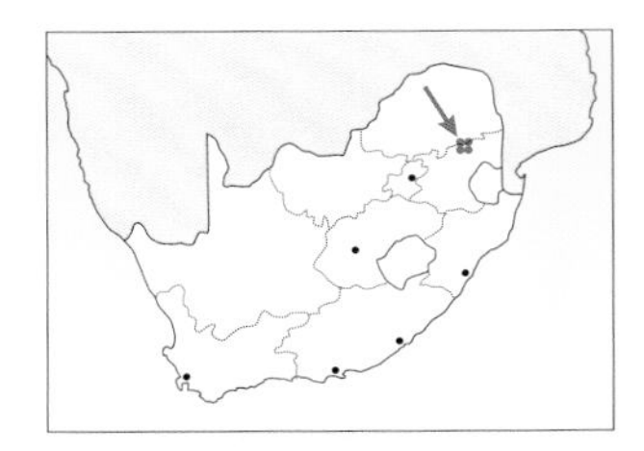

| J | F | M | A | M | J | J | A | S | O | N | D |
|---|---|---|---|---|---|---|---|---|---|---|---|

**STATUS** Rare

**DESCRIPTION** **Plant** 30–50cm. **Corm** obconic, with papery tunics becoming fibrous with age. **Cataphylls** up to 8cm above ground, green or brown. **Leaves** 5 linear blades, ±3mm wide; margins and midrib heavily thickened and raised forming 2 narrow grooves along the entire leaf. **Spike** minutely scabrid, with 8–30 flowers. **Bracts** margins rusty brown, slightly transparent, dry at flowering. **Flowers** unusual combination of dark red spots on a cream background; lower tepals flushed yellow-green and tips of tepals flushed bright pink. Tepals are unequal in length with 'windows' between the dorsal tepal and the lateral upper tepals. Perianth tube up to 9mm. **Anthers** olive-green. **Pollen** white. **Capsules** oblong, 12mm. **Seeds** light brown with darker seed body, unevenly winged. **Scent** unscented.

**DISTRIBUTION** Endemic to the Ehlanzeni district, Mpumalanga, *G. rufomarginatus* is found only between Mashishing and Ohrigstad.

**POLLINATORS** Bees such as species of *Amegilla* and flies belonging to the Nemestrinidae (tangle-veined flies) including members of the genus *Stenobasipteron*, which have short tongues and feed on nectar.

**ECOLOGY & NOTES** The area between Mashishing and Ohrigstad consists of dry hills and valleys on the western side of the Drakensberg escarpment. The most common trees in the area are *Senegalia caffra*. Plants are found in grassland, either in full sun or light shade, in soils of stony shale. They have also been found growing in the cracks of shale pavements and outcrops. Although it seems that such brightly coloured flowers would be easy to see, they are not, as they blend into the grass which, in most years, is already dormant and brown by late April.

**SIMILAR SPECIES** *G. rufomarginatus* has distinctive leaves, rusty-coloured bracts and such unusually coloured flowers that it is unmistakable. Its late flowering season also distinguishes it from other species.

G. rufomarginatus *is known only from Ehlanzeni district, Mpumalanga.*

*Note the red tepal margins and 'window' between dorsal and lateral tepals.*

*Leaves are very narrow and long.*

*Lanceolate tepals have pink tips.*

*Spikes bear rust-coloured bracts.*

*Short-tubed flowers are adapted for pollination by bees.*

G. rufomarginatus *often blends into the dry grasses of its habitat.*

# *Gladiolus vernus*

***vernus*** **= of spring, the flowering period of the species.**

*Gladiolus vernus* has a scattered range, from KwaZulu-Natal to Limpopo. The pale to deep pink flowers appear in July and August.

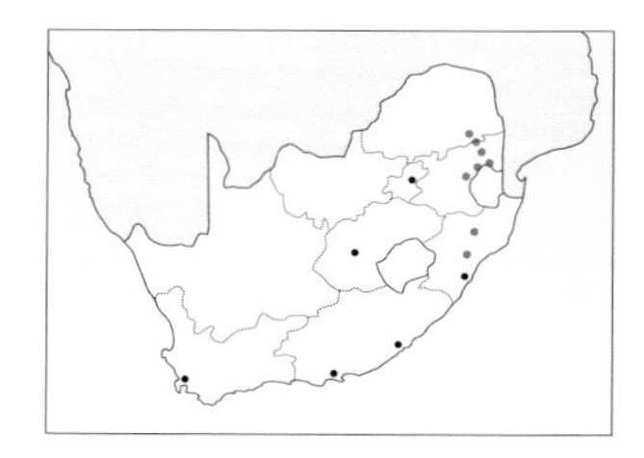

| J | F | M | A | M | J | J | A | S | O | N | D |
|---|---|---|---|---|---|---|---|---|---|---|---|

**DESCRIPTION** **Plant** 45–75cm. **Corm** globose, with coarsely fibred tunics. **Cataphylls** up to 4cm above ground, persisting and accumulating into a thick, fibrous neck at the base of the stem. **Leaves** 6, 3mm wide; lower 4 produced during summer before flowering, persisting to the following spring, usually brown and dry; upper 2 sheath the flowering spike; margins and midrib raised to form 2 grooves on each leaf surface. **Spike** simple or branched, 12–18 flowers. **Bracts** flushed pink, partly dry at flowering. **Flowers** pale to deep pink; lower tepals yellow in the middle and speckled with pink. Tepals of unequal sizes with 'windows' between the dorsal and lateral upper tepals. Perianth tube short, up to 12mm. **Anthers** purplish. **Pollen** white. **Capsules** obovoid, 12mm. **Seeds** light brown with well-developed wing. **Scent** unscented.

*This wide-ranging species has only been collected and recorded from a few sites.*

**DISTRIBUTION** The range extends from Greytown in KwaZulu-Natal northwards along the Drakensberg escarpment to the Wolkberg mountain range in Limpopo, but the species is only recorded from a few isolated sites within this range.

**ECOLOGY & NOTES** Although this part of the escarpment is well watered, *G. vernus* usually grows in drier areas on well-drained slopes among rocks. Plants have been found on both south- and north-facing slopes near the Blyde River, together with other bulbous plants such as *Merwilla plumbea*, *Ledebouria* species and watsonias. This area burns regularly. Flowering plants have only been found in unburned vegetation; this is presumably due to their growth cycle. Growth commences in spring or early summer, the leaves die back in autumn and the plants flower in late winter. If the veld burns in June or July, the *Gladiolus* plants growing there do not flower.

**POLLINATORS** Probably long-tongued bees.

**SIMILAR SPECIES** Unlikely to be confused with any other species as no other *Gladiolus* species flower in this area in July and August.

*Inclined spikes may bear up to 18 alternating flowers.*

*Cataphylls accumulate to form a neck around the base of the stem.*

*A gap between the dorsal and lateral tepals forms a 'window'; note the clawed yellowish lower tepals.*

*It flowers late in the season, capsules mature in spring.*

G. vernus *usually grows in the drier areas of the Drakensberg escarpment, as near the Blyde River Canyon, Mpumalanga.*

# Gladiolus kamiesbergensis

**From the Kamiesberg in Namaqualand.**

*Gladiolus kamiesbergensis* is restricted to the Kamiesberg range of Namaqualand in the Northern Cape. Pale lilac flowers spotted with purple are produced in October.

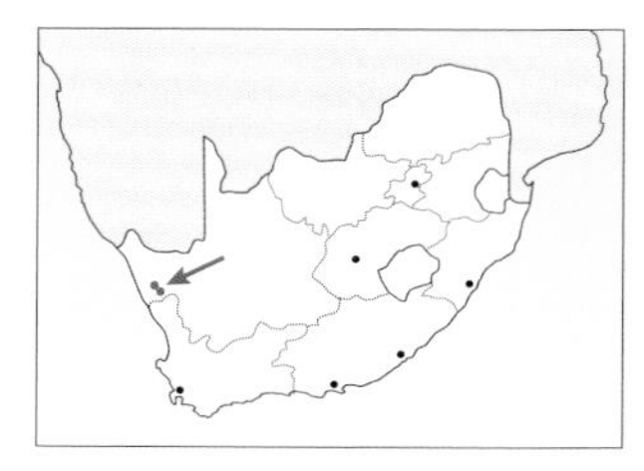

| J | F | M | A | M | J | J | A | S | O | N | D |
|---|---|---|---|---|---|---|---|---|---|---|---|

**STATUS** Rare

**DESCRIPTION** **Plant** 45–90cm. **Corm** globose, with leathery tunics. **Cataphylls** above ground, accumulating with leaf bases to form a thick neck at the base of the stem, rich red-brown. **Leaves** 4, lower 2 have terete blades; with midrib and margins raised and thickened, forming 4 hairline grooves; upper 2 sheathing the stem. **Spike** unbranched, 4–9 flowers, sometimes more. **Bracts** pale green with hyaline margins. **Flowers** pale violet or blueish white with minute purple dots on both sides of the tepals; lower lateral tepals with yellow markings; 'windows' between upper and lower tepals. Lower lateral tepals narrow to 'claws' where they join the upper tepals, before widening abruptly. Perianth tube 12mm. **Anthers** lilac. **Pollen** white. **Capsules** obovoid. **Seeds** ovate, broadly and evenly winged. **Scent** apple scented.

G. kamiesbergensis *is known only from Kamiesberg in Namaqualand, Northern Cape.*

**DISTRIBUTION** Known only from a few localities in the granite hills of the Kamiesberg in Namaqualand, Northern Cape.

**ECOLOGY & NOTES** Found at an altitude of about 1,100–1,200m on both north- and south-facing slopes. The area in which plants have been found receives about 300mm of rain per annum, mostly in winter and spring – a significantly higher rainfall than the surrounding lower-altitude areas. Plants grow among rocks in shrubby, arid fynbos. They seem to flower in burned and unburned veld, although more flowering plants have been seen in the first year after a fire. We have seen the plants in flower several times, flowering best after good late-spring rains.

**POLLINATORS** Probably long-tongued bees including *Anthophora diversipes*.

**SIMILAR SPECIES** There is no other *Gladiolus* species in Namaqualand bearing any resemblance to *G. kamiesbergensis*. It is, however, thought to be closely related to *G. vernus, G. marlothii* and *G. mostertiae*, all of which have the same narrowly clawed lower tepals that abruptly expand to form broad limbs.

*The lower lateral tepals have yellow markings.*

*Upper leaves sheath the stem.*

*Bracts are pale green with translucent margins.*

*Note the 'windows' between the dorsal and lateral tepals.*

*The dorsal tepal is strongly arched, almost bell-like, and lateral tepals are oblique.*

*Lower tepals are 'clawed' before becoming spoon shaped.*

G. kamiesbergensis *grows among granite rocks at relatively high altitudes in a very low-rainfall area.*

# *Gladiolus marlothii*

**Named after Rudolf Marloth (1855–1931), a German-born South African botanist who worked on the Cape flora.**

Endemic to the Roggeveld escarpment in the Karoo, *Gladiolus marlothii* has pale lilac flowers that appear from mid-September to early October.

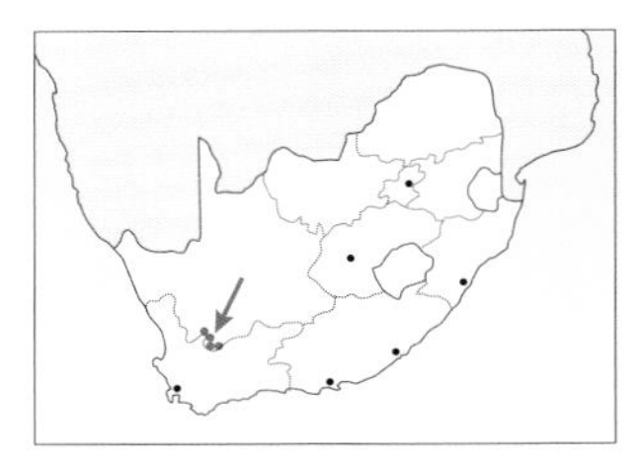

| J | F | M | A | M | J | J | A | S | O | N | D |
|---|---|---|---|---|---|---|---|---|---|---|---|

**DESCRIPTION** **Plant** 45–60cm. **Corm** globose, with coarsely textured tunics. **Cataphylls** upper reaching 10cm above ground; often purple and velvety when young, dry at flowering. **Leaves** 4, 3–4mm wide; with midrib raised and forming flanges on either side of the blade (making it x-shaped in section); lower 2 leaves as long as the spike, upper 2 sheathing the stem and very hairy. **Spike** unbranched, 3–5 flowers. **Bracts** green and membranous. **Flowers** pale blue-lilac, bell shaped; lower tepals with a yellow band surrounded by dense purple speckles, narrow at one end and broad at the other. Perianth tube 10mm. **Anthers** lilac. **Pollen** cream. **Capsules** obovoid. **Seeds** light brown with paler, well-developed wing. **Scent** unscented.

*The species bears up to five blue-lilac flowers on unbranched spikes.*

**DISTRIBUTION** Known only from the Roggeveld escarpment of the Northern Cape, in the areas around the Gannaga and Ouberg passes.

**ECOLOGY & NOTES** Plants are found in loamy soils and usually among rocks, at an altitude of ±1,800m. *G. marlothii* has been found among grass and shrubs in an area close to the escarpment edge with high rainfall, occasionally giving rise to spectacular flower displays. The area has an average annual rainfall of 400mm and experiences extreme climatic conditions. Winters are characterised by snow and freezing temperatures while summers are hot and dry. Associated plants include *Dimorphotheca cuneata*, *Euryops lateriflorus* and many bulbous species such as *Bulbinella latifolia*, *B. cauda-felis*, *Geissorhiza heterostyla* and *Romulea tetragona*. The area rarely burns, but in 2013 after a fire the previous summer, hundreds of *G. marlothii* plants flowered in an area where they had never been seen before, suggesting that the species is not as rare as suspected. However, in most years, unless they have received adequate winter rain, few plants flower.

**POLLINATORS** Probably long-tongued bees, including *Anthophora diversipes*.

**SIMILAR SPECIES** *Gladiolus marlothii* has several distinctive features and a very limited distribution range. The flowers are similar to those of *G. kamiesbergensis*, but the latter is not known from the Roggeveld. *G. mostertiae* has similarly shaped flowers, but it is differently coloured and the species are unlikely to be confused.

*Extremely hairy leaves have raised midribs and flanged margins.*

*Lower tepals are longer than the upper ones and form a bell shape on the spike.*

*Yellow bands on the lower tepals are edged with purple speckles.*

G. marlothii *flowering after fire near Agterkop, Northern Cape.*

# Gladiolus mostertiae

**Named after Aletta Mostert of Nieuwoudtville, who sent the first scientifically recorded specimen to Kirstenbosch in 1920.**

*Gladiolus mostertiae* grows in the Northern Cape along the edge of the Bokkeveld escarpment. Its pale to deep pink flowers appear between mid-November and mid-December.

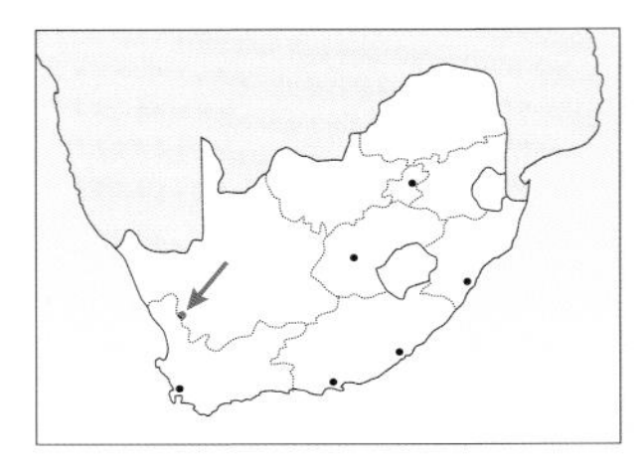

J F M A M J J A S O N D

**STATUS** Endangered

**DESCRIPTION** **Plant** 20–50cm. **Corm** globose, with papery to leathery tunics becoming fibrous. **Cataphylls** uppermost reaching up to 2cm above ground, purple to brownish. **Leaves** 3 or 4 linear blades up to 4mm wide, hairy, with veins strongly thickened. **Spike** occasionally branched, 4–10 flowers. **Bracts** bright green or flushed purple. **Flowers** pale to deep pink; upper tepals darker towards the tips, lower tepals white to cream with yellow-and-green markings, dorsal tepal larger than lower ones and arched over the stamens. Perianth tube 11mm. **Anthers** lilac. **Pollen** whitish. **Capsules** very clearly divided into 3 chambers. **Seeds** pale broad wing with much darker seed body. **Scent** unscented or may have a slight metallic odour.

*Plants grow on the Bokkeveld escarpment near Nieuwoudtville, Northern Cape.*

**DISTRIBUTION** Known from only a few localities close to Nieuwoudtville on the Bokkeveld escarpment in the southwestern parts of the Northern Cape.

**ECOLOGY & NOTES** Plants grow in sandy soil in seasonally wet, marshy places among fynbos. Most of its habitat is highly threatened by invasive vegetation and unsustainable farming practices, including the cultivation of rooibos. Plants have only been found in one area, which has been invaded by pine trees; consequently the plants are unlikely to survive much longer. They do not flower every year as it is often too dry by November and December.

*The dorsal tepal arches over pronounced purple anthers, and the lower tepals have striking yellow markings.*

**POLLINATORS** Probably long-tongued bees including *Amegilla obscuriceps*.

**SIMILAR SPECIES** The flower shape is similar to that of *G. kamiesbergensis* and *G. marlothii*, with the lower tepals being almost spoon shaped. However, the flowers of *G. mostertiae* have a distinctive colour, which cannot be mistaken for any other species. It is the only *Gladiolus* species to flower in November and December in this area.

*Narrow, hairy leaves have thickened veins.*

*Erect spikes bear up to ten delicate flowers.*

*Bright green bracts may be flushed with purple.*

*The heart-shaped yellow-and-green markings are distinctive.*

*Flowering en masse in November near Nieuwoudtville.*

Nikon

*Rachel Saunders focuses intently on* G. lapeirousioides, *Kliprand, Namaqualand.*

# Gladiolus leptosiphon

***leptosiphon* = with a slender perianth tube.**

*Gladiolus leptosiphon* grows in the Swartberg and Baviaanskloof mountains of the Western and Eastern Cape provinces. Its unusual flowers, with a long, pointed dorsal tepal and narrow lower tepals, are cream to pale yellow with red streaks. It flowers from mid-September to October.

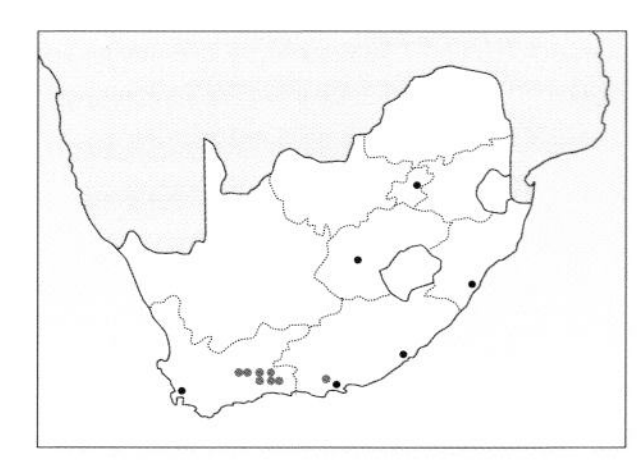

| J | F | M | A | M | J | J | A | S | O | N | D |
|---|---|---|---|---|---|---|---|---|---|---|---|

**STATUS** Vulnerable

**DESCRIPTION** **Plant** 25–45cm. **Corm** coarsely fibred tunics. **Cataphylls** upper may be purple. **Leaves** 6 linear blades, ±2mm wide; midrib thickened. **Spike** sometimes branched, flexed outwards above the leaf, 6–9 flowers. **Bracts** pale green. **Flowers** cream to pale yellow; tepals unequal, lower 3 tepals very narrow with a median dark red streak, and median tepal longer. Perianth tube long, up to 50mm. **Anthers** purple. **Pollen** purple. **Capsules** obovoid. **Seeds** broad reddish wings with black seed body. **Scent** unscented.

*Spikes bear up to nine flowers arranged in two ranks.*

**DISTRIBUTION** Occurs from Oudtshoorn northwards to the Swartberg Mountains and eastwards to the Baviaanskloof Mountains.

**ECOLOGY & NOTES** Plants grow in dry habitats in renosterveld and arid fynbos, in stony shale-derived soils. We found a large population on a dry, pebbly north-facing slope near Oudtshoorn, among *Dodonaea angustifolia*, *Gasteria brachyphylla*, *Rhodocoma arida* and many plants of *Tritonia bakeri*, with which we assume it must share a pollinator, as the flowers are very similar.

**POLLINATORS** Thought to be long-tongued flies, which often pollinate whitish unscented flowers with long tubes and reddish nectar guides.

*The lower median tepal is much longer than the lower lateral tepals.*

**SIMILAR SPECIES** Not likely to be confused with any other species. While the flowers are similar to those of *G. involutus*, the tepal markings differ and the perianth tube in *G. leptosiphon* is more than double the length of that of *G. involutus*. The localities of the two species overlap slightly, but *G. leptosiphon* generally grows in much drier areas.

*Stems flex away from the leaf sheath.*

*Perianth tubes can reach up to 50mm in length.*

G. leptosiphon *grows in dry areas as here near Oudtshoorn, Western Cape.*

# Gladiolus loteniensis

**Named after the Loteni Valley in the Drakensberg, where the first scientific collection of this species was made and recorded.**

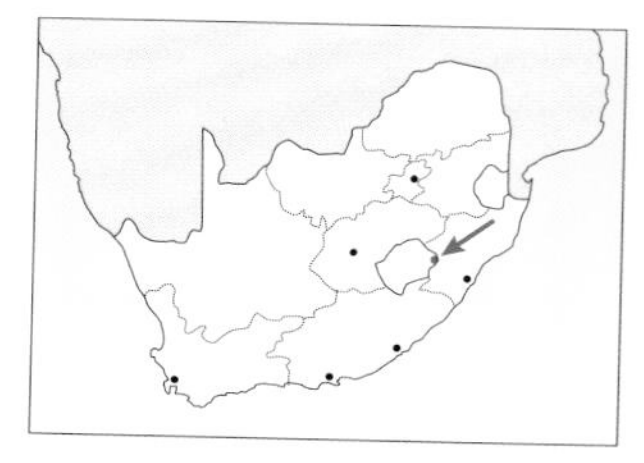

J F M A M J J A S O N D

**STATUS** Critically Endangered

*Gladiolus loteniensis* is known only from the upper Loteni Valley in the Drakensberg. Deeply hooded lilac flowers streaked on the lower tepals appear between December and mid-January.

**DESCRIPTION** **Plant** 30–60cm. **Corm** globose, with fine stolons and cormlets, and papery tunics. **Cataphylls** ±1cm above ground, purple. **Leaves** 4–7, lanceolate, up to 8mm wide; with midrib thickened and margins hyaline. **Spike** may have 1 branch, 3–12 flowers per spike arranged in 2 ranks ±60° apart. **Bracts** green on the lower surface, purplish on the upper surface. **Flowers** pale lilac, the lower 3 tepals with feathery streaks of darker violet on a cream to yellowish background; tepals unequal, dorsal tepal arched over the lower tepals, the other 2 upper tepals twisted upwards, and the middle lower tepal much longer than the other 2. Perianth tube very short, ±5mm. **Anthers** violet. **Pollen** pale lilac. **Capsules** ±1cm long, almost round. **Seeds** pale brown, evenly winged. **Scent** unscented.

*The flowers are deeply hooded, with violet anthers.*

**DISTRIBUTION** This species has only been found in the Loteni Valley, KwaZulu-Natal, at an altitude of 1,800–1,900m.

**ECOLOGY & NOTES** Plants grow in an area of sandstone rock in black peaty soil along the banks of the river. When we visited the area in 2014, the west side of the river had burned the previous winter whereas the east side had not. We only found flowering plants in the burned area, on slightly north-facing slopes, and all at an altitude of 1,800–1,900m. Since finding these plants we have searched in similar habitats in other valleys close to Loteni, in both December and January, but we have not found any other populations. The plants were growing among grass and other members of the Iridaceae such as *Hesperantha baueri* and *Moraea inclinata*; there were also herbaceous plants such as *Anisodontea julii* and various species of *Berkheya*, *Senecio* and *Wahlenbergia*.

*Lower tepals are streaked with dark violet; note the twisted upper lateral petals.*

**POLLINATORS** Thought to be long-tongued bees.

**SIMILAR SPECIES** With such a distinctive locality, *Gladiolus loteniensis* is unmistakable. The small lilac flowers with beautiful feathery streaks on the lower tepals are not similar to any other species in the area. The closest ally to this species is thought to be *G. involutus*, a species found mainly in the southwestern Cape. The two have similarly shaped flowers with unequal tepals and soft-textured corm tunics, and both produce stolons with cormlets at the ends.

*Lanceolate leaves have thickened midribs.*

*Deep purple bracts conceal short perianth tubes.*

*Capsules are almost round.*

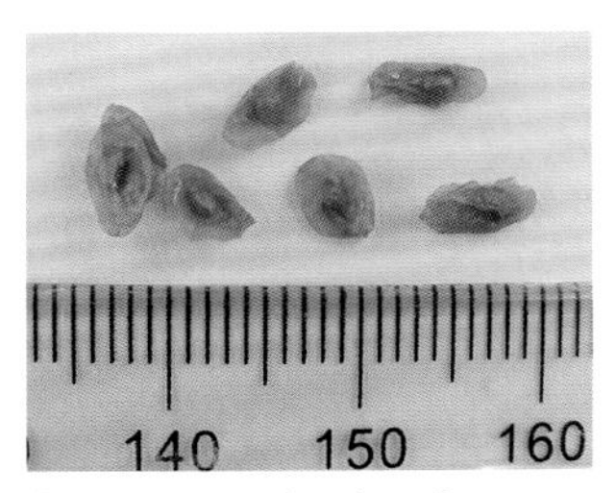


*Seeds are evenly winged, ±6mm.*

G. loteniensis *grows above 1,800m in the Loteni River Valley, Drakensberg, KwaZulu-Natal.*

# Gladiolus involutus

***involutus* = curving inwards, referring to the margins of the lower tepals.**

Fairly common in the southern Cape, *Gladiolus involutus* has white to pink flowers with greenish-yellow nectar guides outlined in pink or purple. Flowers appear from August to September.

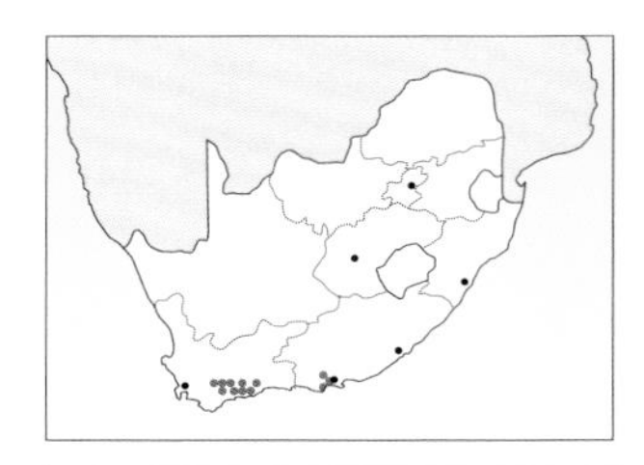

| J | F | M | A | M | J | J | A | S | O | N | D |
|---|---|---|---|---|---|---|---|---|---|---|---|

**DESCRIPTION** **Plant** 20–50cm. **Corm** depressed-globose, with papery tunics and large corms on slender stolons. **Cataphylls** purple above ground, mottled with white. **Leaves** 7–9, as long as the flower spike, ±3mm wide; with thickened midrib and hyaline margins. **Spike** simple or branched, 4–7 flowers. **Bracts** pale green flushed with purple on the upper surface. **Flowers** white, flushed pink towards the tepal bases, lower 3 tepals green edged with a pink line; tepals unequal with the dorsal being the largest, the other 2 upper tepals twisted and the middle lower tepal longer than the other 2; margins of the lower tepals curve upwards making them channelled. Perianth tube up to 18mm. Old flowers turn pink. **Anthers** purple. **Pollen** pale orange. **Capsules** oblong, ±5mm long. **Seeds** light brown with a blackish seed body, evenly winged. **Scent** unscented.

*The flowers turn pink with age.*

**DISTRIBUTION** Found along the Langeberg and Outeniqua mountains from Heidelberg in the west to just east of Gqeberha (Port Elizabeth). Plants have mainly been found on the southern side of the mountain ranges in the coastal areas, but have also been recorded in the region of Montagu, on the north of the mountains.

**ECOLOGY & NOTES** Plants grow in both heavy clay and lighter soils. We have seen plants on the road verge among grass, and after a fire in renosterveld on the northern side of the Langeberg. This species produces long stolons from the base of the corm, with each stolon ending in a small corm. This is an effective method of propagation, which safeguards the plants from destruction by predators such as porcupines. If the parent corm is eaten, the small cormlets on the ends of the stolons are dispersed and will grow during the following winter.

*Lateral tepals create a striking winged shape; lower tepals are marked and channelled.*

**POLLINATORS** Thought to be long-tongued bees, which can reach the large amounts of nectar in the lower part of the tube.

**SIMILAR SPECIES** The upturned margins of the lower tepals of *G. involutus* together with the unusually coloured nectar guides mean that this species is unlikely to be confused with any other. The species has a complex history and has had several names since it was first scientifically illustrated in 1758.

*Perianth tubes emerge from between long bracts.*

G. involutus, *flowering near Herbertsdale along the Garden Route, Western Cape.*

# *Gladiolus vandermerwei*

**Named after JS van der Merwe of Bonnievale, who collected the plants that became the type specimen.**

*Gladiolus vandermerwei* occurs along the Breede River in the Western Cape. Its bright red, unscented flowers with yellow lower tepals are unmistakable.

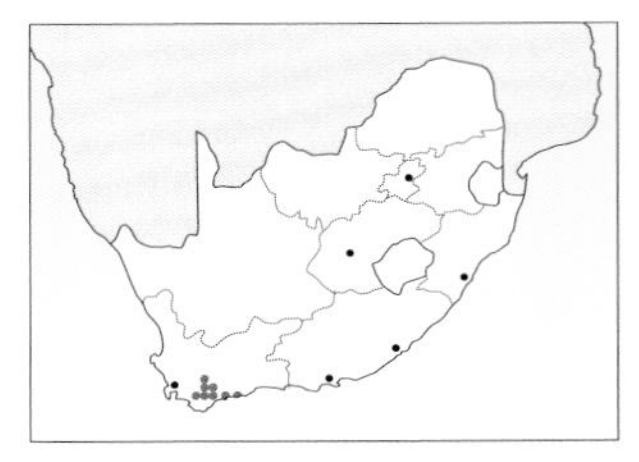

| J | F | M | A | M | J | J | A | S | O | N | D |
|---|---|---|---|---|---|---|---|---|---|---|---|

**STATUS** Endangered

**DESCRIPTION** **Plant** 30–60cm. **Corm** globose, forms cormlets on stolons, with papery tunics. **Cataphylls** up to 10cm above ground, purple mottled with white. **Leaves** 5–7, linear blades 2–5mm wide; margin not thickened, midrib lightly raised. **Spike** simple or branched, 3–8 flowers. **Bracts** dull green to purplish. **Flowers** bright scarlet with lower 3 tepals yellow with scarlet tips; tepals unequal with small linear lower tepals at right angles to the tube. Perianth tube long, up to 45mm, narrow and cylindrical. **Anthers** dark purple. **Pollen** pale yellow. **Capsules** ovoid, up to 20mm. **Seeds** evenly or unevenly winged, brown with darker seed body. **Scent** unscented.

*The species' natural habitat has been fragmented by farming.*

**DISTRIBUTION** The habitat of this endemic to the southern parts of the Western Cape has been severely fragmented by crop cultivation. The species is known from around Bonnievale in the west along the Breede River Valley to the sea at Witsand. There are also records from Bredasdorp and Napier, and from a site near Montagu.

**ECOLOGY & NOTES** The plants that we photographed were near Napier on a rocky hill among boulders, on a southwest-facing slope. The soil type is shale and the area is fairly dry.

**POLLINATORS** Sunbirds are attracted to the bright red colour. The flowers produce nectar with low sugar concentration, a feature usually associated with bird pollination; the wide perianth tube accommodates these avian pollinators.

*Lower tepals are yellow with scarlet tips; anthers are purple.*

**SIMILAR SPECIES** *G. cunonius*, *G. splendens*, *G. saccatus* and *G. abbreviatus* (section *Homoglossum*) all have similar unusually shaped flowers. The most noticeable difference between them is that *G. vandermerwei* has distinct lower lateral tepals even though they are small, whereas the other species have greatly reduced lower lateral tepals that are not easily visible. *G. splendens* and *G. saccatus* can be separated by their distribution (further west and north), and *G. cunonius* is generally found in coastal scrub and on sand dunes close to the sea. All of these species, with their peculiar flower structure, were originally classified in the genus *Antholyza*.

*The lower tepals are narrow and angled away from the perianth.*

G. vandermerwei, *in flower near Bredasdorp, Western Cape.*

# Gladiolus cunonius

**Named after JC Cuno (1708–1780), an 18th-century Dutch botanist.**

*Gladiolus cunonius* grows along the coast from the Cape Peninsula to Knysna. Its bright red, unscented flowers with unusual tepal formations appear from September to mid-November.

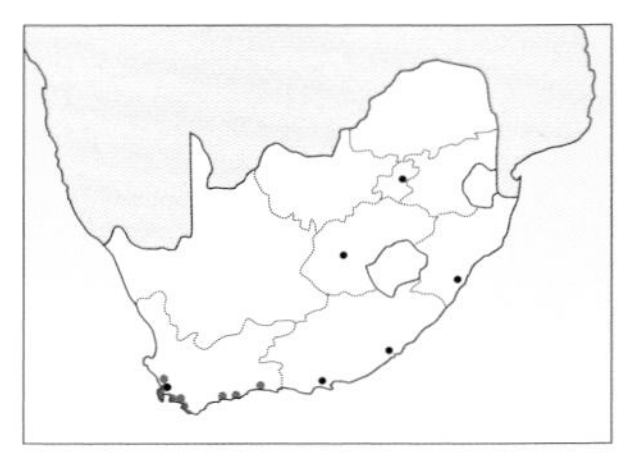

| J | F | M | A | M | J | J | A | S | O | N | D |
|---|---|---|---|---|---|---|---|---|---|---|---|

**DESCRIPTION** **Plant** 30–70cm. **Corm** globose, with papery tunics; forms cormlets on stolons. **Cataphylls** purple above ground. **Leaves** 6–8, blades 5–12mm wide; midrib evident but not thickened; uppermost leaf channelled. **Spike** strongly inclined or horizontal, simple or branched, 5–8 flowers. **Bracts** green or grey-purple. **Flowers** bright red, lower tepals and tube are green fading to yellow; tepals unequal, concave dorsal tepal reaches up to 30mm long; the 2 lateral tepals twist upwards, and the short, scale-like lower tepals curve up to partially block the mouth of the tube. Perianth tube up to 15mm. **Anthers** red, with long tails; lower filament has a small horny tooth at the base. **Pollen** yellow. **Capsules** oblong, up to 30mm. **Seeds** golden brown with a large even wing. **Scent** unscented.

G. cunonius *grows in dunes and coastal scrub.*

**DISTRIBUTION** Extends along the southwestern Cape coast from Blaauwberg in the west to Knysna in the east.

**ECOLOGY & NOTES** Plants grow on sand dunes close to the sea and in coastal scrub. They seem to flower best after fires or in clearings where the vegetation is lower and less dense. The plants produce many stolons from the base of the corm, with small cormlets at the end of each stolon. This ensures that predators such as mole-rats do not destroy the entire population of plants.

**POLLINATORS** *G. cunonius* is pollinated by sunbirds, like many other species with long-tubed red flowers, which have plenty of nectar.

**SIMILAR SPECIES** *G. cunonius*, *G. splendens* and *G. saccatus* share a similar flower structure and are closely related. They all have anthers with deeply forked bases and produce cormlets on stolons, as does *G. vandermerwei*.

| | Distribution | Spikes | Flowers |
|---|---|---|---|
| ***G. cunonius*** | southern Cape, coastal sand | strongly inclined to horizontal, few branches | upper lateral tepals point upwards, lower tepals curve upwards |
| ***G. splendens*** | western Karoo, mainly along rivers | erect, few branches | upper tepals point upwards, lower tepals point down |
| ***G. saccatus*** | Vanrhynsdorp heading north into Namibia | slightly inclined, many branches | perianth tube has sac-like spur at base, upper lateral tepals do not point upwards |

*Leaves are plane and soft textured with midrib and margins not thickened.*

*The spoon-shaped dorsal tepal is longer than the twisting lateral tepals.*

G. cunonius, *here flowering on the Cape Peninsula, Western Cape.*

# Gladiolus splendens

**_splendens_ = brilliant or shining, so named for the bright red flowers.**

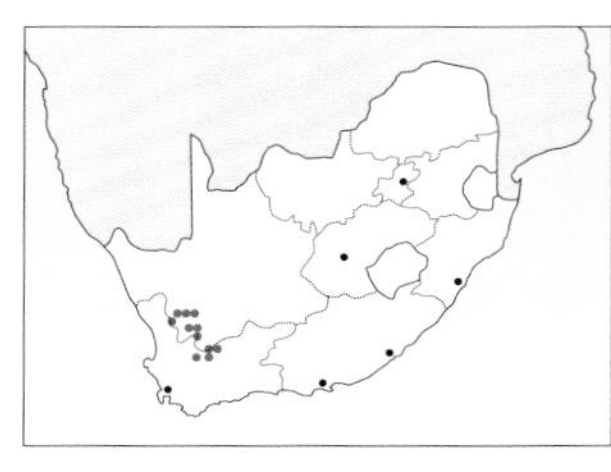

| J | F | M | A | M | J | J | A | S | O | N | D |
|---|---|---|---|---|---|---|---|---|---|---|---|

*Gladiolus splendens* grows in the western Karoo in the winter-rainfall area. Its ranked scarlet, unscented flowers appear between October and mid-November. The flowers are similar to *G. cunonius*, but the two species occur in different areas.

**DESCRIPTION** **Plant** 50–110cm. **Corm** globose, with papery tunics; forms cormlets on fine stolons. **Cataphylls** upper is purple mottled with white. **Leaves** 5–7, basal leaves up to 9mm wide; midrib lightly raised. **Spike** erect, usually branched, sometimes purplish, 8–14 flowers in 2 ranks separated by at least 90°. **Bracts** purplish. **Flowers** bright scarlet, lower tepals and tube are green fading to red; tepals unequal with the spoon-shaped dorsal tepal extending horizontally and the lateral tepals directed upwards; lower tepals very small, extended forwards and downwards. Perianth tube up to 18mm, narrow and cylindrical, abruptly expanding in the upper part. **Anthers** red with sterile tails, lower filament has enlarged horny 'tooth' at the base. **Pollen** yellow. **Capsules** ellipsoid. **Seeds** broadly winged. **Scent** unscented.

G. splendens *has erect spikes with purple bracts and brilliant red two-ranked flowers.*

**DISTRIBUTION** Found in the arid western Karoo from Laingsburg in the south, along the Roggeveld escarpment, to Calvinia in the north.

**ECOLOGY & NOTES** Most often found in sun or light shade on the banks of rivers where water is available for several months. We have always found the plants growing among rocks or shrubs, often thorny, where they are offered some protection from grazing and predators such as porcupines. Stolons extend from the corms, each with a cormlet at the end, further protecting them from mole-rats and porcupines.

**POLLINATORS** Adapted for pollination by sunbirds.

**SIMILAR SPECIES** *G. cunonius*, *G. splendens* and *G. saccatus* share a similar strange flower structure and are closely related. They all have anthers with deeply forked bases, and all produce cormlets on stolons, as does *G. vandermerwei*.

| | Distribution | Spikes | Flowers |
|---|---|---|---|
| ***G. splendens*** | western Karoo, mainly along rivers | erect, few branches | upper tepals point upwards, lower tepals point down |
| ***G. cunonius*** | southern Cape, coastal sand | strongly inclined to horizontal, few branches | upper lateral tepals point upwards, lower tepals curve upwards |
| ***G. saccatus*** | Vanrhynsdorp heading north into Namibia | slightly inclined, many branches | perianth tube has sac-like spur at base, upper lateral tepals do not point upwards |

*Leaves have a slightly raised midrib.*

*The elongate lower median tepal extends downwards; upper lateral tepals are vertical.*

*Flowering spikes are longer than the leaves.*

G. splendens, *encountered here along the Gannaga Pass, Tankwa Karoo, Northern Cape.*

# Gladiolus saccatus

***saccatus* = having a sac or pouch, a reference to the spur at the base of the tube.**

*Gladiolus saccatus* has bright red flowers in two ranks. Its range extends from the northern parts of the Western Cape, through the Northern Cape and further north into central and northern Namibia. Unscented flowers appear from July to mid-August.

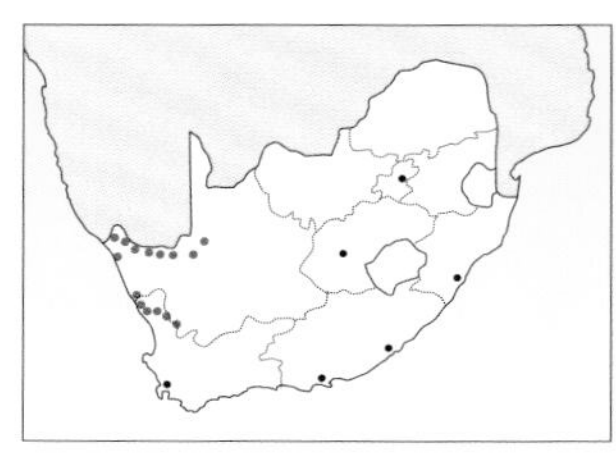

| J | F | M | A | M | J | J | A | S | O | N | D |
|---|---|---|---|---|---|---|---|---|---|---|---|

*There is a yellowish sac-like spur at the base of each perianth.*

**DESCRIPTION Plant** 40–120cm. **Corm** globose to conic, with papery tunics; forms large cormlets on short stolons. **Cataphylls** dark green-brown with white mottling. **Leaves** 5 or 6, basal leaves up to 15mm wide; midrib and 1 other vein thickened. **Spike** dark purplish green, mottled with white at the base, inclined, up to 6 branches, 8–12 flowers in 2 opposed ranks. **Bracts** slightly succulent, brownish purple. **Flowers** bright red, lower tepals and spur bright green at first, then changing to yellow or orange; tepals very unequal with the spoon-shaped dorsal tepal largest, extending horizontally to 45mm, lateral upper tepals pointing downwards and tiny scale-like lower tepals. Perianth tube up to 20mm with a sac-like spur at the lower base. **Anthers** bright red with tails, lower filament with enlarged horny tooth at the base. **Pollen** yellow. **Capsules** ovoid. **Seeds** seed body black, oblong and broadly winged. **Scent** unscented.

**DISTRIBUTION** Occurs in both winter- and summer-rainfall areas. Found from Vanrhynsdorp northwards into Namibia, and eastwards to Upington. *G. saccatus* grows in winter and flowers early in spring (July to mid-August).

**ECOLOGY & NOTES** Usually favours rocky sites, but we have seen it growing on road verges and banks. Plants are often in small groups among other vegetation; they are only visible when in flower as the bright red tepals are very conspicuous.

**POLLINATORS** As for the other bright red, long-tubed species, *G. saccatus*, with its large volumes of nectar, is adapted to sunbird pollination; both Malachite and Dusky sunbirds have been observed visiting.

**SIMILAR SPECIES** *G. cunonius*, *G. splendens* and *G. saccatus* share a similar strange flower structure and are closely related. They all have anthers with deeply forked bases, and all produce cormlets on stolons, as does *G. vandermerwei*.

| | Distribution | Spikes | Flowers |
|---|---|---|---|
| ***G. saccatus*** | Vanrhynsdorp heading north into Namibia | slightly inclined, many branches | perianth tube has sac-like spur at base, upper lateral tepals do not point upwards |
| ***G. cunonius*** | southern Cape, coastal sand | strongly inclined to horizontal, few branches | upper lateral tepals point upwards, lower tepals curve upwards |
| ***G. splendens*** | western Karoo, mainly along rivers | erect, few branches | upper tepals point upwards, lower tepals point down |

*Cataphylls have white mottling.*

*Lower tepals change colour from green to yellow or orange as the flower ages.*

*The dorsal tepals are spoon shaped, lower tepals scale-like.*

G. saccatus, *in flower near Loeriesfontein, Northern Cape.*

# *Gladiolus permeabilis*

***permeabilis* = permeable, for the 'window' between the tepals.**

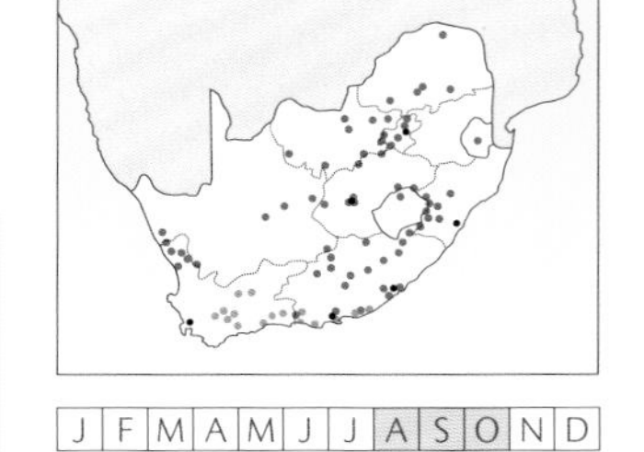

| J | F | M | A | M | J | J | A | S | O | N | D |
|---|---|---|---|---|---|---|---|---|---|---|---|

*Gladiolus permeabilis* is very widespread in summer- and winter-rainfall areas. Viewed in profile, the somewhat dull flowers have a distinctive 'window' between the upper tepals. There are two subspecies.

**DESCRIPTION** **Plant** 15–60cm. **Corm** globose to conic, tunics coarsely fibred. **Cataphylls** uppermost green or flushed purple. **Leaves** 4–7, basal, fanned, rigid, with narrow linear blades; midrib and margins thickened. **Spike** inclined, 4–8 flowers. **Bracts** pale grey-green, sometimes flushed purple. **Flowers** white to cream, dull blue-grey, purple or yellow-brown; upper lateral tepals with brown-purple longitudinal stripe, lower tepals with a yellow section outlined in a darker colour; there is a 'window' between dorsal and lateral tepals. Perianth tube up to 13mm. **Anthers** cream to purple. **Pollen** white. **Capsules** ovoid-ellipsoid. **Seeds** ovate, broadly winged. **Scent** usually scented.

**SUBSPECIES** *G. permeabilis* has two subspecies.
***G. permeabilis* subsp. *permeabilis* (orange on map)** Flowers are blue-grey, purple or brown; tepals spade shaped, not drawn into tails at the tip. Similar to *G. venustus*.
***G. permeabilis* subsp. *edulis* (green on map)** ***edulis* = edible, a reference to the corm.** Flowers are whitish to cream, grey or mauve; tepals unequal, spade shaped, with the ends drawn into twisted tails. Similar to *G. sekukuniensis*.

**DISTRIBUTION** Subsp. *permeabilis* is found in the winter-rainfall area of the southern Cape from Caledon in the west, to Makhanda (Grahamstown) in the east, a summer-rainfall area. Plants usually grow on north-facing slopes in stony shale in renosterveld, but we have seen a large population on a grassy road verge near Gqeberha (Port Elizabeth). Subsp. *edulis* is widespread in the summer-rainfall areas around Oudtshoorn and through the Free State, Mpumalanga and Limpopo, often in drier situations in grass or karoo scrub, in clay soils or among rocks, growing in full sun and in partial shade.

**ECOLOGY & NOTES** In general, populations in winter-rainfall areas are strongly scented, while those in summer-rainfall areas are seldom scented.

**POLLINATORS** Long-tongued anthophorid bees.

**SIMILAR SPECIES** *G. wilsonii* and *G. inandensis* are closely related; *G. uitenhagensis* has a similar 'window' between upper and lower tepals.

| | Flowers | Scent | Region |
|---|---|---|---|
| ***G. permeabilis*** | variable | varies | widespread |
| ***G. uitenhagensis*** | white or mauve; large | faint, sweet scent | Great Winterhoek and Baviaanskloof mountains |
| ***G. wilsonii*** | white to cream; no 'window' | sweetly scented | mainly coastal Eastern Cape |
| ***G. inandensis*** | white to cream | unscented | KwaZulu-Natal |

*Subsp.* permeabilis*; both subspecies have variable colours.*

*Corms of both subspecies may wedge into rock cracks.*

*Subsp.* permeabilis*; dorsal and lateral tepals in both subspecies form a 'window'.*

*Spade-shaped tepals with distinctive markings on subsp.* permeabilis.

*Subsp.* edulis*, Naude's Nek Pass, Eastern Cape.*

*Ends of tepals are twisted on subsp.* edulis.

*Subsp.* edulis*, near Kaapsehoop, Mpumalanga.*

# Gladiolus sekukuniensis

**Named after Sekhukhune, Limpopo, where this species occurs.**

*Gladiolus sekukuniensis* is known only from Sekhukhune in Limpopo. Long-tubed, whitish flowers have red streaks on lower tepals and a characteristic 'window' between upper and lower tepals when viewed in profile. It flowers from March to April.

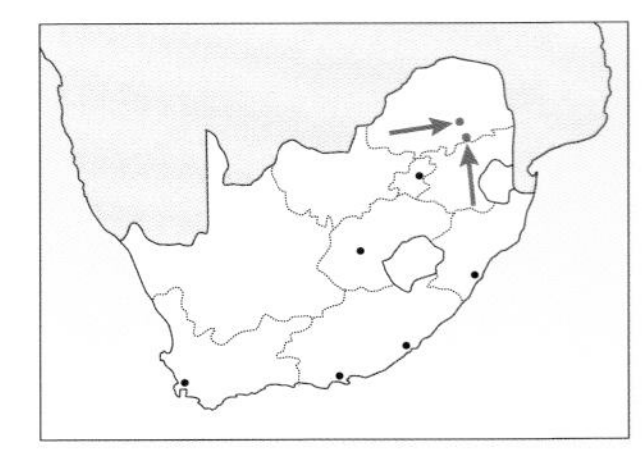

| J | F | M | A | M | J | J | A | S | O | N | D |
|---|---|---|---|---|---|---|---|---|---|---|---|

**STATUS** Vulnerable

**DESCRIPTION** **Plant** 60–110cm. **Corm** globose to conic, with cormlets on suckers, fine tunics. **Cataphylls** uppermost dark green or dry. **Leaves** 5–7, linear, 3mm wide; midrib thickened and raised, margins lightly thickened. **Spike** usually unbranched, 8–17 flowers. **Bracts** cream to pale grey-green. **Flowers** white or cream to pale salmon-pink; tepals unequal, spade shaped, all ending in long twisted tips, with a striking dark red median streak, and a 'window' between the upper dorsal and lateral tepals. Perianth tube 25–35mm. **Anthers** dull blue. **Pollen** cream. **Capsules** ovoid. **Seeds** large dark seed body and light brown wing. **Scent** unscented.

*The flowers are cream with red median stripes.*

**DISTRIBUTION** Known only from a few populations in Sekhukhune, where mining and grazing potentially threaten its habitat.

**ECOLOGY & NOTES** Has only been found on alkaline soils, in lumps of calcrete in banded ironstone, or on norite, an igneous rock high in calcium. The plants grow among rocks in tall grass in open woodland; associated plants include *Senegalia caffra*, *Bauhinia tomentosa* and *Kirkia wilmsii*. The plants are difficult to see even when flowering, as they are slender and the flowers blend in with the vegetation.

*A 'window' between the dorsal and lateral tepals is clearly visible in profile.*

**POLLINATORS** Thought to be long-proboscid flies of the Nemestrinidae family, probably *Stenobasipteron wiedemannii*.

**SIMILAR SPECIES** There are two species in this area that could be confused with *G. sekukuniensis*.

| | Habitat | Flowers | Anthers |
|---|---|---|---|
| ***G. sekukuniensis*** | alkaline soils, Sekukune | 'window' as below, perianth tube 25–35mm | not tailed |
| ***G. permeabilis* subsp. *edulis*** | throughout SA in summer-rainfall areas, not known from alkaline soils | 'window' between upper dorsal and lateral tepals, perianth tube 9–13mm | not tailed |
| ***G. macneilii*** (sect. *Densiflori*) | dolomite, Abel Erasmus Pass | no 'window', perianth tube 40–45mm | tailed at the base |

*Leaves are very narrow.*

*Long perianth tubes are cupped by pale bracts.*

G. sekukuniensis *is known from only three sites. Here it grows north of Burgersfort, Limpopo.*

# *Gladiolus uitenhagensis*

**Named after the area of Kariega (Uitenhage), where it was found and described as new to science.**

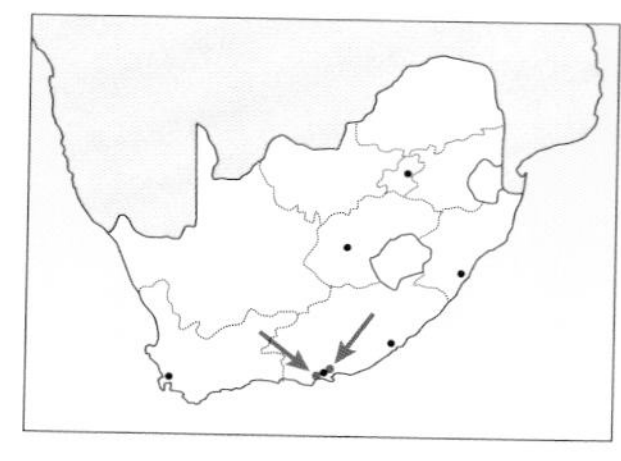

| J | F | M | A | M | J | J | A | S | O | N | D |
|---|---|---|---|---|---|---|---|---|---|---|---|

**STATUS** Vulnerable

Known only from three populations in the mountains of the Eastern Cape, *Gladiolus uitenhagensis* can be confused with *G. permeabilis*, but its flowers are bigger and corm tunics darker.

*Editor's note: This was the only species Rod and Rachel Saunders had not found before their deaths in February 2018. They had searched for it for several years. In 2019, flowering plants were found by Ellie and Rudi Goossens, photographed by Tony Dold, and their identity confirmed by Peter Goldblatt and John Manning. Photographs with kind permission of Tony Dold.*

*The 'window' between the base of the dorsal and lateral tepals appears to be a pair of eyes.*

**DESCRIPTION** **Plant** 35–60cm, some populations up to 75cm. **Corm** orange-yellow, short stolon with single cormlet. **Cataphylls** not known. **Leaves** 3 or 8, up to 2mm wide; midrib strongly thickened. **Spike** unbranched, 5–8 flowers. **Bracts** grey-green. **Flowers** mauve to white; each lower tepal with a yellow spear-shaped mark outlined in purple (Groendal population) or creamish white with fine purple markings, but often with no markings (Baviaanskloof population). Tepals unequal: dorsal largest and inclined over the stamens, with a 'window' between base of dorsal and upper lateral tepals. Perianth tube 28–35mm. **Anthers** yellow or blue. **Pollen** yellow or purple. **Capsules** ovoid-ellipsoid. **Seeds** ovate, broadly and unevenly winged. **Scent** light, sweet scent in the evening.

**DISTRIBUTION** Recorded only from three sites in the Great Winterhoek Mountains: Groendal Reserve near Kariega (Uitenhage), north of Gqeberha (Port Elizabeth); the Elandsberge; and the Baviaanskloof.

**ECOLOGY & NOTES** Plants grow on well-drained stony sandstone-derived soils among fynbos; they apparently flower only one or two years after fire. The colour and markings of plants in the recorded sites are slightly different, with those from Groendal being mauve with yellow markings and with yellow anthers and pollen, and those from the Elandsberge and Baviaanskloof being cream to white with blue anthers and purple pollen. The descriptions for corm, capsules, seeds and scent (above) are for the Baviaanskloof population. This information was not available for the plants found in Groendal.

**POLLINATORS** Thought to be long-tongued anthophorid bees.

**SIMILAR SPECIES** Most similar to *G. permeabilis* especially subsp. *permeabilis*. Differences include size of flower and perianth tube, and texture of corm.

| | Flower size | Habitat |
|---|---|---|
| ***G. uitenhagensis*** | 50–65mm long, perianth tube 28–35mm | sandstone soils, highly restricted locality |
| ***G. permeabilis*** | 30–50mm long, perianth tube 13mm | clay soils |

*Long-tubed flowers alternate up erect spikes.*

*Capsules are ovoid-ellipsoid.*

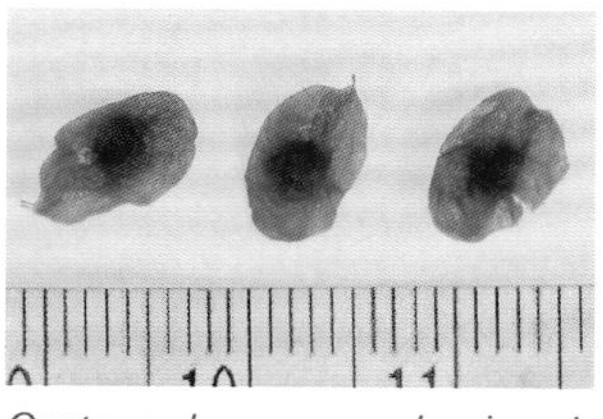

*Ovate seeds are unevenly winged.*

*Flowers resemble those of* G. permeabilis *but are larger.*

G. uitenhagensis *has been recorded in only three sites; these plants were found in the Baviaanskloof near Steytlerville, Eastern Cape.*

# Gladiolus acuminatus

***acuminatus*** **= acuminate tepals, with narrow tapering ends.**

*Gladiolus acuminatus* grows at low elevations in the southern parts of the Western Cape. It flowers from mid-August to late September. Pale yellow to pale green flowers flushed with brown are lightly scented.

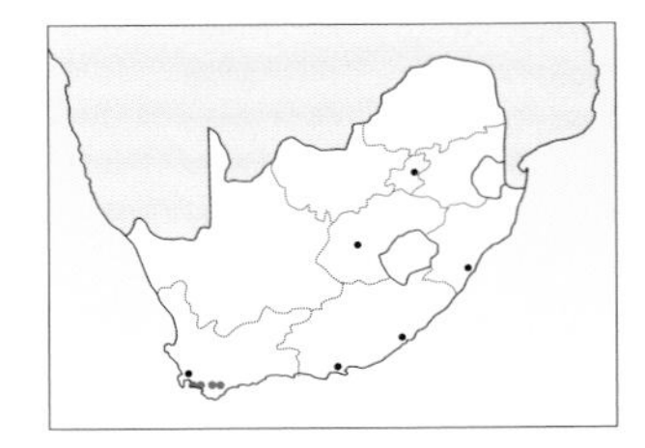

| J | F | M | A | M | J | J | A | S | O | N | D |
|---|---|---|---|---|---|---|---|---|---|---|---|

**STATUS** Endangered

**DESCRIPTION** **Plant** 25–50cm. **Corm** globose, with papery tunics becoming fibrous with age. **Cataphylls** green or purple above ground. **Leaves** 5 or 6 or more if spike is branched, up to 1.5mm wide; midrib slightly thickened. **Spike** simple or branched, erect or inclined, 3–10 flowers. **Bracts** pale green. **Flowers** small, cream to pale yellow or greenish; tepals up to 20mm, backs flushed brown or purple, tepal margins undulate and apices lightly twisted. Perianth tube ±20mm. **Anthers** white-cream. **Pollen** white-cream. **Capsules** small, 10–15mm. **Seeds** oval and small, unevenly winged, seed body dark and large. **Scent** lightly scented.

**DISTRIBUTION** Range extends from Onrus (near Hermanus) to Bredasdorp in the southern parts of the Western Cape.

**ECOLOGY & NOTES** Plants grow on stony shale flats and north-facing slopes at low altitudes. The vegetation type in these areas is renosterveld, and much of its former habitat is now under wheat or grazed by sheep. The species is poorly recorded and although it has been assumed to be rare, it may also be due to the small size and the cryptic colour of the flowers. There is concern about loss of pollinators due to farming practices. The plants that we photographed were on the bank of a road verge and were very difficult to see, even though they were not in dense vegetation.

G. acuminatus *is endemic to the southern coast of the Western Cape.*

**POLLINATORS** Thought to be moths as the flowers have a sweet scent. The perianth tube is long and the flowers are a pale greenish cream; both of these features are attractants of moths.

**SIMILAR SPECIES** Plants in flower cannot be mistaken for any other species. The perianth tube is long in relation to the size of the flower and encloses the anthers. The flowers are an unusual colour and the almost-equal tepals taper at the ends. However, in leaf, the plants can be confused with those of *G. permeabilis* and *G. stellatus*. The known range of *Gladiolus stellatus* is further east (Swellendam and eastwards), but the ranges of *G. permeabilis* and *G. acuminatus* overlap.

*Cream flowers are flushed with brown and may have brown median markings.*

*Tepals undulate and twist with brown-flushed backs.*

*Long perianth tubes emerge from pale green bracts.*

*Leaves are narrow; cataphylls are green or purple above ground.*

*Even in flower, plants can be difficult to see. Here, near Bredasdorp, Western Cape.*

# *Gladiolus karooicus*

***karooicus*** **= from the Karoo.**

The sweetly scented yellow flowers of *Gladiolus karooicus* appear between mid-August and late September. Long-tubed with erect lateral tepals, it is unlikely to be confused with any other plant when in flower.

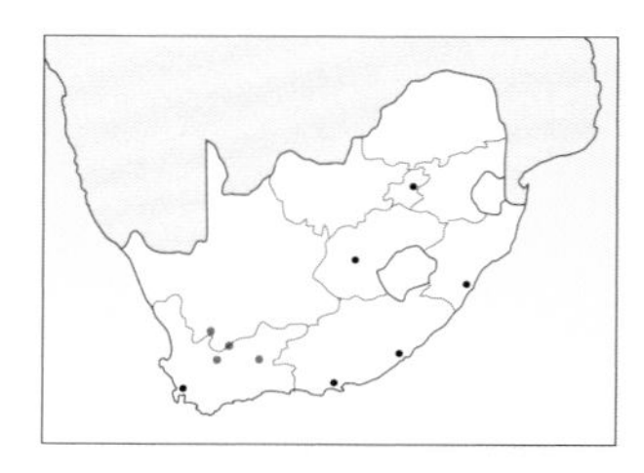

J | F | M | A | M | J | J | A | S | O | N | D

**DESCRIPTION** **Plant** 15–50cm. **Corm** conical, with papery tunics. **Cataphylls** dull purple above ground. **Leaves** usually 4, blades linear, 2mm wide, slightly twisted; midrib slightly thickened. **Spike** inclined, sometimes branched, 2–5 flowers. **Bracts** grey-green, sometimes purplish or dry at tips. **Flowers** yellow. Clawed tepals unequal and narrow, with windows between upper dorsal and upper lateral tepals. Upper tepals with grey veins outside, lower tepals deep yellow fading to grey with age. Perianth tube 12mm. **Anthers** light purple. **Pollen** whitish. **Capsules** obovoid, 20mm. **Seeds** golden brown with darker seed body, ovate, broadly winged. **Scent** sweetly violet scented.

*Erect upper lateral tepals give the flowers a characteristic 'surrender' shape.*

**DISTRIBUTION** This species has only been found on the Klein Roggeveld (Komsberg), the foot of the Witteberg near Matjiesfontein, and north of the Great Swartberg near Prince Albert, Western Cape.

**ECOLOGY & NOTES** Plants are normally found in damp areas such as along streambanks and in gullies, and in decomposed shale soils. The population that we photographed were all growing among karoo bushes and in grass clumps where they were protected from both sun and grazing.

*Lower tepals are strongly clawed.*

**POLLINATORS** Thought to be long-tongued bees.

**SIMILAR SPECIES** In flower, this species is unlikely to be confused with any other. The bright yellow flowers are very different from those of *G. venustus*, which grows in the same area.

*Narrow twisted leaves have thickened midribs.*

*The short perianth tubes are concealed by green bracts.*

*Inclined spikes bear up to five flowers, sometimes more.*

*Before flowering, this species can be difficult to see.*

# *Gladiolus stellatus*

***stellatus*** **= star-like, as the flowers resemble small stars.**

The radially symmetrical, star-shaped flowers of *Gladiolus stellatus* make it distinctive. Growing along the southern parts of the Western Cape, between Swellendam and Gqeberha (Port Elizabeth), it flowers in the morning only, from August to October.

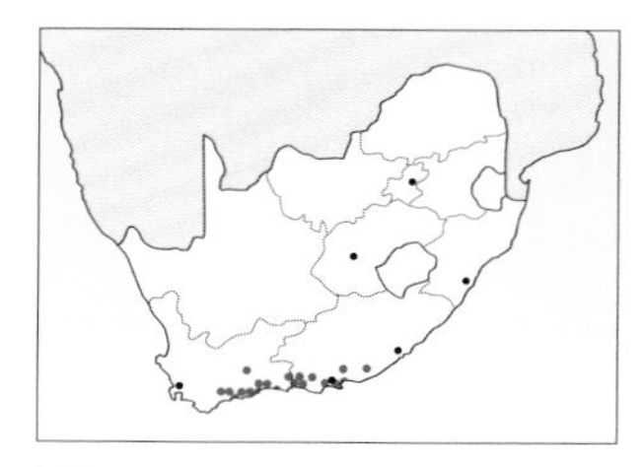

J F M A M J J A S O N D

**DESCRIPTION** **Plant** 30–60cm. **Corm** globose, often producing stolons at base, fine- to coarse-fibred tunics. **Cataphylls** up to 30mm above ground, green or purple. **Leaves** usually 6–9, up to 3mm wide; midrib thickened. **Spike** inclined, usually with 1 or 2 branches, 5–12 flowers. **Bracts** pale green but brownish when the first flowers open. **Flowers** white to cream to pale lilac with a darker median streak on each tepal; star-shaped and radially symmetrical with all tepals equal. Perianth tube short, up to 7mm. Flowers only open in the morning and close around midday. **Anthers** pale yellow. **Pollen** yellow. **Capsules** ellipsoid. **Seeds** light brown, ±5mm, large dark seed body. **Scent** very sweetly scented.

G. stellatus *bears up to 12 flowers on branched spikes.*

**DISTRIBUTION** *G. stellatus* is known from the southern Cape, from Swellendam in the west to Gqeberha (Port Elizabeth) in the east, and inland into the Little Karoo. Most commonly found on the coastal plain and south of the Langeberg and Outeniqua mountains, it has also been recorded from the Langkloof and the Little Karoo north of the mountains.

**ECOLOGY & NOTES** Often found on dry sites in stony grassland and renosterveld, on the tops of hills or north-facing slopes. Flowers profusely after fire, but fire is not a prerequisite. The small flowers do not look like a species of *Gladiolus*.The flowers are only open in the morning; they close at midday and then re-open the following day. The population of plants that we photographed were growing together with *Moraea polyanthos*. The flowers of these two species are very similar, but the *Moraea* flowers only open at midday and remain open in the afternoon, when the *Gladiolus* flowers are closed.

**POLLINATORS** Thought to be generalists such as bees and butterflies. The flowers offer both pollen and nectar as a reward, and their sweet scent perfumes the air as soon as the flowers open early in the morning. It is thought that *G. stellatus* and *Moraea polyanthos* share pollinators, with the flowers opening at different times of the day.

*This is one of a few gladioli with radially symmetrical flowers.*

**SIMILAR SPECIES** In flower, this species cannot be confused with any other *Gladiolus* species. The actinomorphic (radially symmetrical) white-lilac flowers are unlike those of any other species, but they could conceivably be mistaken for *Moraea polyanthos*.

*Leaves are narrow with a thickened midrib.*

*Tepals have dark median streaks.*

G. stellatus *flowering on Robinson Pass, Western Cape.*

# *Gladiolus wilsonii*

**Named after either John Wilson from Scotland – who grew the plants Baker described in 1886 – or his brother Alexander Wilson, who collected the plants for John near Gqeberha (Port Elizabeth).**

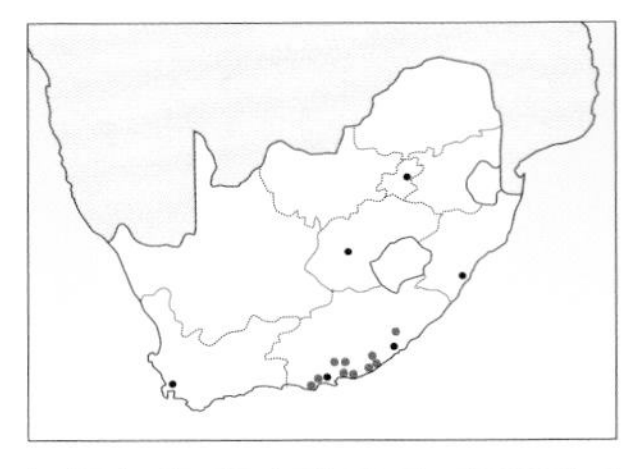

| J | F | M | A | M | J | J | A | S | O | N | D |
|---|---|---|---|---|---|---|---|---|---|---|---|

Although it grows in the summer-rainfall areas in the Eastern Cape, *Gladiolus wilsonii* is unusual because it flowers from October to November, after scant autumn and winter rainfall. Its pale flowers tinged with lilac are sweetly scented. It can be mistaken for *G. inandensis* when flowering, but their leaves differ.

**DESCRIPTION** **Plant** 30–60cm. **Corm** globose, with small stolons, fine- to coarse-fibred tunics. **Cataphylls** uppermost green or dry. **Leaves** 6–8, blades linear, up to 3mm wide; midrib thickened. **Spike** inclined, simple or branched, 4–12 flowers. **Bracts** green, may be flushed purple. **Flowers** white to cream with pink to purple on the reverse of the 3 upper tepals, lower tepals may have mauve streaks near the base; tepals are unequal with the dorsal being the largest, upper tepals are not narrowed below, so there are no 'windows' between upper and lower tepals. Perianth tube up to 13mm. **Anthers** purple. **Pollen** cream. **Capsules** ±15mm. **Seeds** light brown, 6mm, unevenly winged. **Scent** sometimes scented.

*The spikes may bear up to 12 pale flowers.*

**DISTRIBUTION** Although a winter-growing plant, *G. wilsonii* occurs in the summer-rainfall areas of the Eastern Cape from Jeffreys Bay in the west to Engcobo in in the east. It is best known from the Makhanda (Grahamstown) and East London areas.

**ECOLOGY & NOTES** Present in open grassland and sometimes flowers en masse in spring. Grows in winter and relies on the late autumn or winter rain that occasionally falls in this area. Although it is reported to be scented, we saw plants in three different populations and could not detect a scent in any of the flowers.

**POLLINATORS** Thought to be bees, as the flowers have short tubes and are sometimes scented.

**SIMILAR SPECIES** There are two similar species: *G. permeabilis* and *G. inandensis*.

| | Flowers | Growth cycle |
|---|---|---|
| ***G. wilsonii*** | no window between upper tepals; colour is white, cream to blueish | winter growing, flowers together with green leaves |
| ***G. permeabilis*** | profile view shows 'window' between bases of dorsal and upper lateral tepals; colour is brownish, purplish, grey or cream | summer or winter growing, flowers with green leaves |
| ***G. inandensis*** | no window between upper tepals; white or cream | summer growing, flowers with dry leaves from previous growth season |

*The narrow upper leaves sheath the stem.*

*The green-purple bracts are as long as the perianth tubes.*

*Dorsal tepals are the largest, with no 'windows'.*

*This white form has delicate mauve streaks.*

*Strongly flushed tepals give a lilac appearance to this flower.*

*Flowering in grasslands near Gqeberha (Port Elizabeth), Eastern Cape.*

# Gladiolus inandensis

**The type collection was made in Inanda, near Durban.**

Growing on quartzitic sandstone soils in the Eastern Cape and KwaZulu-Natal, *Gladiolus inandensis* has pale, unscented flowers that appear from late July to September.

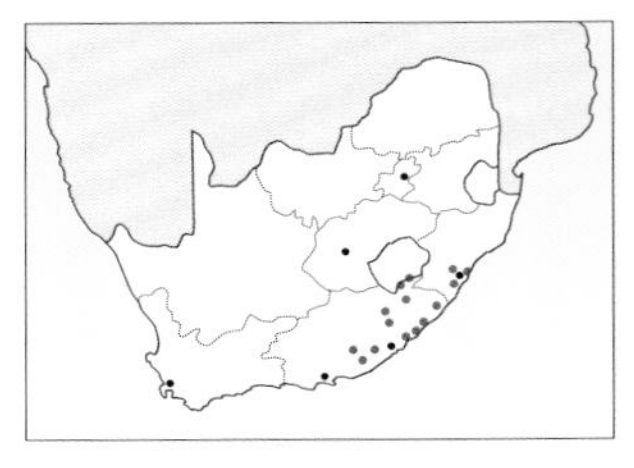

| J | F | M | A | M | J | J | A | S | O | N | D |
|---|---|---|---|---|---|---|---|---|---|---|---|

**DESCRIPTION** **Plant** 20–45cm. **Corm** globose, with large cormlets. **Cataphylls** leathery and dark green above ground, forming a fibrous neck around base of stem. **Leaves** on flowering stem are dry or absent. At the end of summer there are 4–6 leaves up to 20cm long, oval in transverse section when fresh, but midrib thickened and prominent when dry, upper leaves entirely sheathing. **Spike** inclined, sometimes with 1 branch, 4–12 flowers. **Bracts** green or flushed purple. **Flowers** white to cream with pink on reverse of 3 upper tepals and sometimes on the midline; unequal tepals. Perianth tube short, up to 12mm. **Anthers** mauve. **Pollen** cream. **Capsules** narrow, up to 15mm long. **Seeds** ovate, broadly and evenly winged, golden brown. **Scent** unscented.

*The plant bears four to 12 pale flowers with unequal tepals.*

**DISTRIBUTION** Found from Komani in the Eastern Cape, northwards to Empangeni north of Durban in KwaZulu-Natal.

**ECOLOGY & NOTES** Plants seem restricted to quartzitic sandstone soils and are most commonly seen near the coast, but have also been recorded as far inland as southern Lesotho. They are found in clumps in rocky grassland on hillsides and flower best after fire, although we photographed some flowering plants in old grassland near Oribi Gorge. This species has an unusual growth cycle: plants flower very early in spring before the new leaves have emerged. If the area burned during the winter, the flower spikes have no leaves. If the grass did not burn, the spikes still have the previous season's leaves attached, but they are dry and dead. After flowering, the new leaves emerge from the same corm, but from a new shoot next to the flowering stem.

**POLLINATORS** Long-tongued bees.

**SIMILAR SPECIES** In flower shape and colour, this species is almost identical to *G. wilsonii*. However, the growth cycles of the two species are very different. In the areas where their distribution overlaps, the plants must be examined very carefully to distinguish them.

| | Leaves | Growth cycle |
|---|---|---|
| ***G. inandensis*** | bladeless, sheathing; new leaves appear at end of flowering season; terete in transverse section | summer growing, flowers with dry leaves from previous growth season |
| ***G. wilsonii*** | plane, linear leaves with raised midrib | winter growing, flowers together with green leaves |

*Pink is seen on reverse of upper three tepals.*

*Short floral tubes allow pollination by bees.*

*Bracts are the same length as the perianth tubes.*

G. inandensis *has no true dormancy period. Encountered near Durban, KwaZulu-Natal.*

# *Gladiolus robertsoniae*

**Named after WM Robertson, who made the first recorded collection in 1916.**

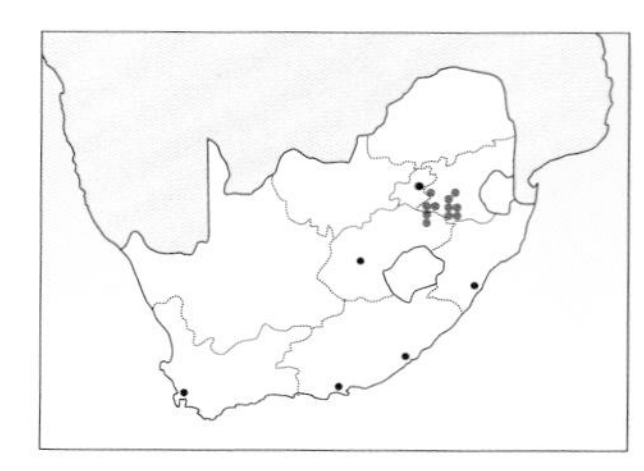

| J | F | M | A | M | J | J | A | S | O | N | D |
|---|---|---|---|---|---|---|---|---|---|---|---|

**STATUS** Near Threatened

*Gladiolus robertsoniae* is endemic to Mpumalanga and the Free State, where it flowers from October to November. White flowers flushed with lilac are scented day and night, and are pollinated by moths.

**DESCRIPTION** **Plant** 20–40cm. **Corm** globose, with medium-textured netted tunics. **Cataphylls** upper reaching 2cm above ground, green or brown. **Leaves** 3 or 4, blades linear to curved, 2–4mm wide; midrib thickened. **Spike** simple or with 1 branch, 4–8 flowers. **Bracts** dark green or flushed purple. **Flowers** white, sometimes flushed with violet towards tepal apices, throat with fine red lines; tepals unequal with dorsal tepal the smallest. Perianth tube up to 44mm, about twice as long as the bracts. **Anthers** white or light mauve. **Pollen** yellow. **Capsules** globose, up to 12mm. **Seeds** ovate, broadly winged. **Scent** carnation scented.

**DISTRIBUTION** Known from a small area of the Highveld of Mpumalanga and the Free State: from Ermelo westwards to Frankfort and Villiers. There are increasing demands on its habitat.

**ECOLOGY & NOTES** Plants grow in rocky areas among dolerite outcrops. As they flower at the end of a long dry season, they are usually found in areas that remain damp for most of the year, such as in seeps or on riverbanks. The soil in this species' habitat is black heavy turf, and the gladioli we saw grew among many other bulbous plants such as *Crinum bulbispermum*, and *Hypoxis*, *Ledebouria* and *Tulbaghia* species. Most of the habitat is heavily grazed by cattle. A large population of *G. robertsoniae* close to the N3 between Johannesburg and Durban has been split down the middle by the national road, and is threatened by grass cutting and developments along the route.

**POLLINATORS** Pollinated by moths, attracted to the long-tubed, scented white flowers, which remain open at night.

**SIMILAR SPECIES** There are no other species flowering in this area in October that can be confused with *G. robertsoniae*.

*Yellow pollen is visible here on the white flowers.*

*Tips of the tepals may be flushed with violet.*

Narrow leaves sheath the base of the stem.

Fine red stripes are visible in the throat.

Perianth tubes are long and slender; note how the dorsal tepal is the shortest.

Inclined spikes straighten as flowers open. Here, growing near Villiers, Free State.

# Gladiolus arcuatus

***arcuatus*** **= arched, referring either to the sickle-shaped leaves or the inclined flower spike.**

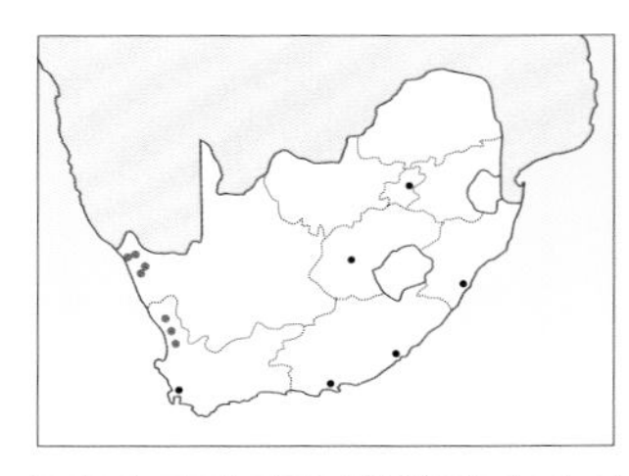

J F M A M J J A S O N D

The dull purple-brown flowers of *Gladiolus arcuatus* appear from June to August at lower elevations and August to September at higher ones. This species is widespread in the drier northern areas of the Western Cape and Namaqualand.

**DESCRIPTION** **Plant** 12–20cm. **Corm** depressed-globose, with papery tunics. **Cataphylls** uppermost green above ground. **Leaves** 5 or 6, 5–10mm wide, curved and sickle shaped, often twisted. Hairy on the sheaths, midrib and margins; margins and midrib may be raised. **Spike** inclined, 4–7 flowers. **Bracts** pale green, up to 35mm. **Flowers** shades of dull grey-purple to purple, green or brownish; unequal tepals with 'windows' between dorsal and lateral tepals and dorsal tepal arching; clawed lower tepals with yellow in the lower two-thirds. Perianth tube 10–15mm. **Anthers** pale grey. **Pollen** cream. **Capsules** ellipsoid. **Seeds** saucer shaped, asymmetrical, dark seed body, paler wing. **Scent** apple scented.

*This species usually has dull-coloured flowers.*

**DISTRIBUTION** Found in the dry northwestern parts of the winter-rainfall area. Recorded from near Steinkopf in the Northern Cape southwards through Namaqualand to the area around Klawer and Vredendal. We have also seen it flowering in southern Namibia, in the Rosh Pinah area.

**ECOLOGY & NOTES** Found in a variety of soils: granite-derived stony soil in Namaqualand and more silty soils in the southern parts of its range. They grow among small shrubs including *Dimorphotheca cuneata* and *Elytropappus rhinocerotis*, and other Iridaceae such as *Lapeirousia silenoides*.

*More brightly coloured flowers are found further north in its range.*

**POLLINATORS** Long-tongued bees such as honeybees and *Anthophora* species. These bees are probably attracted to the flowers by their sweet scent.

**SIMILAR SPECIES** *G. arcuatus* may be confused with *G. scullyi*.

| | Corm | Leaves |
|---|---|---|
| ***G. arcuatus*** | depressed-globose to discoid, soft papery corm tunics | pubescent to velvety, curved |
| ***G. scullyi*** | globose, orange to red, tunics of hard almost woody layers | fan of grey-green to glaucous leaves |

*Velvety leaves are often twisted; note how the spikes are inclined.*

*In profile the flowers have 'windows' between the upper and lower tepals.*

G. arcuatus, *found in the Kamiesberg, Northern Cape.*

# *Gladiolus viridiflorus*

***viridiflorus* = green-flowered, although in many of the plants, green is not the dominant colour.**

*Gladiolus viridiflorus* has dull greenish-brown flowers, making it inconspicuous. Growing in the arid areas of the Western Cape and Namaqualand, its flowering cycle is early, from May to July, and very short.

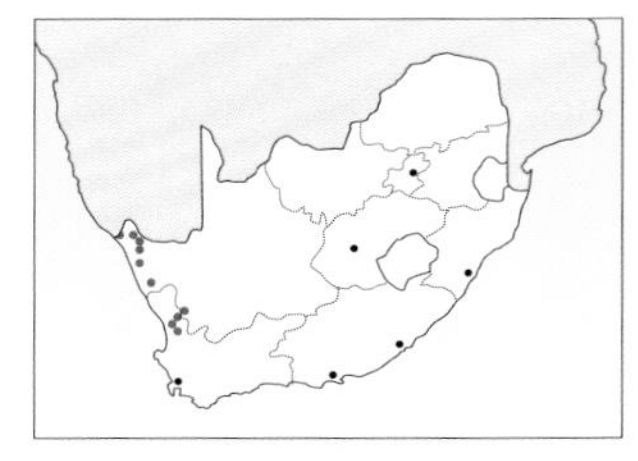

| J | F | M | A | M | J | J | A | S | O | N | D |
|---|---|---|---|---|---|---|---|---|---|---|---|

**DESCRIPTION** **Plant** 8–15cm. **Corm** globose to conic, with hard, woody tunics. **Cataphylls** mottled purple and white, above ground. **Leaves** usually 4, 3–6mm wide, blades loosely twisted or coiled; with prominent midrib. **Spike** inclined 30–60°, not straight but scalloped between the flowers, sometimes branched, 3–8 flowers. **Bracts** green and slightly glaucous, margins transparent. **Flowers** upper tepals greenish grey to yellowish; tepals unequal with the lower 3 longer than the upper 3 and marked with a broad yellow transverse band outlined with purple. Perianth tube shorter than bracts. **Anthers** greenish. **Pollen** pale yellow. **Capsules** broadly ellipsoid. **Seeds** large (± 10mm), broadly and evenly winged, yellowish brown with reddish centre. **Scent** freesia scented.

*This species has distinctive twisted leaves.*

**DISTRIBUTION** Recorded from the banks of the Orange River east of Alexander Bay in the Northern Cape, southwards through Namaqualand, to Clanwilliam in the south.

**ECOLOGY & NOTES** Most of the localities of this species are in arid areas that may receive as little as 150mm of winter rain. Plants grow among low bushes on soils derived from granite or sandstone, either sandy or heavier clay. The plants flower early, in autumn or early winter, and can complete their flowering cycle before the rainy season finishes. In years of good rains, the plants will continue to grow into early summer, and in these years we have noticed that the seeds take several months to ripen. This is unusual as most *Gladiolus* seeds are ripe ±4–6 weeks after flowering. Although the species is recorded as having very fragrant flowers, we were unable to detect any scent at all in the three populations that we photographed. It is possible that this was due to the fact that we visited the plants quite early in the morning when the air temperature was still low.

**POLLINATORS** Long-tongued bees.

**SIMILAR SPECIES** The closest relatives of *G. viridiflorus* are thought to be *G. arcuatus*, *G. deserticola*, *G. scullyi* and *G. venustus*. However, the distinctive mottled cataphylls, coiled leaves, strongly inclined spike and greenish flowers with yellow and purple markings on the lower tepals make *G. viridiflorus* easy to identify.

*Mottled cataphylls grow above ground.*

*Greyish-green bracts have transparent margins.*

*The lower tepals are the longest.*

*Note the narrow 'windows' between the greenish dorsal and lateral tepals.*

G. viridiflorus, *here growing near Soebatsfontein, Namaqualand, Northern Cape.*

# Gladiolus deserticola

***deserticola* = living in a desert landscape, wilderness-lover.**

Restricted to the Richtersveld area of the Northern Cape, *Gladiolus deserticola* has star-shaped blue flowers, which appear from mid-August to mid-September. They are weakly scented.

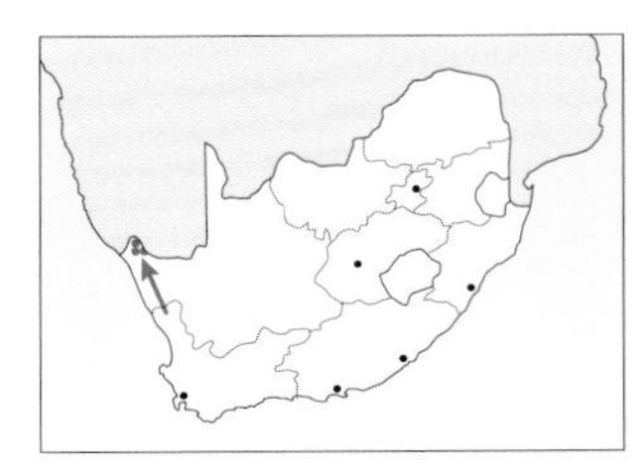

| J | F | M | A | M | J | J | A | S | O | N | D |
|---|---|---|---|---|---|---|---|---|---|---|---|

**STATUS** Vulnerable

**DESCRIPTION** **Plant** 10–20cm. **Corm** conic, with firm tunic. **Cataphylls** uppermost up to 8cm above ground, purple mottled with white. **Leaves** 4 or 5, linear blades ±2mm wide, erect or trailing and often twisted; midrib raised. **Spike** erect or inclined, sharply flexed above the leaves and scalloped between the flowers, simple or branched, 4–6 flowers. **Bracts** green. **Flowers** dark blue, actinomorphic except for asymmetrical cream markings on the 2 lower lateral tepals; each tepal with a darker marking on the midline. Perianth tube 11mm. **Anthers** pale yellow. **Pollen** pale yellow. **Seeds** ±5mm, light brown, with well-developed wing. **Scent** gently rose scented.

*The plant's radially symmetrical dark blue flowers are unlikely to be confused with those of any other* Gladiolus *species.*

**DISTRIBUTION** Known only from the Richtersveld in the Northern Cape. It has been seen at both high and lower altitudes.

**ECOLOGY & NOTES** Plants seem to favour cooler south-facing slopes where they grow in heavier clay soils in protected places such as in the shade of rocks or shrubs. The Richtersveld is extremely hot and arid, and during some years it receives almost no rain at all. Presumably the plants do not flower in these years.

**POLLINATORS** Long-tongued bees, which forage on nectar.

**SIMILAR SPECIES** The closest relative is thought to be *G. viridiflorus* with similar mottled cataphylls, corm tunics, twisted leaves and a strongly inclined scalloped flower spike. However, the very limited distribution range of this species, together with the highly distinctive star-shaped, dark blue flowers, make it unmistakable.

*Spikes are scalloped between the flowers.*

*Actinomorphic flowers are uncommon in the genus.*

*Plants seem to favour protected places; encountered here in the Richtersveld, Northern Cape.*

# Gladiolus scullyi

**Named after the Magistrate of Namaqualand in the 1890s, WC Scully (1855–1943), who had a keen interest in the natural history of the area.**

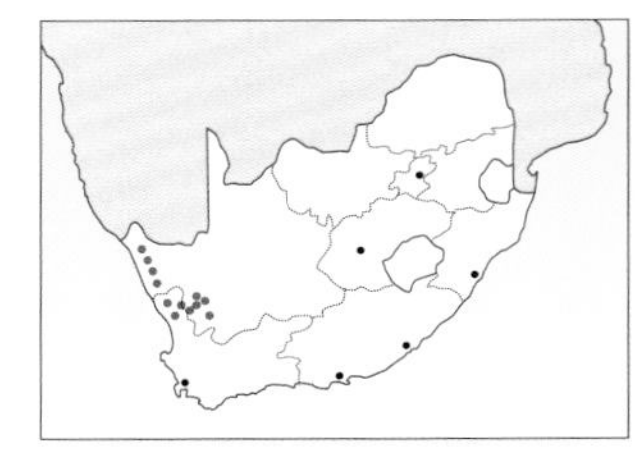

| J | F | M | A | M | J | J | A | S | O | N | D |
|---|---|---|---|---|---|---|---|---|---|---|---|

Common in the dry western parts of the Northern Cape, *Gladiolus scullyi* has dull greenish to brownish flowers with yellow lower tepals tipped with purple. They are strongly scented.

**DESCRIPTION** **Plant** 20–35cm. **Corm** globose-conic, with many cormlets at the base, orange to reddish, with hard, woody tunics. **Cataphylls** pale green or purple above ground. **Leaves** ±6 in distichous fan, 3–5mm wide, lanceolate, grey-green; midrib lightly raised. **Spike** erect or inclined, may be branched, 5–8 flowers. **Bracts** pale green, translucent on the veins. **Flowers** muted greenish cream to yellow-brown; tepals unequal, windowed between dorsal and upper lateral tepals, lower tepals narrowed into claws and narrowly channelled; upper tepals with darker stripe along midline and sometimes flushed purple, lower tepals streaked in throat. Perianth tube 14mm. **Anthers** light purple to cream. **Pollen** cream. **Capsules** broadly obovoid. **Seeds** dark red-brown, saucer shaped, seed body asymmetrically placed. **Scent** strongly scented.

*Spikes bear up to eight strongly scented flowers.*

**DISTRIBUTION** Occurs from Namaqualand southwards to the Calvinia district in the Northern Cape. These arid areas receive a low winter rainfall.

**ECOLOGY & NOTES** Plants grow in granite and shale-derived soils. In the southern end of their distribution range, they can be found together with *G. venustus*.

**POLLINATORS** Long-tongued bees such as *Anthophora diversipes*, which are attracted by the strong scent.

*Flowers are similar to those of* G. venustus, *but are more muted in colour.*

**SIMILAR SPECIES** *G. scullyi* can be confused with *G. arcuatus* but is more likely to be confused with *G. venustus*.

| | Flowers | Upper tepals | Lower tepals |
|---|---|---|---|
| ***G. scullyi*** | muted colours, strongly scented | 'window' between base of dorsal and upper lateral tepals | claws narrowly channelled |
| ***G. venustus*** | strongly contrasting colours, lightly scented | no gap between dorsal and upper lateral tepals, which are twisted upwards and lie partly behind the dorsal tepal | claws deeply channelled with prominent ear-like lobes |

*Lanceolate leaves have lightly raised midribs.*

*Lower tepals narrow into claws.*

G. scullyi *flowering en masse in Namaqualand, Northern Cape.*

# Gladiolus venustus

**_venustus_ = beautiful, for the brightly coloured flowers.**

*Gladiolus venustus* is widely spread across the interior of the winter-rainfall area. Purple, pink and sometimes yellow flowers are sweetly scented and appear from August to mid-September.

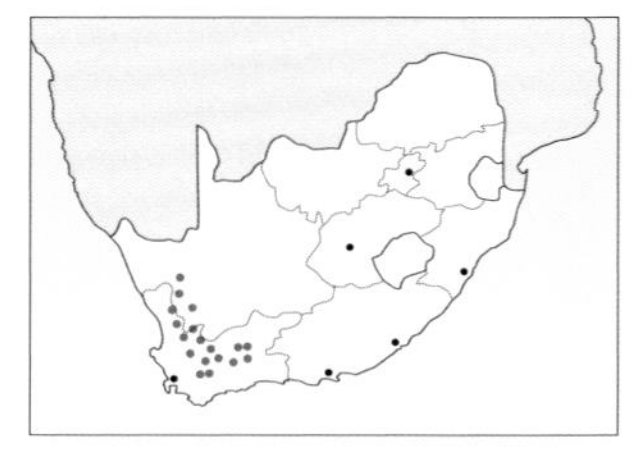

| J | F | M | A | M | J | J | A | S | O | N | D |
|---|---|---|---|---|---|---|---|---|---|---|---|

**DESCRIPTION** **Plant** 12–35cm. **Corm** conic, with tunics of hard, woody layers. **Cataphylls** uppermost green. **Leaves** 5 or 6, linear blades ±4mm wide; midrib lightly raised. **Spike** inclined, usually branched, 5–8 flowers. **Bracts** pale green, with transparent veins. **Flowers** shades of blue-purple to deep pink with yellow on lower tepals; tepals unequal, upper laterals pointing upwards and lying at least partly behind the dorsal tepal; lower tepals with deeply channelled claws. Perianth tube long, 12–17mm. **Anthers** greyish to yellow. **Pollen** yellow. **Capsules** globose, broad and squat. **Seeds** reddish brown, saucer shaped, asymmetrical, with seed body closer to one side. **Scent** delicately rose scented.

*This species is colourful when flowering en masse.*

**DISTRIBUTION** Found in the southwestern Cape, from the Clanwilliam area eastwards through the Cederberg to the Swartberg, extending through the Tankwa Karoo and southwards to Worcester. Plants have been recorded as far north as the Nieuwoudtville area and southern Namaqualand.

**ECOLOGY & NOTES** Usually found on dry slopes in heavier clay and shale soils, but also sometimes in sandstone. Intermediates between *G. venustus* and *G. scullyi* have been found in the Nieuwoudtville and southern Namaqualand areas.

**POLLINATORS** Long-tongued bees foraging for nectar.

*Some forms can be difficult to distinguish from* G. scullyi.

**SIMILAR SPECIES** *G. scullyi* and *G. venustus* are easily confused. The upper tepals are different. Flower colour can be a distinguishing factor but this is not always the case.

| | Flowers | Upper tepals | Lower tepals |
|---|---|---|---|
| ***G. venustus*** | strongly contrasting colours, lightly scented | no gap between dorsal and upper lateral tepals, which are twisted upwards and lie partly behind the dorsal tepal | claws deeply channelled with prominent ear-like lobes |
| ***G. scullyi*** | muted colours, strongly scented | gap between base of dorsal and upper lateral tepals (window) | claws narrowly channelled |

*Linear blades with raised midribs surround the spike.*

*Strongly inclined spikes bear up to eight flowers.*

*A purple form contrasts with the dry soil.*

*Pale green bracts complement this yellow-pink form.*

*Bracts have transparent veins.*

G. venustus *flowering in the Tankwa Karoo, Northern Cape.*

# *Gladiolus salteri*

**Named after Captain TM Salter (1883–1969), an amateur botanist who collected the plants used for the description.**

*Gladiolus salteri* occurs only in the Namaqualand region of the Northern Cape. When not in flower, it is hard to distinguish from *G. scullyi*, but its pale pink flowers with darker markings are distinctive.

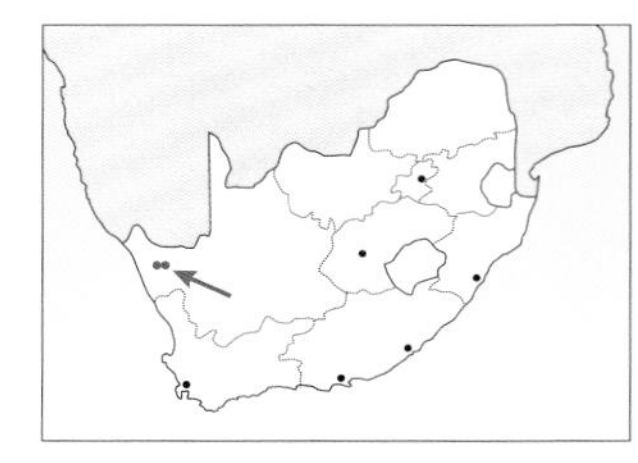

| J | F | M | A | M | J | J | A | S | O | N | D |
|---|---|---|---|---|---|---|---|---|---|---|---|

**STATUS** Rare

**DESCRIPTION** **Plant** 10–20cm. **Corm** globose-conic, with coarsely fibred tunics. **Cataphylls** not above ground. **Leaves** 4, curved to one side, 4–7mm wide, grey-green. **Spike** inclined to nearly horizontal, may be branched. **Bracts** pale grey-green. **Flowers** pale pink, lower tepals cream with deep pink markings; tepals unequal, dorsal longest, narrowed below into a slender claw, arching to upright; lower tepals narrow and pointing forward. Perianth tube long, 18–22mm. **Anthers** cream. **Pollen** pale yellow. **Capsules** broadly ovoid. **Seeds** saucer shaped, seed body darker than the wing and asymmetrically placed, 5mm. **Scent** delicately rose scented.

*The recurved, clawed dorsal tepal gives the flower an unusual shape.*

**DISTRIBUTION** This species is known only from the Northern Cape, close to Springbok.

**ECOLOGY & NOTES** Plants grow in coarse, gritty decomposed granite, or in red, wind-blown sand, usually on the edges or in the middle of drainage lines. Despite searching among many granite hills, we only found one population of plants, which flowered and set seed satisfactorily in a year of good rainfall. This population of plants responded to late rain by flowering well into October. We found a large population of the tiny *Brunsvigia namaquana* in the same area as this *Gladiolus*.

*Lower tepals have delicate deep pink markings.*

**POLLINATORS** Thought to be long-tongued bees.

**SIMILAR SPECIES** When plants are in flower, this species is unmistakable. Out of flower, it is possible to mistake *G. salteri* for *G. scullyi*, although the leaves of *G. salteri* are generally shorter and more arched than those of *G. scullyi*.

*The distinctive flower shape is clear in profile.*

*Sickle-shaped leaves arch to one side.*

G. salteri *is known only from near Springbok, Northern Cape.*

# Gladiolus lapeirousioides

***lapeirousioides* = resembling a species from the genus *Lapeirousia*, particularly *L. fabricii*.**

Known only from Loeriesfontein in the Northern Cape, *Gladiolus lapeirousioides* has creamy, unscented flowers with yellow and red markings, and appears from late August to early October.

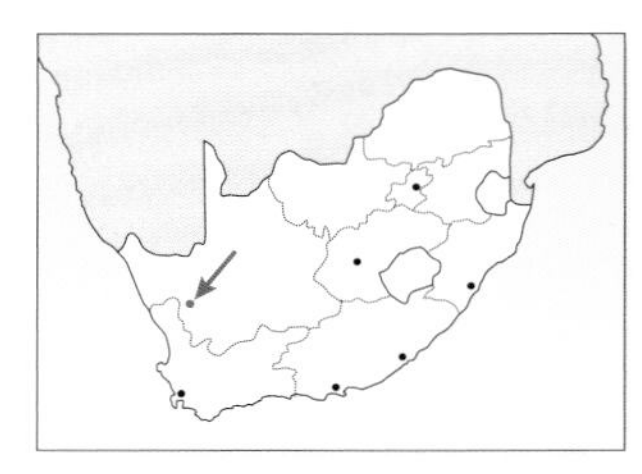

| J | F | M | A | M | J | J | A | S | O | N | D |
|---|---|---|---|---|---|---|---|---|---|---|---|

**STATUS** Vulnerable

**DESCRIPTION** **Plant** 8–14cm. **Corm** conic, with cormlets around the base on short stolons, tunics bright yellow becoming brown with age. **Cataphylls** upper green or purple, to just above the ground. **Leaves** 4–6 forming a fan, curved, 3–5mm wide; midrib and margins lightly thickened. **Spike** nearly horizontal, may be branched, with 5–12 flowers. **Bracts** green to greyish green. **Flowers** creamy white; tepals unequal and spreading at right angles to the tube; lower tepals with yellow spot surrounded by red. Perianth tube long, up to 40mm. **Anthers** purplish. **Pollen** yellow. **Capsules** broadly obovoid. **Seeds** saucer shaped, seed body asymmetrical, broadly winged. **Scent** unscented.

*Horizontal spikes bear up to 12 flowers.*

**DISTRIBUTION** It has only been recorded close to the Northern Cape town of Loeriesfontein.

**ECOLOGY & NOTES** The plants that we photographed were growing among shale, with their corms buried beneath rocks where they are protected from desiccation and against predators such as porcupines. This area normally receives less than 200mm of annual rain, but some winters are extremely dry with little to no rain falling. In these dry years it appears that the plants do not even break dormancy, and flowers are only seen in years of good rain. It is possible that the plants are more widespread, but good wet winters are rare. We only found four plants in flower in 2012, a year when little rain was recorded in this area, and we have not seen these plants in flower since then.

*The delicate flower may easily be mistaken for a species in the genus* Lapeirousia.

**POLLINATORS** Probably long-tongued horse flies (family Tabanidae), which are often involved in pollination of cream to pale pink long-tubed flowers with red nectar guides.

**SIMILAR SPECIES** The extremely long perianth tube and the creamy white flowers with yellow and red markings are unlike those of any other *Gladiolus* species in this area. As its name implies, this gladiolus is similar to a *Lapeirousia* species when in flower, and they could easily be confused.

*Leaves are sickle shaped.*

*Surrounded by shale; note the long perianth tubes.*

G. lapeirousioides *grows in arid terrain near Loeriesfontein, Northern Cape.*

# *Gladiolus orchidiflorus*

***orchidiflorus*** **= referring to the orchid-like appearance of the flower.**

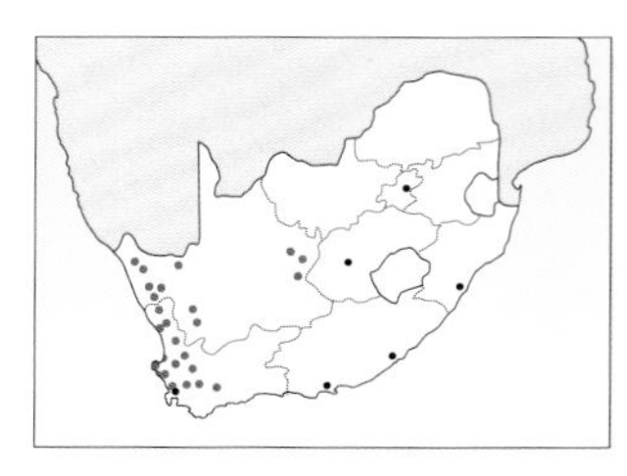

J F M A M J J A S O N D

Widely distributed in the Western and Northern Cape, and extending eastwards to Kimberley, *Gladiolus orchidiflorus* flowers from August to September, and later in its northern range. Its dull green to purple flowers have yellow-and-purple lower tepals and are sweetly scented.

*The bracts stand erect over the arched tepals.*

**DESCRIPTION** **Plant** 18–45cm. **Corm** globose to conic, with many cormlets, tunics coarsely fibred to leathery. **Cataphylls** uppermost green or purplish. **Leaves** up to 8, sword shaped to linear, up to 7mm wide (or wider), soft textured; midrib and margins lightly thickened. **Spike** inclined, often branched, 5–12 flowers in 2 ranks 60° apart. **Bracts** green to purplish grey. **Flowers** grey-green to dull purple; dorsal tepal narrow and strongly arched over the stamens, upper laterals spade shaped, lower tepals arching outwards and down. Dorsal tepal darker on the outside, upper lateral tepals with a dark midline, lower tepals with a bright yellow band outlined in darker purple towards the tip. Perianth tube up to 14mm, with rough ridges in the throat of the tube. **Anthers** purple. **Pollen** cream. **Capsules** ellipsoid. **Seeds** round, broadly and evenly winged with darker seed body. **Scent** intensely sweet, violet-like scent.

*Note the flower's very narrow dorsal tepal.*

**DISTRIBUTION** A widespread species, extending from the Cape Peninsula in the southwestern Cape northwards to Namaqualand in the Northern Cape, and into southern Namibia.

**ECOLOGY & NOTES** Plants generally grow in sandy soils usually derived from sandstone or from granite, but they have also been found in light clay soils. The plants we photographed were in deep sand near the western coastal plain, and in coarser granitic sand on the Kamiesberg. In both areas the plants were growing among shrubs, probably deriving some protection from them as both areas were grazed by sheep. We also found plants flowering in the year after a fire on the West Coast north of Cape Town. The corms of this species have numerous small cormlets around the base, which are easily dislodged and serve as protection against porcupines (which only eat the large mother corm).

**POLLINATORS** Bees from the families Anthophoridae (long-tongued bees) and Apidae (including honeybees, carpenter bees).

**SIMILAR SPECIES** The flower shape of *G. orchidiflorus* is similar to that of several other species such as *G. watermeyeri*, *G. virescens*, *G. ceresianus* and even *G. alatus*, but there are several distinguishing features: the leaves of *G. orchidiflorus* are up to 10mm wide and soft textured, with lightly thickened veins and margins. The flowers are grey-green or purplish grey and the dorsal tepal is extremely narrow (only 3–6mm).

*Leathery, linear leaves have thickened midribs.*

*Inclined spikes bear up to 12 flowers.*

*Lower tepals have a bright yellow band outlined in purple.*

*Pale purple form.*

*Narrow, arched dorsals give an undulating effect to inclined spikes. Seen here near Kamieskroon, Northern Cape.*

# *Gladiolus watermeyeri*

**Named after EB Watermeyer, who collected corms that were cultivated and subsequently flowered at Kirstenbosch in 1917.**

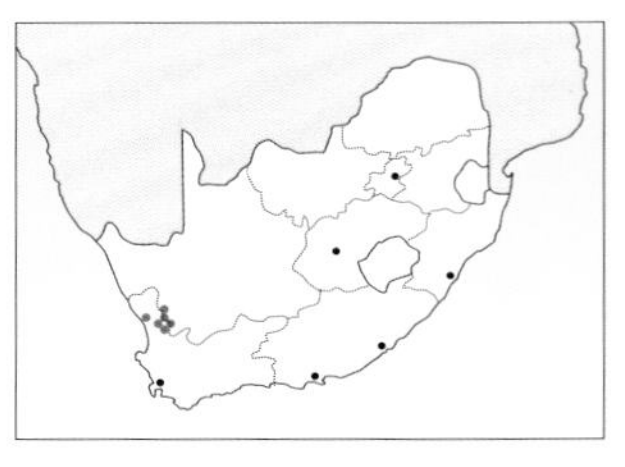

| J | F | M | A | M | J | J | A | S | O | N | D |
|---|---|---|---|---|---|---|---|---|---|---|---|

*Gladiolus watermeyeri* occurs in the area surrounding Nieuwoudtville around the Western Cape–Northern Cape border. The lateral tepals of its green to cream flowers are boldly veined and the flowers, appearing from August to mid-September, are strongly scented.

**DESCRIPTION** **Plant** 16–35cm. **Corm** globose to depressed-globose, with papery tunics. **Cataphylls** usually purple above ground. **Leaves** 4 or 5, sword shaped to linear blades, strongly 2–5-ribbed; margins thickened and raised. **Spike** inclined, 2–6 flowers. **Bracts** bright green, transparent along the veins. **Flowers** unequal with the dorsal tepal hooded over the filaments; upper lateral tepals creamish and heavily veined with maroon or pink, dorsal tepal translucent grey, lower tepals olive-green to yellow. Perianth tube 14–16mm. **Anthers** dark green. **Pollen** yellowish. **Capsules** broadly ellipsoid, large (up to 30mm) and inflated. **Seeds** reddish, ovate, evenly winged. **Scent** strongly sweetly scented.

*The flower is distinctive with its yellow lower tepals and dark-veined lateral tepals.*

**DISTRIBUTION** Known from the Nieuwoudtville area of the Northern Cape, southwards to the Gifberg and Wuppertal areas in the Western Cape. It is best known from west and south of Nieuwoudtville along the edge of the Bokkeveld escarpment.

**ECOLOGY & NOTES** Plants grow in sandstone-derived sandy soils, and flower best after a fire (not common in the Nieuwoudtville area) or after the vegetation has been cleared. Despite the bizarre coloration of the flowers, the plants are difficult to see, and they are often first detected by their scent.

**POLLINATORS** Large anthophorid bees, such as *Anthophora diversipes*, a common pollinator of short-tubed *Gladiolus* species.

**SIMILAR SPECIES** Although similar in coloration to other species in section *Hebea*, it is unlikely to be confused with them.

| | Leaves | Upper dorsal tepal |
|---|---|---|
| ***G. watermeyeri*** | broad, ridged leaf blades | translucent and strongly hooded |
| ***G. virescens*** | linear to terete, 1–3mm wide | erect and feathered, margins may be recurved |
| ***G. ceresianus*** | terete and 4-grooved | erect and feathered, margins recurved |
| ***G. uysiae*** | narrow, unridged leaves, short and curved | usually erect, reddish purple and feathered |

*Leaf blades are strongly ribbed.*

*Bracts protect seed capsules.*

*The lowermost leaf is longer than the spike; the dorsal tepal is typically narrow and translucent grey.*

*Plants can be difficult to see; often the first indication of their presence is their sweet smell.*

# Gladiolus virescens

***virescens*** **= greenish, referring to the colour of the flowers of the plants collected by CP Thunberg (1743–1828), who gave the species its scientific name.**

Occurring in the southern half of the winter-rainfall area, *Gladiolus virescens* has strongly scented, pale, boldly veined flowers that appear from mid-August to September and sometimes much later.

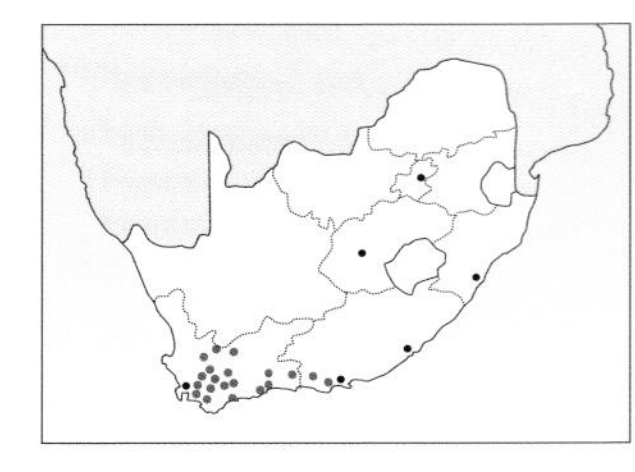

| J | F | M | A | M | J | J | A | S | O | N | D |
|---|---|---|---|---|---|---|---|---|---|---|---|

*This species has variable flower colours. The flowers of this form, from Boswaarmoed, Western Cape, show a cream throat and dark pink veining.*

**DESCRIPTION** **Plant** short, 12–20cm. **Corm** globose, with soft-papery tunics that do not persist. **Cataphylls** not usually visible above ground. **Leaves** 3, linear to terete, 1–3mm wide, oval to round in section; margins raised and arched towards the midrib making the leaf 2- or 3-grooved on each surface. Leaves sheath the stem, but do not conceal it, with the internode between the upper 2 leaves always exposed. **Spike** flexed at the base, inclined, 3–7 flowers. **Bracts** green with transparent veins, inner bract almost enveloped by outer bract. **Flowers** yellow to yellow-green or pink, upper 3 tepals feathered in various colours, lower 3 each with a yellow band across the limb; dorsal tepal straight and inclined, margins sometimes recurved. Perianth tube ±12mm. **Anthers** pale or dark. **Pollen** yellow. **Capsules** oblong-ellipsoid. **Seeds** roundish, broadly and evenly winged, light translucent yellow-brown. **Scent** variable: usually strongly violet scented but pink variants are unscented.

**DISTRIBUTION** Found from Bot River in the Western Cape to Humansdorp in the Eastern Cape, inland in the Little Karoo near Montagu, on the Witteberg Mountains south of Matjiesfontein, in the Ceres area and in the mountains northeast of Ceres, including along the Riet River in the Swartruggens Mountains.

**ECOLOGY & NOTES** Usually found in heavier clay soils and on stony shale slopes, but has also been recorded in sandy areas. On the Waboomsberg, plants were growing in heavy sand among fynbos; in the Riet River population, they were growing in deep alluvial sand. Interestingly, the plants with strangely coloured and shaped flowers from the outlying areas northeast of Ceres flowered much later – in December as opposed to August and September. Natural hybrids with *G. alatus* have been recorded near Rawsonville and Bot River.

**POLLINATORS** Long-tongued bees, mostly of the genus *Anthophora*.

**SIMILAR SPECIES** This species is very variable in colour and form and is not always easy to identify. The species that is most similar, and for which it is most commonly mistaken, is *G. ceresianus*.

| | Corm | Leaves | Flower colour |
|---|---|---|---|
| ***G. virescens*** | globose with papery tunics | stem visible between upper 2 leaves | usually yellow, yellow-green or pink |
| ***G. ceresianus*** | coarse blackish fibres | stem completely covered between upper 2 leaves | purple-brown to greenish grey |

*Pink feathering is striking on the upper three tepals of this form.*

*Inclined spikes bear up to seven flowers.*

*Yellow form, Bot River.*

*Yellow-green flushed with mauve, Bot River.*

*Dorsal tepals are erect.*

*Orange form, Van der Stel Pass.*

*Strong veining and pale anthers, Waboomsberg.*

G. virescens *flowering in Suurbraak, Western Cape.*

# Gladiolus ceresianus

***ceresianus* = from Ceres, where the type collection was made.**

Widespread in the Western Cape interior, *Gladiolus ceresianus* has purple-brown to greenish flowers that are strongly veined and feature yellow markings on the lower tepals. The strongly scented flowers appear from mid-August to September.

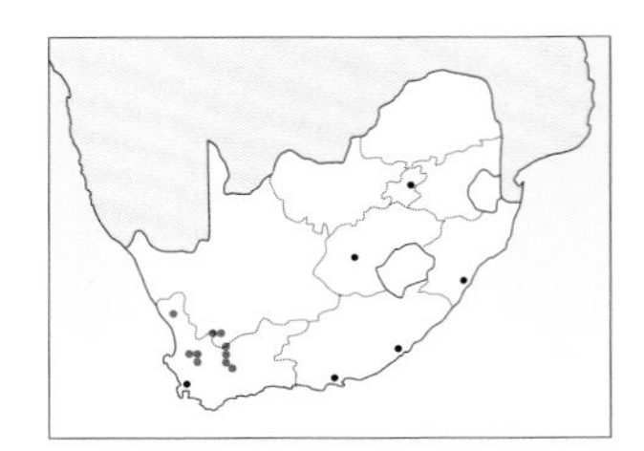

| J | F | M | A | M | J | J | A | S | O | N | D |
|---|---|---|---|---|---|---|---|---|---|---|---|

**DESCRIPTION** **Plant** short, 10–20cm. **Corm** conic, with tunics of coarse blackish fibres. **Cataphylls** uppermost dark green to reddish. **Leaves** 3, blade terete and 4-grooved, 1–2mm diameter. All leaves overlap and completely conceal the stem. **Spike** inclined, unbranched, 2–4 flowers. **Bracts** green or flushed reddish. **Flowers** mostly purple-brown or greenish grey; upper 3 tepals with purple veins, lower tepals clear yellow in the lower two-thirds; dorsal tepal erect with recurved margins, upper lateral tepals spade shaped, all tepals narrowed into claws. Perianth tube 12–16mm. **Anthers** light purple. **Pollen** yellow. **Capsules** elongate, ±1cm. **Seeds** almost round, reddish in colour. **Scent** intensely sweetly scented.

*This plant can be confused with some forms of* G. virescens.

**DISTRIBUTION** Recorded from the Kouebokkeveld north of Ceres, eastwards to Matjiesfontein and north to the Roggeveld escarpment near Middelpos, with an outlying population near Vredendal, west of Vanrhynsdorp.

**ECOLOGY & NOTES** Usually found on rocky slopes and flats in clay soils.

**POLLINATORS** Solitary bees.

*Seed capsules maturing; note their elongate form.*

**SIMILAR SPECIES** May be confused with some forms of *G. virescens* and *G. uysiae*.

| | Corm | Leaves |
|---|---|---|
| ***G. ceresianus*** | conic, coarsely textured, black fibrous tunics | leaf bases overlap, sheathing the stem completely, blade terete and 4-grooved, strongly raised midrib |
| ***G. virescens*** | globose, soft, papery tunics which do not persist | stem internode between upper 2 leaves is exposed, blade linear or terete, 2- or 3-grooved, raised midrib |
| ***G. uysiae*** | flattened, softly membranous tunics, long stolons with cormlets | plane, leathery almost fleshy leaves, falcate, no raised veins |

*Seeds ready for wind dispersal.*

*Green bracts encase short perianth tubes.*

*Dorsals have recurved margins and narrow into claws; upper laterals are spade shaped.*

*Inclined spikes bear few flowers.*

*Dorsal tepals are erect.*

G. ceresianus, *Kouebokkeveld, Western Cape.*

# *Gladiolus uysiae*

**Named after GJ Uys from Nieuwoudtville, who provided the plants scientifically described by L Bolus.**

Very similar in flowering appearance to *Gladiolus ceresianus*, the tiny *G. uysiae* is widespread in the interior of the Western and Northern Cape. Sweetly scented flowers of purplish brown or grey-green appear from mid-August to September.

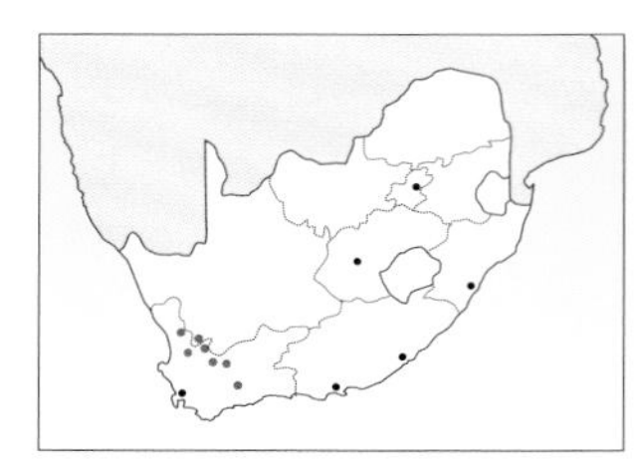

| J | F | M | A | M | J | J | A | S | O | N | D |
|---|---|---|---|---|---|---|---|---|---|---|---|

*The plant has very short leaves and upright dorsal tepals.*

**DESCRIPTION** **Plant** short, 7–18cm. **Corm** depressed-globose, producing long stolons with cormlets, membranous to papery tunics. **Cataphylls** pale. **Leaves** 3, curved, 3–6mm wide; no thickened veins or midrib. **Spike** flexed at the base, inclined, 1–3 flowers. **Bracts** green with transparent margins, large and look inflated, the inner bract enveloped by the outer. **Flowers** reddish purple, upper tepals feathered with purple, lower tepals yellow with dark yellow-green band; dorsal tepal erect, upper laterals spear shaped, all tepals clawed. Perianth tube 10–12mm. **Anthers** purple. **Pollen** yellow. **Capsules** obovoid. **Seeds** translucent brown, round, evenly winged. **Scent** strongly sweetly scented.

**DISTRIBUTION** Known only from the Western Cape, *G. uysiae* has been recorded from Calvinia in the Northern Cape, west towards Nieuwoudtville and south across the Roggeveld escarpment to the Ceres district.

**ECOLOGY & NOTES** Plants usually grow in renosterveld and are often found on shale banks; however, we have also photographed them in sandy soils south of Nieuwoudtville. They are almost always found in dense colonies – a result of the number of stolons and cormlets produced by each parent plant. We have noticed that plants do not flower every year, and sometimes whole colonies do not flower, whereas an adjacent colony does. Plants are inconspicuous and extremely difficult to see; they are usually detected by their fragrance, which is strong and carried by the wind! Even after detecting the scent, it sometimes takes a concerted effort to find the plants as they seem to blend into the background.

**POLLINATORS** Large, long-tongued bees such as *Anthophora diversipes*.

**SIMILAR SPECIES** The two species most likely to be confused with *G. uysiae* are *G. ceresianus* and *G. virescens*. *G. watermeyeri* has similarly coloured flowers to *G. uysiae*, but has ribbed leaves, and the dorsal tepal of *G. watermeyeri* is arched whereas that of *G. uysiae* is erect.

| | Corm | Leaves |
|---|---|---|
| ***G. uysiae*** | flattened, softly membranous tunics, long stolons with cormlets | plane, leathery almost fleshy leaves, falcate, no raised veins |
| ***G. ceresianus*** | conic, coarsely textured, black fibrous tunics | leaf bases overlap, sheathing the stem completely, blades terete and 4-grooved, strongly raised midrib |
| ***G. virescens*** | globose, soft, papery tunics which do not persist | stem internode between upper 2 leaves is exposed, blades linear or terete, 2- or 3-grooved, raised midrib |

*Bracts have transparent margins.*

*All the tepals are clawed.*

*Note the recurved dorsal tepal margins.*

*Lower tepals have a yellow-green band.*

G. uysiae *growing en masse, Rooiwal, Northern Cape.*

# *Gladiolus equitans*

***equitans* = riding, referring to the way sheaths of the broad leaves clasp the base of the leaf above them.**

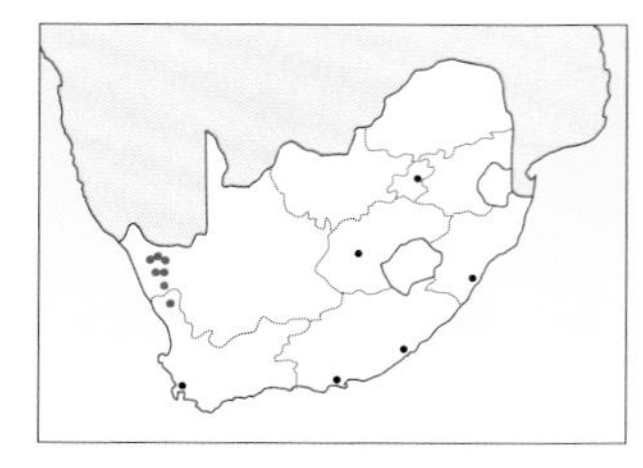

J F M A M J J A S O N D

*Gladiolus equitans* has orange to scarlet flowers with yellow lower tepals tipped in red. It grows only in the Namaqualand region of the Northern Cape, where it flowers from late August to mid-September, and sometimes earlier or later at higher elevations. Its broad leaves are distinctive.

*Note the unusual wide leaves.*

**DESCRIPTION** **Plant** 15–30cm. **Corm** globose, with membranous to leathery tunics. **Cataphylls** not really visible. **Leaves** 4, up to 4cm wide, forming an almost rigid fan; margins heavily thickened and often reddish. **Spike** flattened and 3-winged, unbranched, 3–7 flowers. **Bracts** green, strongly keeled, with keels and margins reddish. **Flowers** orange to scarlet, lower lateral tepals yellow with red apices; dorsal tepal concave and hooded over the stamens, upper laterals broadly heart shaped. Perianth tube up to 15mm. **Anthers** cream. **Pollen** cream. **Capsules** oblong. **Seeds** dark reddish brown, ovate, evenly winged. **Scent** lightly scented.

**DISTRIBUTION** Known only from Namaqualand in the Northern Cape. It has been recorded from the Knersvlakte in the south to Okiep in the north, but it is best known from the Kamiesberg.

**ECOLOGY & NOTES** Plants grow only in granite soils and although they are occasionally found in sandy areas, their preferred habitat is rock cracks. It is likely that these are the only places that offer protection from predators such as porcupines, mole-rats and baboons.

**POLLINATORS** Large bees of the Anthophoridae family.

**SIMILAR SPECIES** *G. equitans*, *G. speciosus*, *G. pulcherrimus*, *G. alatus* and *G. meliusculus* have similarly coloured and shaped flowers. *G. equitans* can be distinguished from them by its broad sturdy leaves with thickened margins and its locality, which it does not share with the other species.

| | Leaf blades | Dorsal tepal | Filaments | Lower tepals |
|---|---|---|---|---|
| ***G. equitans*** | broad (40mm), margins thickened | hooded | glabrous | yellow with red apices |
| ***G. speciosus*** | plane, narrow (7mm) | hooded | glabrous | yellow markings (upper tepals) in some populations |
| ***G. pulcherrimus*** | plane (10mm) | erect | hairy | yellowish green in lower ⅔ |
| ***G. alatus*** | ribbed, narrow (6mm) | erect | hairy | yellowish green in lower ⅔ |
| ***G. meliusculus*** | ribbed, strongly raised veins | erect | hairy | dark reddish at edge of yellow |

*Lower leaves form a two-ranked fan.*

*Inclined spikes bear up to seven flowers.*

*Bright flowers have wide lateral tepals with median stripes, and deep yellow bands on clawed lower tepals.*

*Keels and margins of bracts are reddish.*

*Plants wedge themselves into cracks in granite rocks in Namaqualand as here in the Kamiesberg, Northern Cape.*

# Gladiolus speciosus

***speciosus* = beautiful, referring to the flowers.**

Broadly extended across the western parts of the Western Cape and crossing into the Northern Cape, *Gladiolus speciosus* has orange flowers with yellow-green lower tepals tipped with orange. Strongly scented flowers appear around September and October.

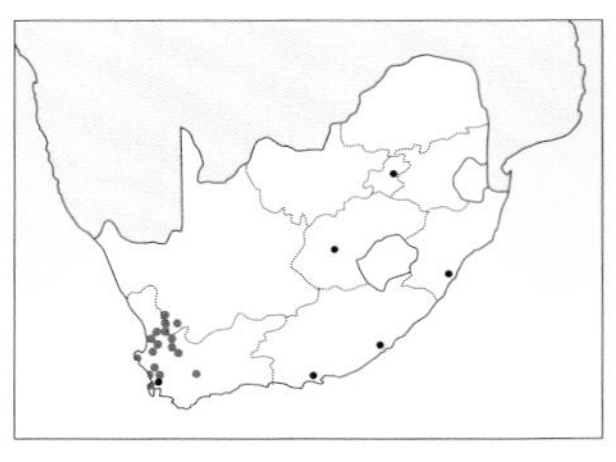

| J | F | M | A | M | J | J | A | S | O | N | D |
|---|---|---|---|---|---|---|---|---|---|---|---|

**DESCRIPTION** **Plant** 12–20cm. **Corm** depressed-globose, with slender stolons terminating in cormlets, papery tunics. **Cataphylls** uppermost dry and grey-brown. **Leaves** 4–6, linear to curved, 3–7mm wide; margins and midrib lightly raised; lower leaves basal, upper 2 or 3 arising on the stem but diverging from it. **Spike** inclined, flexed outwards above the sheathing part of the top leaf, compressed and 3-sided and winged above, 2–5 flowers. **Bracts** green, keeled, red on the keel and margins. **Flowers** dorsal tepal nearly horizontal and concave, hooded over the stamens; upper 3 tepals mainly orange although some populations have large yellow markings on the outside of the lateral tepals; lower tepals yellow with orange apices. Perianth tube up to 13mm. **Anthers** greenish yellow. **Pollen** yellow. **Capsules** ellipsoid, sometimes inflated and pithy. **Seeds** large, evenly winged, light translucent brown, rarely found. **Scent** strongly scented.

*Note the hooded, unicoloured dorsal tepal.*

**DISTRIBUTION** Extends from Mamre close to Cape Town, northwards to Lambert's Bay and eastwards to the Bokkeveld escarpment. A small population of plants is also known from the Swartruggens Mountains east of the Kouebokkeveld.

**ECOLOGY & NOTES** The small plants grow in deep, well-drained sand among annuals and fynbos plants such as restios. They are found along coastal sandveld as well as inland in drier habitats. An interesting feature of this species is that the plants rarely make seeds. We know of several large populations of plants that flower regularly, and we have only once found capsules and a few seeds. It is not possible that the flowers are never pollinated, so we can only assume that they rely on the stolons and cormlets for reproduction. Most of the plants we know of grow in large colonies, suggesting that they are all derived from a central mother plant.

**POLLINATORS** Probably long-tongued Anthophoridae bees.

**SIMILAR SPECIES** This species is most likely to be confused with *G. alatus*, which has a more or less erect dorsal tepal and ribbed leaves, whereas *G. speciosus* has a very hooded dorsal tepal and leaves with no raised veins. Some populations of *G. speciosus* have distinctive flowers with yellow markings on the outside of the lateral tepals.

*Lateral tepals are orange inside and green-yellow on the reverse.*

*Bracts are keeled and edged in red. Note plane leaves.*

*Tepals have distinctive outer coloration in some populations.*

G. speciosus *grows in well-drained sand. Found here near Graafwater, Western Cape.*

# Gladiolus pulcherrimus

***pulcherrimus*** **= most beautiful, referring to the flowers.**

*Gladiolus pulcherrimus* grows in the northwestern parts of the Western Cape. Salmon to orange flowers have lime-green lower tepals and are rose scented. They appear from August to October and can be confused with *G. alatus*.

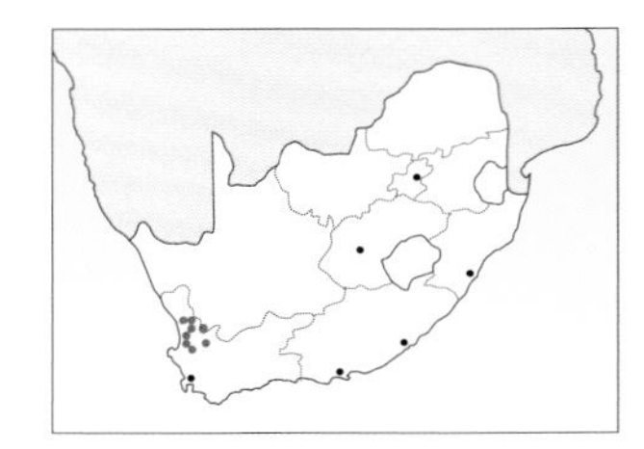

| J | F | M | A | M | J | J | A | S | O | N | D |
|---|---|---|---|---|---|---|---|---|---|---|---|

**DESCRIPTION** **Plant** 20–40cm, sometimes taller. **Corm** depressed-globose, with membranous tunics. **Cataphylls** uppermost dry and brown. **Leaves** 5–7, grey-green with a waxy bloom, 6–12mm wide, sword shaped, plane; margins lightly thickened, midrib hardly evident. **Spike** inclined, compressed and 3-angled, angles winged, usually branched, with 5–8 flowers. **Bracts** grey-green, keeled. **Flowers** salmon coloured; upper tepals feathered with red, lower tepals greenish with salmon apices, dorsal tepal erect and sometimes whitish inside. Perianth tube ±11mm. **Stamens** hairy. **Anthers** greenish yellow. **Pollen** yellowish. **Capsules** ellipsoid, up to 4cm. **Seeds** light brown, ovate to fiddle shaped. **Scent** delicately rose scented.

*The salmon-coloured flowers have yellow bands on the lower tepals, and greenish-yellow anthers.*

**DISTRIBUTION** Known from the Olifants River Valley near Clanwilliam, the Lambert's Bay area and the Cederberg, Gifberg and Nardouw mountains.

**ECOLOGY & NOTES** Plants grow in deep sandy soils in dry fynbos among proteas, leucadendrons, and in restio clumps. They have been recorded together with *G. speciosus*, but plants seem to flower at different times. *G. pulcherrimus* is drought tolerant.

**POLLINATORS** Probably long-tongued Anthophoridae bees.

*Branched spikes bear up to five or more flowers.*

**SIMILAR SPECIES** The species most likely to be confused with *G. pulcherrimus* is *G. alatus*, which has almost identical flowers, but the former usually has plane leaves with no venation, whereas the latter has strongly ribbed leaves. *G. pulcherrimus* is a much more robust species, sometimes reaching 60cm in height, whereas *G. alatus* is usually shorter than 20cm; however, if growing in shade, the latter can become much taller, making it more difficult to tell the two species apart, and a close study of the leaves is required to distinguish between them. Further differences between these and other plants in the section *Hebea* are described on page 264.

*Soft leaves form a basal fan.*

*Plants in the Paleisheuwel area have plane leaves with visible venation.*

*Keeled bracts support open, showy flowers.*

G. pulcherrimus *is drought resistant. Flowering here near Clanwilliam, Western Cape.*

# Gladiolus alatus

***alatus*** **= winged, referring to the wings on the stems.**

*Gladiolus alatus* is widespread in the winter-rainfall region. Scented, bright orange-red flowers with yellow-green lower tepals appear from August until October. They can be confused with *G. pulcherrimus* and *G. speciosus*, but unlike these two species, *G. alatus* has strongly ribbed leaves.

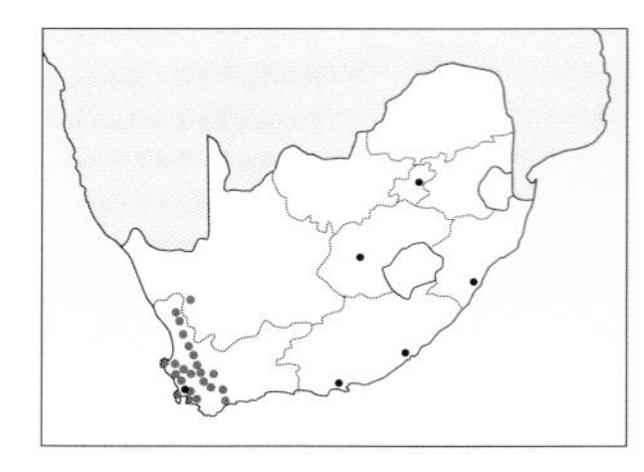

| J | F | M | A | M | J | J | A | S | O | N | D |
|---|---|---|---|---|---|---|---|---|---|---|---|

This is a common plant in the Western Cape.

**DESCRIPTION** **Plant** 15–20cm. **Corm** depressed-globose, with membranous tunics. **Cataphylls** not visible above ground. **Leaves** 3 or 4, lanceolate to curved, 2–6mm wide, strongly ribbed and often lightly scabrid on the veins. **Spike** flexed outwards, rarely branched, often with small cormlets in the leaf axils (below the ground), 2–5 flowers. **Bracts** strongly keeled with margins and keels purple. **Flowers** orange to scarlet; lower tepals yellow-green with orange apices, dorsal tepal almost erect. Perianth tube 10–14mm. **Stamens** long, hairy. **Anthers** dull green to yellowish. **Pollen** yellowish. **Capsules** 2cm. **Seeds** light glossy brown, ovate to round. **Scent** sweetly scented.

**DISTRIBUTION** *G. alatus* has been recorded from a large area of the Western Cape, from the Bokkeveld escarpment in the Nieuwoudtville area in the Northern Cape, southwards to the Cape Peninsula and eastwards to Bredasdorp.

Dorsal tepals are almost erect.

**ECOLOGY & NOTES** Found in deep sand as well as stony soil derived from granite or sandstone. Plants appear to be quite tolerant of drought and can colonise areas that dry out quickly, such as road verges and old disturbed farmlands. They often grow in huge colonies, possibly owing to the cormlets, which are produced in the leaf axils. Natural hybrids with *G. virescens* have been recorded near Rawsonville and Bot River.

**POLLINATORS** Thought to be *Anthophora diversipes*, which collects nectar from the flowers and pollinates them at the same time.

**SIMILAR SPECIES** *G. alatus* is similar to *G. speciosus*, *G. pulcherrimus*, *G. meliusculus* and *G. equitans*. The latter's range is further north.

| | Leaves | Lower tepals |
|---|---|---|
| ***G. alatus*** | strongly ribbed | yellow-green with small orange tips |
| ***G. speciosus*** | plane | yellow markings, red apices |
| ***G. pulcherrimus*** | plane – no ribs | greenish with small orange tips |
| ***G. meliusculus*** | strongly ribbed | predominantly salmon with small yellow portion outlined with dark red or purple |

*The common name in Afrikaans is* Kalkoentjie.

*Margins and keels of bracts are purple.*

G. alatus *has flexed spikes; the cataphylls are not visible above ground.*

# *Gladiolus meliusculus*

***meliusculus* = a little better, possibly implying that it is more attractive than *G. alatus*.**

Pink-reddish or orange, the flowers of *Gladiolus meliusculus* appear from mid-August to October on the Western Cape's coastal plains. It is pollinated by monkey beetles.

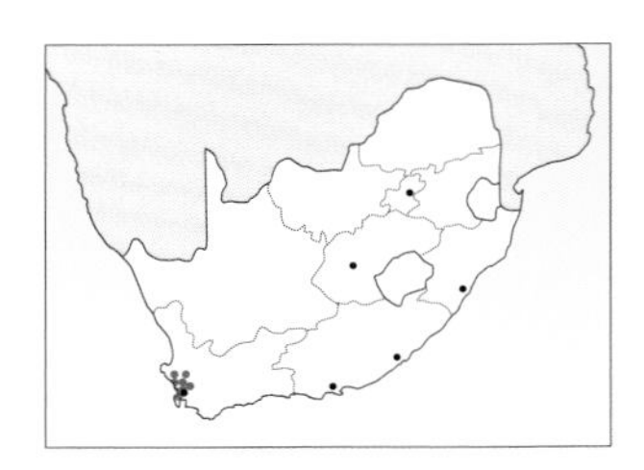

| J | F | M | A | M | J | J | A | S | O | N | D |
|---|---|---|---|---|---|---|---|---|---|---|---|

**STATUS** Near Threatened

**DESCRIPTION** **Plant** 12–20cm. **Corm** depressed-globose, producing stolons with cormlets, membranous tunics. **Cataphylls** uppermost purple above ground. **Leaves** 4, up to 5mm wide; strongly ridged; cormlets form in axils below ground. **Spike** inclined, unbranched, compressed, with ridges or wings, 3–6 flowers. **Bracts** green, keeled; keels and margins reddish. **Flowers** upper tepals salmon-pink to brick-red to orange, lower tepals mainly salmon coloured with yellow or greenish area edged in purple towards the inside of the flower, dorsal tepal more or less erect. Perianth tube up to 11mm. **Stamens** short, hairy. **Anthers** purple to buff. **Pollen** cream to yellowish. **Capsules** obovoid. **Seeds** light pinkish brown, broadly winged. **Scent** honey scented.

*Spikes bear up to six flowers.*

**DISTRIBUTION** Found in the Western Cape in an area around Malmesbury, Darling, Hopefield and Porterville. It used to occur on the Cape Peninsula but is no longer found there.

**ECOLOGY & NOTES** The plants grow in sandy soils that remain damp in the growing season. Unfortunately much of the habitat has been destroyed and this species is becoming very rare outside protected areas, occurring mostly on farms in Darling. It used to be fairly common along the N7 national road in the Malmesbury area, but many sites were destroyed when the road was widened. The species is also threatened by agriculture (ploughing as well as grazing), housing developments and sand mining. Plants are usually found in large colonies, probably because they produce copious numbers of seeds; they also multiply by cormlets on the ends of stolons and leaf axils.

*A Shaggy monkey beetle on* G. meliusculus.

**POLLINATORS** Thought to be monkey beetles, which are known to pollinate flowers with similar markings such as *Romulea obscura* and *R. eximia*.

**SIMILAR SPECIES** *Gladiolus meliusculus* is most likely to be confused with *G. alatus*. The main difference is the colour of the lower tepal: in *G. meliusculus* it is predominantly salmon with a small yellow portion outlined with dark red or purple, whereas in *G. alatus* it is yellow-green with a small orange tip. While their geographical ranges overlap, *G. meliusculus* flowers a little later than *G. alatus*. For more differences between these species, see the table on page 270.

*Leaves are strongly ridged.*

*Bracts have reddish margins and keels.*

G. meliusculus *flowering among* Heliophila *near Malmesbury, Western Cape.*

*Rachel Saunders records* G. taubertianus *at Heuningvlei in the Cederberg, Western Cape.*

# Gladiolus atropictus

***atropictus*** **= darkly painted, referring to the dark lines on the lower tepals.**

Known only from the southwestern Cape, the delicate blue-violet flowers of *Gladiolus atropictus* are strongly scented, appearing from mid-July to August.

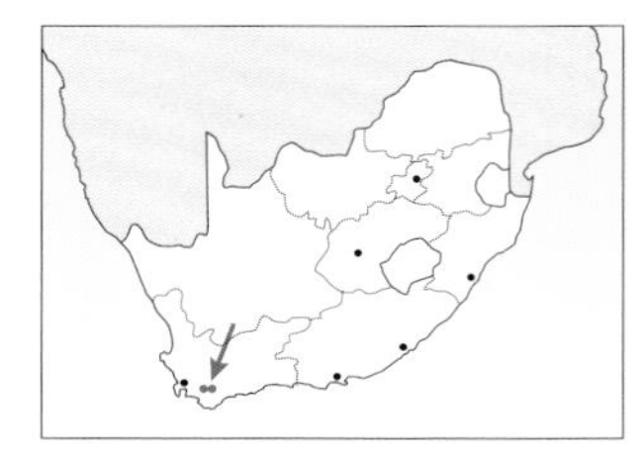

| J | F | M | A | M | J | J | A | S | O | N | D |
|---|---|---|---|---|---|---|---|---|---|---|---|

**STATUS** Vulnerable

**DESCRIPTION** **Plant** 30–60cm. **Corm** elongate-conical, with papery tunics. **Cataphylls** uppermost green or purple. **Leaves** 4, 2 basal, 2 sheathing, up to 2mm wide; margins and midrib raised. **Spike** erect, unbranched, 1 or 2 flowers. **Bracts** greenish grey with broad, membranous margins. **Flowers** violet, lower tepals white to pale yellow with dark violet feathering and are elongate, curving downwards. Perianth tube ±13mm. **Anthers** pale cream. **Pollen** white. **Capsules** oval. **Seeds** pale brown with darker seed body, evenly winged. **Scent** strong freesia-like scent.

**DISTRIBUTION** Known only from a tiny area of the southwestern Cape on the northern slopes of the Riviersonderend Mountains close to Greyton and McGregor. Only a few populations are known and the plants are rarely seen.

**ECOLOGY & NOTES** Grows on rocky sandstone slopes in dry fynbos among restios. Plants are scattered and flower well in mature vegetation, suggesting that they do not require a recent fire to stimulate flower production.

**POLLINATORS** Long-tongued bees.

**SIMILAR SPECIES** Thought to be closely related to *G. violaceolineatus* and *G. comptonii*. All three species have a similar corm shape and elongate tapering tepals. However, the restricted locality of this species and its distinctive dark blue-violet flowers with feathery markings make it unlikely to be mistaken for any other species.

*The plant was first collected for scientific purposes in 1978.*

*Sweetly scented flowers are pollinated by long-tongued bees.*

*The outer bracts are longer than the perianth tube.*

*Flowers are dark violet with feathery markings.*

*Spikes bear one or two flowers.*

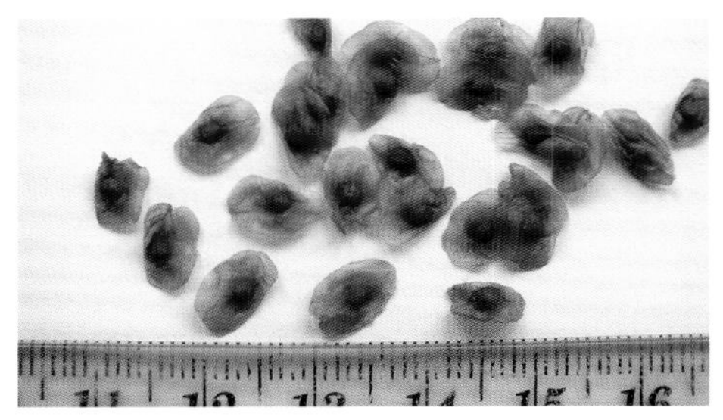

*Evenly winged seeds are ±8mm at their longest point.*

*The upper tepals are larger than the lower ones.*

G. atropictus *is known from only a few sites in the Riviersonderend Mountains in the Western Cape.*

# Gladiolus violaceolineatus

***violaceolineatus*** **= lined or streaked with violet.**

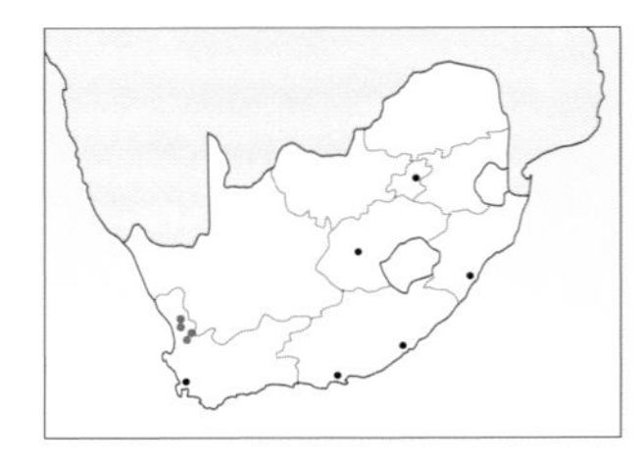

| J | F | M | A | M | J | J | A | S | O | N | D |
|---|---|---|---|---|---|---|---|---|---|---|---|

*Gladiolus violaceolineatus* has a restricted range in the northwestern parts of the Western Cape. The scented pale blue flowers appear from July to mid-August, and later at higher elevations. It is unlikely to be confused with other species as the violet lines on the tepals are distinctive.

*The violet streaks for which the species is named are distinctive.*

**DESCRIPTION** **Plant** 30–50cm. **Corm** conical, cartilaginous to papery tunics. **Cataphylls** uppermost green or purple, sometimes hairy. **Leaves** 4, 1 or 2 basal and 2 sheathing, each 1–2mm wide, 3-angled; the margin on one side of the leaf has raised edges, the other margin and the midrib are not raised. **Spike** erect, unbranched, with 1–4 flowers. **Bracts** green to grey-green with membranous margins. **Flowers** pale blue-lilac, all tepals streaked with dark violet; lower tepals white to yellow below, with the tube and base of tepals sometimes with red markings. Perianth tube up to 15mm. **Anthers** pale grey. **Pollen** whitish. **Capsules** unknown. **Seeds** light brown and ovate, broadly winged. **Scent** sweetly scented.

**DISTRIBUTION** Recorded from mountains in the northwestern interior of the Western Cape. Populations are known from the Citrusdal area northwards through the Cederberg to the Gifberg.

*This species is seldom seen.*

**ECOLOGY & NOTES** On rocky sandstone slopes or flats, often in shade and usually on south-facing slopes that are slightly more moist and cool. The plants favour the cooler sites in the areas in which they grow, which can be very hot and dry in spring and summer. They seem to flower better after a fire than in mature vegetation, although we know of a small population that flowers every year as long as there is sufficient rain. It is not a common species; plants are rarely seen.

**POLLINATORS** Thought to be long-tongued bees.

**SIMILAR SPECIES** No similar species in this area of the Western Cape. It has distinctive leaves and heavily streaked and spotted flowers with long tapering tepals.

*Very narrow leaves are three-angled.*

*Tepals are elongate and curve outwards.*

*Bracts are green and longer than the perianth tube.*

G. violaceolineatus *grows on sandstone, as here in the Gifberg, Western Cape.*

# *Gladiolus comptonii*

**Named after Professor RH Compton (1886–1979), the founder of the Compton Herbarium at Kirstenbosch National Botanical Garden.**

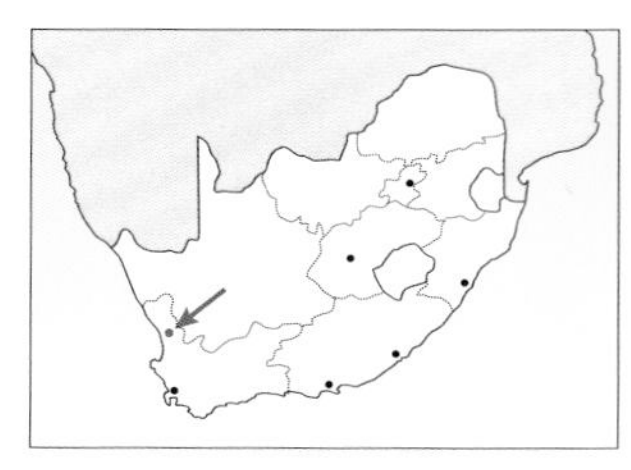

| J | F | M | A | M | J | J | A | S | O | N | D |
|---|---|---|---|---|---|---|---|---|---|---|---|

**STATUS** Critically Endangered

*Gladiolus comptonii* occurs only in the northwestern region of the Western Cape. Its sweetly scented, bright yellow flowers appear in July.

**DESCRIPTION** **Plant** 30–45cm. **Corm** globose-ovoid, with leathery tunics becoming fibrous with age. **Cataphylls** uppermost green or purple and minutely hairy. **Leaves** 4, 1 or 2 basal and the others sheathing, 2–3mm wide; with margins and midrib lightly raised. **Spike** erect and unbranched, with 1–3 flowers. **Bracts** green to grey-green with membranous margins. **Flowers** bright yellow, all tepals with reddish streaks and spots; tepals elongate and taper towards the apex. Perianth tube ±12mm. **Anthers** yellow. **Pollen** yellow. **Capsules** ellipsoid. **Seeds** light brown, ovoid, evenly winged. **Scent** slightly sweetly scented.

*The flowers of this plant are always bright yellow and streaked.*

**DISTRIBUTION** Known from only one locality in the Olifants River Mountains, at the Heerenlogement, which is a large cave.

**ECOLOGY & NOTES** The plants that we photographed were scattered throughout mature, dry fynbos in rocky sandstone soil at an altitude of ±700m above sea level. The plants were on a south- or southwest-facing aspect of the mountain, among *Protea nitida* and fairly dense clumps of restios. This species has not been seen in any of the other nearby mountains and seems to be a highly localised endemic. On our way up the hill we noticed some plants of a blue-flowered form of *Gladiolus carinatus* also flowering in the mature, thick fynbos.

**POLLINATORS** Thought to be long-tongued bees.

**SIMILAR SPECIES** The bright yellow flowers are unlikely to be mistaken for any other species in this area.

*Bracts are longer than the perianth tube.*

*All tepals are elongate and curved.*

G. comptonii *is known only from one location on the slopes of the Olifants River Mountains, Western Cape.*

# Gladiolus roseovenosus

***roseovenosus* = red-veined, named for the markings on the lower tepals.**

*Gladiolus roseovenosus* is known only from the Outeniqua Mountains. This species is a late-summer bloomer, despite growing in a winter-rainfall area. Its unscented cream-pink flowers appear from mid-February to April.

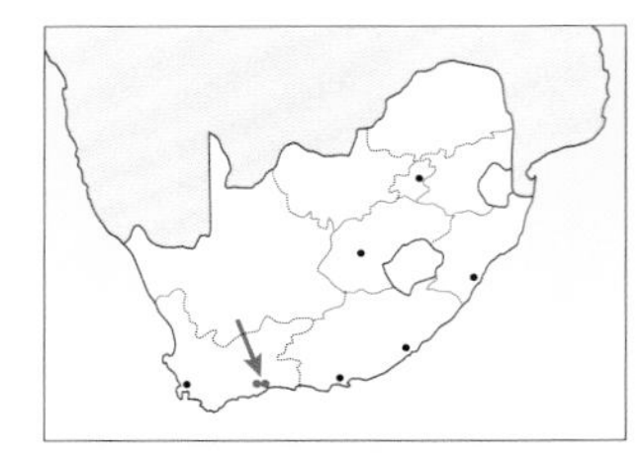

| J | F | M | A | M | J | J | A | S | O | N | D |
|---|---|---|---|---|---|---|---|---|---|---|---|

**STATUS** Critically Endangered

**DESCRIPTION** **Plant** 30–50cm. **Corm** globose, with papery tunics. **Cataphylls** uppermost dark green or purple, sometimes mottled with white. **Leaves** 4, mainly sheathing, the short blade is 1.5–2mm wide; with margins and midrib lightly thickened. **Spike** erect and unbranched, with 2–4 flowers. **Bracts** pale green. **Flowers** large, white to pale pink with dense, red feathery streaks mainly on the lower tepals; pink on reverse of tepals. Perianth tube long, up to ±40mm. **Anthers** purple. **Pollen** cream. **Capsules** not known. **Seeds** not known. **Scent** unscented.

**DISTRIBUTION** Known only from one small area in the south of the Western Cape. Two localities are known, both of which are in the Outeniqua Mountains: one near George, the other on the Robinson Pass.

**ECOLOGY & NOTES** Grows in peaty sandstone-derived soil either in flat areas or on gentle slopes. We photographed plants in old, dense fynbos, but they were growing mainly along the edge of a path, so it is likely that they would respond to fire. The plants are close to pine plantations and much of their original habitat has been lost to afforestation and alien invasive species. We attempted to hand-pollinate the flowers but could find little pollen. When we returned about six weeks later, we discovered that none of the plants had set seed, but to our surprise we found another five or six plants in flower! The plants seem to flower over a fairly long period. The area where they grow receives rain more or less all year round, with higher rainfall in winter.

*This species is endangered by alien invasives and afforestation.*

**POLLINATORS** Possibly long-tongued flies.

**SIMILAR SPECIES** The flowers of this plant are distinctive, and it is therefore unlikely to be mistaken for any other species.

*Flowers are large and unscented.*

*Leaves are narrow, linear and plane.*

*Note the long perianth tube.*

*Bracts are pale green; note the purple anthers.*

G. roseovenosus *grows in the Outeniqua Mountains, here on Robinson Pass.*

# Gladiolus carinatus

***carinatus* = with a keel, referring to the raised midvein on the leaf blades.**

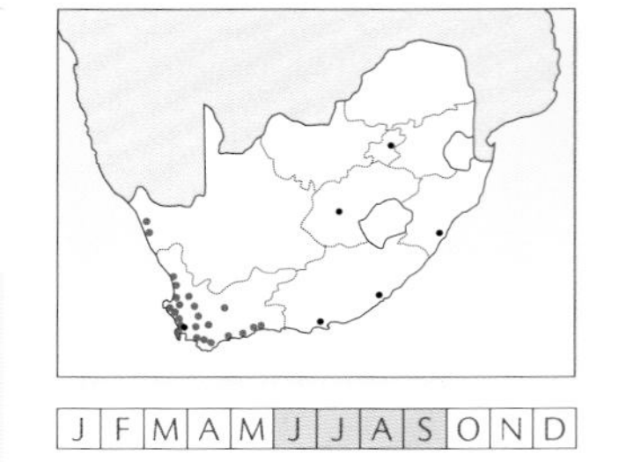


*Gladiolus carinatus* is widespread in the winter-rainfall region of the Western Cape. Appearing in winter and spring, the flowers are shades of blue, yellow and pink.

**DESCRIPTION** **Plant** 20–50cm or taller. **Corm** globose, with leathery to woody tunics. **Cataphylls** the uppermost usually purple and mottled with white. **Leaves** 3, with linear blade 3–9mm wide; midrib (but not margins) thickened. **Spike** erect and unbranched, usually 4–6 flowers. **Bracts** green or greyish or purplish green, with transparent margins. **Flowers** large and variable, blue to grey to purple or pink, or shades of yellow; lower tepals have a yellow band across that may be outlined above and/or below with reddish to purple streaks. Perianth tube ±16mm. **Anthers** cream. **Pollen** cream. **Capsules** oblong. **Seeds** light yellow-brown, broadly winged. **Scent** usually highly scented.

**DISTRIBUTION** A commonly seen species in the Northern and Western Cape. Found from the Cape Peninsula northwards into Namaqualand and eastwards to Knysna.

**ECOLOGY & NOTES** Most often found in well-drained sandy soils, but also occurs on stony mountain slopes up to 1,000m above sea level. Blue- to pink-blue-flowered plants are most common along the coast and on the Cape Peninsula. Further inland, in places such as the Kouebokkeveld, Ceres, Worcester and the Cederberg, the most common shade is pale yellow or buff. Populations in the Kouebokkeveld grow in clumps and are very short (15–20cm) with yellowish-brown to cream flowers. Despite examining many cataphylls, we never found a spotted one! It is possible that these plants are diverging and are on their way to forming a new species. Blue-flowered plants have been seen near Nieuwoudtville and along the N7 near Klawer, so there are no fixed rules about colour!

**POLLINATORS** Long-tongued bees of the family Anthophoridae, and *Apis mellifera*, the honeybee.

**SIMILAR SPECIES** The only species similar in appearance to *G. carinatus* is *G. griseus*, also found along the coastal plain. *G. carinatus* is not always easy to identify. This is because there are many atypical plants within the species that have unscented, mainly yellowish flowers, cataphylls that are not noticeably spotted, and extremely narrow leaves.

| | Habitat | Flowers | Flowering time, scent |
|---|---|---|---|
| ***G. carinatus*** | acidic pH sandstone soils in rocky areas or deep sand | blue, pinkish mauve or yellow, up to 54mm long (tube 15mm), 4–6 flowers | spring, usually highly scented |
| ***G. griseus*** | high pH calcareous sand near the sea | mauve to grey, up to 36mm long (tube up to 10mm), 6–12 flowers | winter, faintly scented |

*Leaves have thickened midribs.*

*Cataphylls are usually mottled.*

*Pink form, Mamre.*

*Pale pink form, Elandsberg.*

*Blue form, Brackenfell.*

*Yellow form, Worcester.*

*Dark mauve form, Heuningvlei.*

G. carinatus *growing in sandy soils; a yellow-flowered form found in the Kouebokkeveld, Western Cape.*

# Gladiolus griseus

***griseus*** **= grey-coloured, named for the flower colour.**

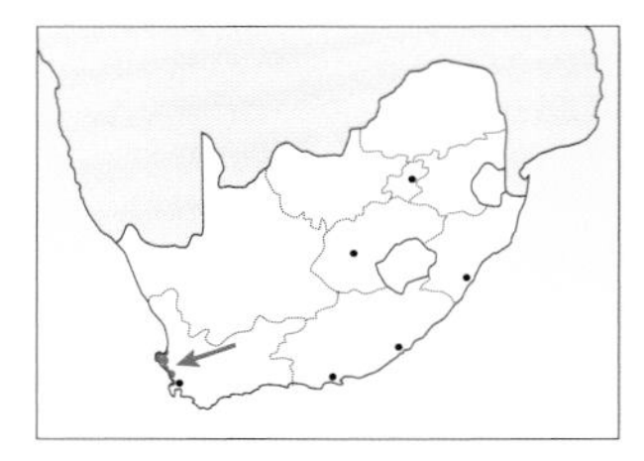

| J | F | M | A | M | J | J | A | S | O | N | D |
|---|---|---|---|---|---|---|---|---|---|---|---|

**STATUS** Critically Endangered

*Gladiolus griseus* is endemic to the West Coast. Its mauve-grey flowers have yellow on the lower tepals and are lightly scented. It flowers in autumn and winter. When not in flower, it can be confused with *G. carinatus*.

**DESCRIPTION** **Plant** 20–80cm. **Corm** globose, with papery tunics. **Cataphylls** uppermost purple mottled with white. **Leaves** 3, 1 basal and 2 sheathing, linear blade 2–7mm wide; midrib (but not margins) moderately thickened. **Spike** erect and unbranched, usually 6–15 flowers. **Bracts** green or greyish or purplish green. **Flowers** mainly shades of mauve to grey with a yellow band on the lower tepals outlined with maroon feathering; dorsal tepal arched. Perianth tube short, up to 10mm. **Anthers** cream. **Pollen** cream. **Capsules** oblong. **Seeds** yellow-brown and ovate, broadly winged. **Scent** faintly sweetly scented.

*This coastal species is known only from the West Coast.*

**DISTRIBUTION** Known only from the West Coast of the Western Cape, from Cape Town to Saldanha.

**ECOLOGY & NOTES** Grows in calcareous soils along the coast, close to the sea. Found in restio clumps and among coastal plants such as *Metalasia muricata*, *Salvia lutea* and *Osteospermum moniliferum*. Much of this plant's habitat has been destroyed by housing developments and roads, making this a species that we rarely see. However, its early flowering time (in winter rather than in spring) and cryptically coloured flowers may also be responsible for it being unobserved.

**POLLINATORS** Probably bees such as *Apis mellifera*. Not many insects are active during the winter months when this species is in flower.

**SIMILAR SPECIES** May be confused with *G. carinatus*.

| | Habitat | Flowers | Flowering time, scent |
|---|---|---|---|
| ***G. griseus*** | high pH calcareous sand near the sea | mauve to grey, up to 36mm long (tube up to 10mm), 6–12 flowers | winter, faintly scented |
| ***G. carinatus*** | acidic pH sandstone soils in rocky areas or deep sand | blue, pinkish mauve or yellow, up to 54mm long (tube 15mm), 4–6 flowers | spring, usually highly scented |

*Leaf midribs are thickened.*

*Spikes bear many small flowers.*

*Veins on bracts are translucent.*

*The lower tepals show yellow bands with maroon feathering.*

*Dorsal tepals are recurved.*

G. griseus *growing on the West Coast of the Western Cape.*

# Gladiolus quadrangulus

***quadrangulus*** **= quadrangular or four-sided; the strongly thickened midrib gives the leaf a four-sided shape in transverse section.**

*Gladiolus quadrangulus* grows from the Cape Peninsula up the West Coast. Its pale pink to white star-shaped (actinomorphic) flowers appear from August to September.

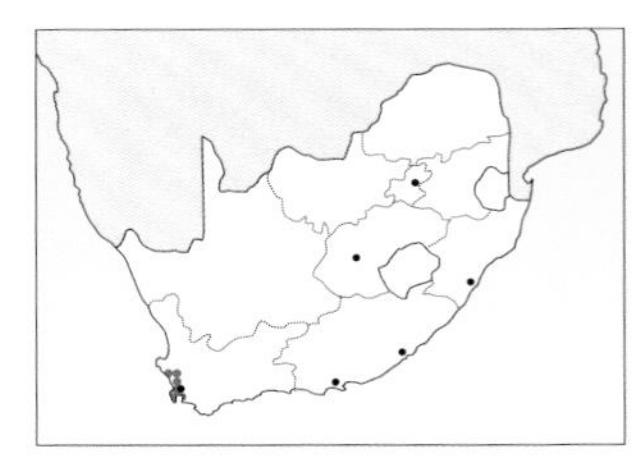

| J | F | M | A | M | J | J | A | S | O | N | D |
|---|---|---|---|---|---|---|---|---|---|---|---|

**STATUS** Endangered

**DESCRIPTION** **Plant** 14–35cm. **Corm** globose, with papery to leathery tunics. **Cataphylls** uppermost purplish to brown with white spots. **Leaves** 3, with 2 sheathing the spike, blade linear, 1–2mm wide; margins not thickened, midrib strongly thickened. **Spike** erect and unbranched, with 2–5 flowers. **Bracts** up to 45mm, green, long and narrow. **Flowers** pale pink, mauve or white, actinomorphic and stellate; tepals with darker veins and almost equal. Perianth tube short, up to 10mm, with stamens erect surrounding a central style that is tightly enclosed in a tube. **Anthers** yellow. **Pollen** yellow. **Capsules** ellipsoid. **Seeds** ovate, evenly winged, small (±4mm). **Scent** lightly narcissus scented.

**DISTRIBUTION** The historical range is from Cape Town to Malmesbury in the southwestern Cape. However, today it is known only from a few isolated sites, most of which are highly vulnerable to development. The biggest threats are housing developments and roads.

**ECOLOGY & NOTES** Grows in sandy soil in areas that are seasonally very wet. The year we photographed them, the plants were almost completely submerged! The water is often slightly saline or brackish, and if the winter is wet, the plants remain waterlogged for several months. Although the natural habitat of this species is highly threatened, the plants set large amounts of seed and they are easily cultivated in pots.

*Spikes bear up to five flowers.*

**POLLINATORS** Thought to be generalists such as bees and other insects. The flowers do not produce nectar; the only reward is pollen, which is very easily accessible on the anthers.

**SIMILAR SPECIES** On first sight it may be more difficult to identify, as species with actinomorphic flowers are rare in *Gladiolus*, and the plants look more like *Geissorhiza* or *Hesperantha* than *Gladiolus*. The only *Gladiolus* species that bears any resemblance to *G. quadrangulus* is *G. stellatus* (section *Hebea*), also with pale mauve to white actinomorphic flowers. However, the habitat of *G. stellatus* is different and the two species do not grow anywhere near one another. They might be confused if grown in a pot, and to distinguish between them one needs to take note of the cataphylls, number and shape of leaves, arrangement of the flowers, etc.

*A lilac-blue form.*

*Long, narrow bracts completely enclose the perianth tube.*

*Tepals are almost equal, and have darker veins.*

*Flowers are actinomorphic (radially symmetrical).*

*Plants resemble species of* Geissorhiza *or* Hesperantha.

G. quadrangulus *growing in seasonally wet soils near Malmesbury, Western Cape.*

# Gladiolus mutabilis

***mutabilis* = changeable, referring to the variation in flower colour.**

*Gladiolus mutabilis* occurs mostly in the Eastern Cape and to a lesser extent in the Western Cape. The flowers vary in colour from blue to mauve to brown or cream, and they appear early, from May to July.

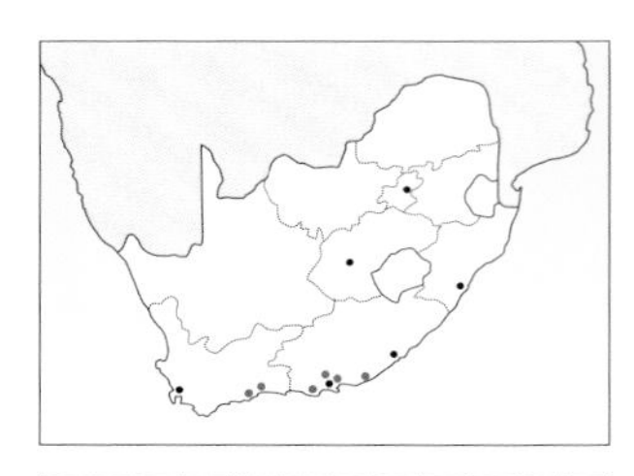



**DESCRIPTION** **Plant** 25–50cm. **Corm** elongate-obconic, with cartilaginous tunics becoming fibrous with age. **Cataphylls** uppermost green and leathery, forming a fibrous neck around the base of the stem. **Leaves** 4, 1 basal, 3 on the stem, fleshy; margins and midrib not raised. **Spike** erect, unbranched, 2–5 flowers. **Bracts** up to 40mm, glaucous-green and long; margins rolled in and slightly twisted. **Flowers** blue, mauve, brownish to cream shaded with brown; the lower 3 tepals with cream or yellow at the base, spotted and streaked with purple to brown, tepal margins undulating. Perianth tube up to 17mm, enclosed in the bracts. **Anthers** lilac to cream. **Pollen** pale yellow. **Capsules** ellipsoid. **Seeds** reddish brown, ovate, evenly winged. **Scent** violet scented.

*The tepal margins are undulate.*

**DISTRIBUTION** Found from Albertinia and Riversdale eastwards to Gqeberha (Port Elizabeth) and Makhanda (Grahamstown).

**ECOLOGY & NOTES** Grows in sandstone-derived soils among rocks; seems to flower best after fire.

**POLLINATORS** Presumed to be long-tongued bees.

**SIMILAR SPECIES** Not a particularly distinctive species in flower; it could be mistaken for several others such as *G. exilis*, *G. vaginatus*, *G. gracilis* and *G. subcaeruleus*.

| | Corm | Leaves |
|---|---|---|
| ***G. mutabilis*** | elongate-obconic, with cartilaginous tunics becoming fibrous | 4, fleshy, no raised midrib or margins |
| ***G. exilis*** | ovoid, leathery tunics becoming fibrous | 3, slightly fleshy, no raised midrib or margins |
| ***G. vaginatus*** | globose, papery tunics becoming fibrous | 2, lower leaf with no blade, second leaf inconspicuous, no foliage leaf after flowering |
| ***G. gracilis*** | globose, hard concentric tunics | 4, margins raised into wings extending at right angles to the blade |
| ***G. subcaeruleus*** | conic, firm papery tunics become fibrous to form a neck | 3, all entirely sheathing, foliage leaf after flowering is oval in section with grooves (hairy) |

*Monkey beetles explore the long bracts.*

*Erect spikes bear up to five flowers.*

*Broad lower tepals are distinctive.*

G. mutabilis *grows on sandstone soils in the Eastern Cape.*

*Mauve form, Swellendam.*

# Gladiolus exilis

***exilis* = small and thin, referring to the size of the plants.**

*Gladiolus exilis* occurs only in the mountainous area of the southwestern Cape. Its flowers are usually pale blue, appearing from April to May. This species can be mistaken for *G. mutabilis*, from which it is differentiated by its succulent leaves and corm.

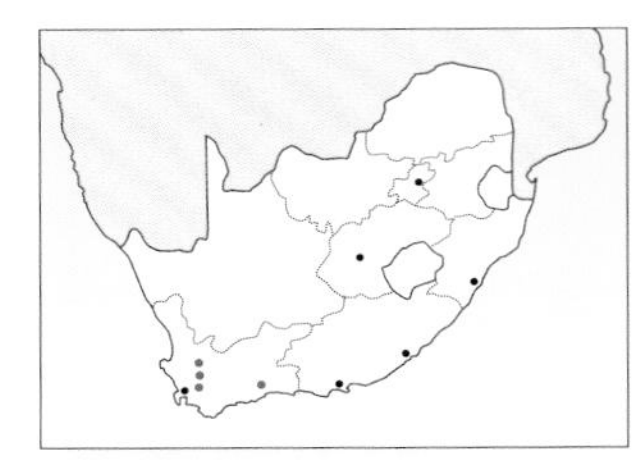

J F M A M J J A S O N D

**STATUS** Near Threatened

**DESCRIPTION** **Plant** 30–60cm. **Corm** ovoid, with leathery tunics becoming fine fibres with age. **Cataphylls** uppermost green, decaying into fibres that make a neck around the stem. **Leaves** 3, 1 basal, 2 on the stem, 1mm wide, slightly succulent. **Spike** slightly flexed above the leaf sheath, unbranched, with 1–4 flowers. **Bracts** grey-green. **Flowers** pale blue-mauve to white; tepals unequal, lower 3 longer than upper in profile, lower lateral tepals have yellow in the throat, streaked and speckled with dark mauve. Perianth tube up to 15mm. **Anthers** pale mauve. **Pollen** whitish. **Capsules** ellipsoid. **Seeds** small, dark brown and ovate. **Scent** lightly scented.

*This species has delicate flowers.*

**DISTRIBUTION** Known only from the Bainskloof Mountains and the Witzenberg Mountains near Tulbagh in the southwest of the Western Cape. An unusual form in the southeast awaits further verification. Alien invasive species, particularly hakea plants, are its biggest threats.

**ECOLOGY & NOTES** Grows in well-drained rocky areas on slopes; sometimes found near the interface between sandstone-derived soils and clay soils. Plants flower best in the first two years after fire, before the fynbos becomes too dense. The flowers are inconspicuous and are easily overlooked. The plants that we photographed were growing on south- and southwest-facing slopes and were in shade for a large part of the day.

*White form, Bainskloof; note the yellow in the throat.*

**POLLINATORS** Long-tongued bees such as *Amegilla fallax*.

**SIMILAR SPECIES** Could be mistaken for species such as *G. mutabilis*, *G. vaginatus* and *G. gracilis*.

| | Corm | Leaves |
|---|---|---|
| ***G. exilis*** | ovoid, cartilaginous tunics becoming fibrous | 3, slightly fleshy, no raised midrib or margins |
| ***G. mutabilis*** | elongate-obconic, cartilaginous tunics becoming fibrous | 4, fleshy, no raised midrib or margins |
| ***G. vaginatus*** | globose, papery tunics becoming fibrous | 2, lower leaf with no blade, second leaf inconspicuous; no foliage leaf after flowering |
| ***G. gracilis*** | globose, hard concentric tunics | 4, margins raised into wings extending at right angles to the blade |

*Leaves are narrow and slightly succulent.*

*Flowering after fire.*

*Long bracts encase short perianth tubes.*

*Lower tepals are longer than the upper ones.*

G. exilis *flowering in Bainskloof, Western Cape.*

# Gladiolus vaginatus

***vaginatus* = enclosed in a sheath, referring to the basal leaf that entirely sheaths the flowering stem.**

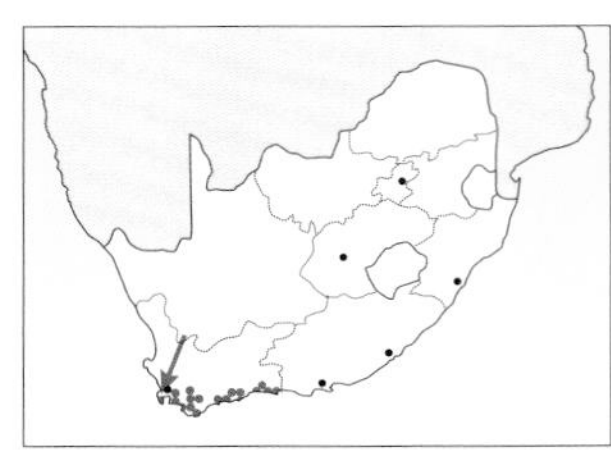

J F M A M J J A S O N D

**STATUS** Vulnerable

*Gladiolus vaginatus* grows in the coastal region of the Western Cape. Its scented, pale blue flowers are easily confused with *G. exilis* and *G. mutabilis*, but it is distinguished from these by having only two leaves. It flowers from March to April.

**DESCRIPTION Plant** 30–50cm or taller. **Corm** globose to depressed-globose, with thin, papery tunics. **Cataphylls** uppermost green or purplish. **Leaves** 2 on flowering stem, lower 1 entirely sheathing with no blade, upper 1 reduced to a scale, also sheathing. Leaves on non-flowering plants are linear, 1–1.5mm wide. **Spike** unbranched, remaining green for several months, 2–6 or more flowers. **Bracts** grey to purplish green. **Flowers** pale blue with unequal tepals; in profile lower tepals longer than upper by up to 10mm, lower tepals streaked and dotted with purple. Perianth tube long, up to 20mm. **Anthers** mauve. **Pollen** pale mauve. **Capsules** broadly ovoid. **Seeds** light yellow-brown, oblong, broadly winged. **Scent** lightly scented, although may be unscented.

*Pale blue flowers appear in late summer to autumn.*

**DISTRIBUTION** Recorded along the coast from the Cape Peninsula eastwards to Knysna.

**ECOLOGY & NOTES** Most records show plants in calcareous sands or limestone, both of which have an alkaline pH. However, plants have been found around Caledon and Elim in gravelly clay, with an intermediate or acid pH. We saw plants growing and flowering in thick, old vegetation, but presumably this species would respond to burning and produce a flush of flowers in the first or second year after fire.

**POLLINATORS** Thought to be anthophorid bees that pollinate other small, short-tubed flowers.

**SIMILAR SPECIES** Although this species is easily mistaken for *G. mutabilis* and *G. exilis*, it can be identified by the presence of only two leaves, both of which lack blades. Its flowers are similar to those of *G. mutabilis* and *G. subcaeruleus*, both of which flower in autumn.

| | Corm | Leaves |
|---|---|---|
| ***G. vaginatus*** | globose, papery tunics becoming fibrous | 2, lower leaf with no blade, second leaf inconspicuous; no foliage leaf after flowering |
| ***G. mutabilis*** | elongate-obconic, cartilaginous tunics becoming fibrous | 4, fleshy, no raised midrib or margins |
| ***G. exilis*** | ovoid, cartilaginous tunics becoming fibrous | 3, slightly fleshy, no raised midrib or margins |
| ***G. subcaeruleus*** | conic, firm papery tunics become fibrous to form a neck | 3, all entirely sheathing, foliage leaf after flowering is oval in section with grooves (hairy) |

*The upper cataphyll is purple-green.*

*Dull-coloured bracts cup the perianth.*

*The lower tepal markings are not always present or easily visible.*

*This species is thought to be pollinated by anthophorid bees.*

*The plants grow on calcium-rich soils near the coast, but also occur in neutral to acidic soils inland; here growing near Napier, Western Cape.*

# Gladiolus maculatus

***maculatus* = speckled, referring to the markings on the flowers.**

Widespread across the winter-rainfall areas, *Gladiolus maculatus* bears yellow to lilac flowers from April to June. It is scented day and night.

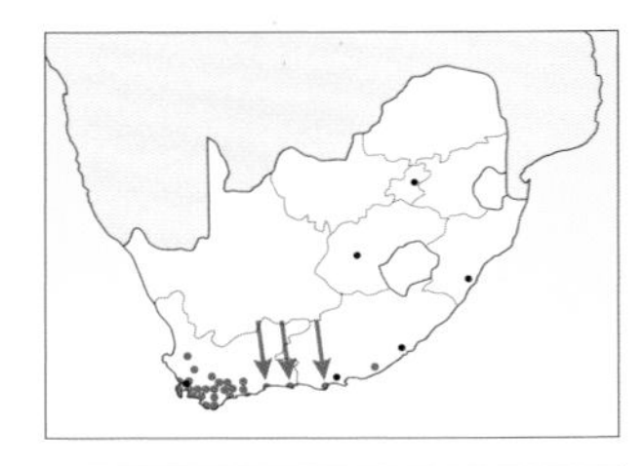

| J | F | M | A | M | J | J | A | S | O | N | D |
|---|---|---|---|---|---|---|---|---|---|---|---|

**DESCRIPTION** **Plant** 30–80cm. **Corm** globose, with papery tunics. **Cataphylls** uppermost green or purple. **Leaves** 3 or 4, a basal leaf sheaths the lower half of the stem, with a fairly fleshy plane blade, other leaves arising on the flower stem; margins and midrib not thickened. **Spike** inclined, unbranched, 1–3 flowers. **Bracts** green or greyish, soft textured. **Flowers** variable, from dull yellow to brown to orange to lilac, all very speckled with brown to purple spots of different sizes; tepals have undulating margins, dorsal tepal is sometimes transparent on either side of the midline. Perianth tube long, up to 35mm. **Anthers** brown to cream. **Pollen** yellow. **Capsules** ovoid. **Seeds** ovate, with golden brown wing. **Scent** strongly sweetly scented.

*Growing after fire in Kleinmond, Western Cape.*

**DISTRIBUTION** Extending northwards from the Cape Peninsula and eastwards to Makhanda (Grahamstown) and Hlanganani (Alexandria) in the Eastern Cape. Mainly found in the coastal areas and slightly inland. More common in the western areas.

**ECOLOGY & NOTES** Grows in heavier soils such as clay in renosterveld, but also found in sandy soils among fynbos, for example on the Cape Peninsula at Silvermine Nature Reserve and in the Betty's Bay area. The flowers, although very colourful, are quite cryptic and are often inconspicuous among the vegetation. These plants respond to fire, flowering profusely in the first few years after a burn, after which flowering becomes sporadic.

*A plant seen flowering after fire, Elandsberg Reserve, Western Cape.*

**POLLINATORS** Thought to be moths as the flowers have a long perianth tube and are strongly scented at night as well as during the day. The sugar concentration of the nectar is relatively high. The moth *Cucullia terensis* has been seen visiting the flowers.

**SIMILAR SPECIES** Flowers are superficially similar to *G. hyalinus*, *G. liliaceus* and *G. recurvus* but tepal forms differ. The leaves of the three species differ considerably.

*Short inner and long outer bracts are clearly visible.*

*Reddish form, Silvermine.*

*Orange form, Silvermine.*

*Brown form, Elandsberg Reserve.*

*Cream-and-lilac form, Table Mountain.*

G. maculatus *growing in the foothills of Table Mountain, Western Cape.*

# Gladiolus albens

***albens* = white, named for the flower colour.**

*Gladiolus albens* is endemic to the Eastern Cape, where it flowers from March to May. Its lightly scented, cream to white flowers with long perianths are adapted for pollination by moths.

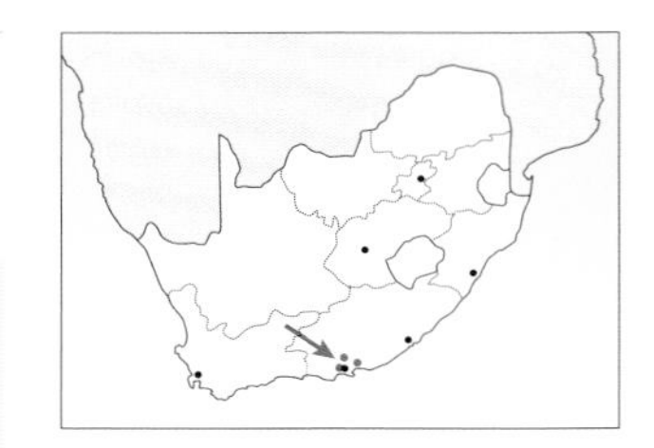

| J | F | M | A | M | J | J | A | S | O | N | D |
|---|---|---|---|---|---|---|---|---|---|---|---|

**DESCRIPTION** **Plant** 25–50cm. **Corm** globose, with papery tunics. **Cataphylls** uppermost 9cm above ground, green or dull purple. **Leaves** 3, lower 2 basal, blade 1–2mm wide, slightly fleshy; with 3 equal veins. **Spike** unbranched, slightly flexed above upper leaf, 1–3 flowers. **Bracts** green to grey. **Flowers** cream to white, sometimes with faint streaks and speckles; tepals ovate and almost equal in size. Perianth tube long and slender, up to 65mm, widening gradually in the upper third. **Anthers** cream. **Pollen** yellow. **Capsules** not known. **Seeds** not known. **Scent** lightly scented.

**DISTRIBUTION** Recorded only in the Eastern Cape in the Albany and Hlanganani (Alexandria) areas, including Makhanda (Grahamstown) and Gqeberha (Port Elizabeth), with one record from the Somerset East area.

**ECOLOGY & NOTES** Grows in fynbos-type vegetation and among grass in the summer-rainfall region, but this species has a winter-growing cycle. Plants grow through the winter months, flowering in the autumn. Although there are a fair number of herbarium records of this species, it is rarely seen, even by people who botanise regularly in its distribution range. The plants that we photographed were next to a railway line and were protected from grazing. Many areas where it previously occurred are heavily grazed by cattle and this has impacted greatly on the species.

*Flowers appear in the autumn.*

**POLLINATORS** Thought to be moths, which are attracted to the white long-tubed flowers and their light fragrance.

**SIMILAR SPECIES** Unlikely to be mistaken for any other species.

*Ovate tepals give the flower its open shape.*

*Slightly fleshy leaves rise from the stem and have short blades.*

*Very long perianth tubes suggest that the flowers are pollinated by moths.*

G. albens *is seldom seen, here encountered near Bathurst, Eastern Cape.*

# Gladiolus meridionalis

***meridionalis* = southern; this species is found along the south coast.**

*Gladiolus meridionalis* has a discontinuous distribution along parts of the coast in the Western and Eastern Cape. Pink flowers with long perianth tubes and high-sucrose nectar are adapted for sunbird pollination. Flowering season is from April to June.

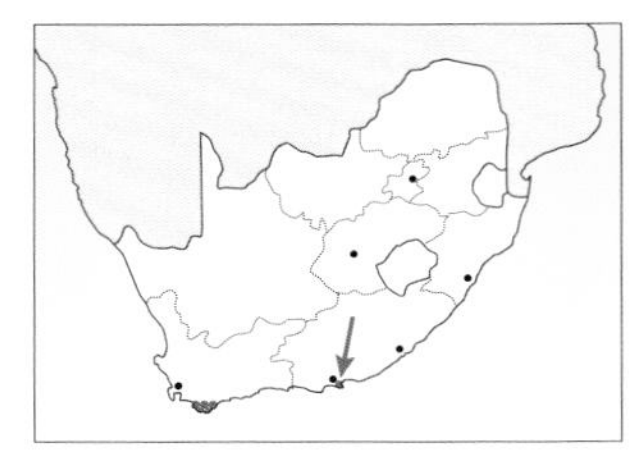

| J | F | M | A | M | J | J | A | S | O | N | D |
|---|---|---|---|---|---|---|---|---|---|---|---|

**DESCRIPTION** **Plant** 30–80cm. **Corm** globose, with papery to leathery tunics. **Cataphylls** uppermost green or purple. **Leaves** 3, lower 2 basal and slightly fleshy; flat blade with no raised margins or midrib. **Spike** unbranched, lightly flexed above the leaves, with 1 or 2 flowers. **Bracts** up to 50mm long, green to dull purple. **Flowers** pale to dark pink (in the west) or occasionally cream to pinkish orange (in the east), often with darker spots and speckles; tepals subequal. Perianth tube narrow and long, up to 50mm. **Anthers** light brown. **Pollen** yellow. **Capsules** ovoid, 12mm. **Seeds** golden brown and ovate, evenly winged. **Scent** faintly sweetly scented.

**DISTRIBUTION** An unusual range: plants have been recorded in the southwestern Cape in the Elim–Gansbaai area, and in the Eastern Cape in the Gqeberha (Port Elizabeth) area, but nowhere in between.

**ECOLOGY & NOTES** Grows in stony sandstone soils among fynbos, both on mountain slopes and flat areas, and usually within a few kilometres of the sea.

**POLLINATORS** *G. meridionalis* is one of a few species in the genus that are adapted to sunbird pollination. These species produce high-sucrose nectar, a feature more commonly associated with insect-pollinated plants than with bird-pollinated ones, but the adaptation to bird pollination makes them unusual.

**SIMILAR SPECIES** Unlikely to be confused with any other species owing to the distribution, flowering time and flower colour.

*The plant has long, narrow bracts and elongate perianth tubes.*

*Flowers are pink in its western range and pinkish orange or cream in its eastern range.*

*Cataphylls reach well above ground.*

*The long perianth tube is an adaptation to sunbird pollination.*

*Pollen speckles the lower tepals.*

*Pinkish-orange eastern form.*

*Pink western form.*

G. meridionalis *grows in sandy soils close to the sea.*

# Gladiolus priorii

**Named after Richard Prior (1809–1902), a medical doctor and amateur botanist who collected plants in South Africa in the 1840s.**

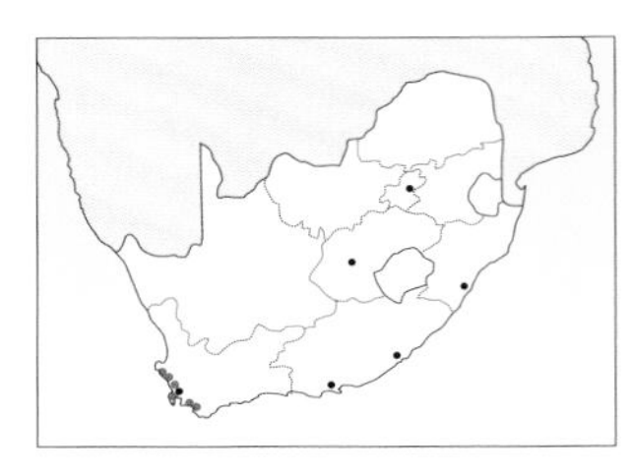

| J | F | M | A | M | J | J | A | S | O | N | D |
|---|---|---|---|---|---|---|---|---|---|---|---|

The range of *Gladiolus priorii* extends from Saldanha down to the Cape Peninsula and across to Hermanus. Its bright red flowers with long perianth tubes appear from May to June and are adapted for sunbird pollination.

**DESCRIPTION** **Plant** 30–50cm. **Corm** ovoid, with papery tunics. **Cataphylls** green to purplish above ground. **Leaves** usually 4, the lowest mostly sheathing the stem with a short blade (10cm), slightly fleshy; no raised margins or midrib. **Spike** unbranched, flexed outwards above the leaf sheaths, with 1–4 flowers. **Bracts** up to 50mm long, green. **Flowers** scarlet with pale yellow or cream base. Perianth tube up to 45mm, slender in the lower part and abruptly expanded in the upper part. **Anthers** pale yellow. **Pollen** yellow. **Capsules** ellipsoid, up to 15mm. **Seeds** golden brown and ovate, evenly winged. **Scent** unscented.

**DISTRIBUTION** Found only in the extreme coastal southwestern Cape. It has been recorded from Saldanha Bay north of Cape Town, south to the Peninsula and east to Hermanus.

**ECOLOGY & NOTES** In most of its range, this species is found on sandstone-derived rocky slopes, but north of Cape Town it grows on granite outcrops. Most of the plants we found were growing with their corms well wedged between rocks, protecting them from porcupine predation.

*The species has scarlet flowers with lanceolate tepals.*

**POLLINATORS** The red flowers, long perianth tube and large volumes of nectar seem likely to be adaptations for pollination by sunbirds.

**SIMILAR SPECIES** Very similar to *G. watsonius* and *G. teretifolius*, two other scarlet-flowered species.

| | Flowering time | Leaves | Flowers |
|---|---|---|---|
| ***G. priorii*** | May–Jun | slightly fleshy, no venation | scarlet, yellow on lower tepals, yellow anthers |
| ***G. watsonius*** | Aug–Sep | midrib thickened, margins strongly thickened forming wings | scarlet, purple anthers |
| ***G. teretifolius*** | Jul–Aug | oval to nearly terete with margins and midrib thickened | scarlet, purple anthers |

*Stems flex away from the sheathing leaves.*

*Bracts are long.*

*Flowers are adapted for pollination by sunbirds.*

*Note the creamy yellow throat within the scarlet tepals.*

G. priorii *flowering near Hermanus, Western Cape.*

# Gladiolus brevitubus

***brevitubus* = short-tubed, referring to the short perianth tube, which is unusual in *Gladiolus*.**

Endemic to the southwestern Cape, *Gladiolus brevitubus* flowers in late October and November. Its orange, radially symmetrical flowers are scented.

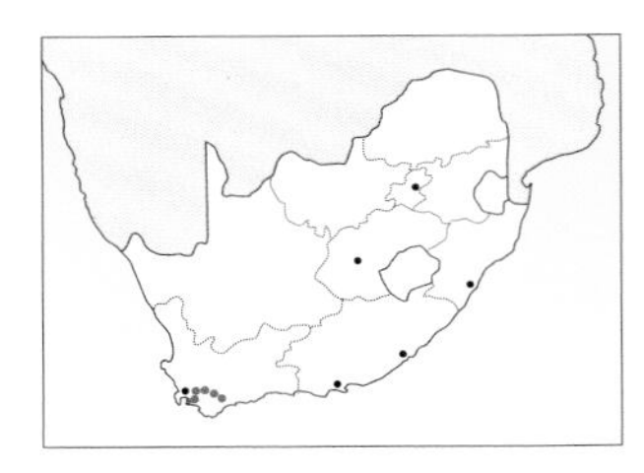

| J | F | M | A | M | J | J | A | S | O | N | D |
|---|---|---|---|---|---|---|---|---|---|---|---|

**DESCRIPTION** **Plant** 12–40cm. **Corm** globose, with tunics of fine- to medium-netted fibre. **Cataphylls** accumulate to form a short fibrous neck around the stem. **Leaves** 3 or 4, 2 of which sheath the stem, blade linear, to 3mm wide; margins and midrib lightly thickened. **Spike** inclined and unbranched, with 2–8 flowers. **Bracts** green with reddish margins. **Flowers** actinomorphic, orange to salmon with yellow in the centre; tepals almost equal, with the outer whorl slightly larger than the inner. Perianth tube very short, 4mm, with small round papillae at the throat. **Anthers** yellow. **Pollen** yellow. **Capsules** obovoid. **Seeds** broadly and evenly winged. **Scent** lightly sweetly scented.

*The flower is so unusually shaped that it doesn't immediately resemble a gladiolus.*

**DISTRIBUTION** Occurs only in the mountains from the Hottentots Holland range eastwards to the Riviersonderend Mountains and south into the Kleinrivier Mountains near Hermanus.

**ECOLOGY & NOTES** Found up to 1,500m above sea level on rocky mountain slopes in sandy soils. Flowers particularly well after fires, but also flowers in fairly old, unburned fynbos. We found plants on south- and east-facing slopes, which are cooler slopes that are kept moist by mist blown in by the southeast wind in summer.

**POLLINATORS** Thought to be small, pollen-collecting bees of the family Halictidae, commonly called sweat bees. They seem to be attracted to the pollen, and also to the papillae at the base of the filaments and in the throat of the perianth tube.

*Inclined spikes bear up to eight flowers.*

**SIMILAR SPECIES** Actinomorphic flowers are rare in the *Gladiolus* genus, and at first sight these flowers are more likely to be mistaken for a *Geissorhiza* or *Hesperantha* than recognised as a *Gladiolus*. However, the plants are unmistakable once they have set seed as the seeds are typical winged *Gladiolus* seeds.

*Narrow leaves have thickened margins and midribs.*

*Actinomorphic flowers have small round papillae at the throat.*

*Note the short, rounded bracts, which are green with reddish margins.*

G. brevitubus *flowering above Hermanus, Western Cape.*

# *Gladiolus rogersii*

**Named after Reverend Moyle Rogers (1835–1920), a clergyman and botanist who made the type collection.**

The sweetly scented, dark blue to purple flowers of *Gladiolus rogersii* appear from August to October. Mainly a winter-rainfall species, it can sometimes appear in autumn. It can be confused with *G. bullatus*. The species' range extends from Agulhas to the Outeniqua Mountains.

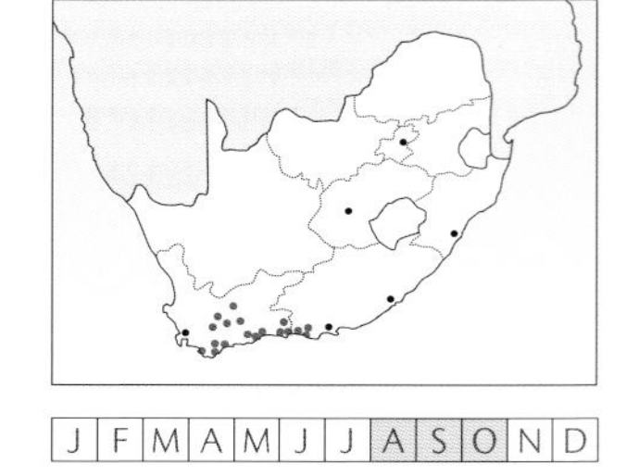

| J | F | M | A | M | J | J | A | S | O | N | D |
|---|---|---|---|---|---|---|---|---|---|---|---|

**DESCRIPTION** **Plant** 30–50cm. **Corm** globose, with finely fibred tunics. **Cataphylls** uppermost purple, mottled or dry and brown. **Leaves** 3 or 4 linear leaves, 1.5–4mm wide; margins and midribs lightly to heavily thickened; leaves sometimes terete, upper leaves sheathing. **Spike** unbranched, usually with 3–6 flowers. **Bracts** up to 20mm long with narrow transparent margins, green or flushed grey to purple, diverging from the axis. **Flowers** inflated and bell-like, blue to purple; lower tepals lightly spotted with purple and with a V-shaped transverse band of white or yellow outlined with dark purple. Perianth tube 12–19mm. **Anthers** blue. **Pollen** cream. **Capsules** not known. **Seeds** prominent darkish seed body with wide paler wing. **Scent** lightly sweetly scented, may be unscented.

**DISTRIBUTION** Occurs from Pearly Beach in the west, along the coastal plain and the mountains of the southern Cape to Humansdorp in the east. It is also found on some of the inland mountain ranges such as the Kammanassie Mountains east of Oudtshoorn, and in the Little Karoo near Barrydale.

**ECOLOGY & NOTES** Grows in a large range of conditions, from limestone flats near the coast to well-drained sandstone slopes at elevations of over 1,000m. Flowering occurs either in autumn or spring, which is somewhat unusual. There are four main population types; the features of three of these types suggest a relationship with the habitat conditions. Populations in the Langeberg and Outeniqua mountains have soft corm tunics, green or purplish cataphylls, and plane leaves with lightly thickened margins and midribs. Coastal limestone populations have coarsely fibrous corm tunics, mottled cataphylls and leaves that are often terete or oval. Plants in drier habitats have a neck of fibres around the stem base and broad leaves, 4–6mm wide. Populations of early flowering plants have coarsely fibrous corm tunics and unscented flowers.

**POLLINATORS** Long-tongued anthophorid bees, which pollinate other short-tubed blue flowers with a sweet scent.

**SIMILAR SPECIES** *G. rogersii* is extremely variable with different markings, leaf profiles and flowering times, making it quite difficult to identify in the veld. It is similar to its close relative *G. bullatus* and also very similar in appearance to *G. patersoniae* and *G. inflatus*, which are thought to be more distantly related. All have flowers that are very similar in colour and shape.

| | Cataphylls | Floral bracts | Flowers |
|---|---|---|---|
| ***G. rogersii*** | sometimes mottled | 15–22mm, smooth | up to 6 per spike |
| ***G. bullatus*** | green or purple | 30mm or longer, striated | 1 or 2 per spike |

*Inclined spikes bear bell-like flowers.*

*Flowers grow alternately on two sides of the spike.*

*Flower markings and colours are variable.*

*Linear leaves have thickened midribs and margins.*

*Unusual pink form, Kammanassie Mountains.*

*Pale lilac form, Nature's Valley.*

G. rogersii *flowering in Nature's Valley, Western Cape. The Tsitsikamma Mountains can be seen in the distance.*

# Gladiolus bullatus

***bullatus* = bubble-like, referring to the shape of the flower.**

*Gladiolus bullatus* is endemic to the southwestern parts of the Western Cape. It has bell-shaped purple or blue flowers marked with yellow. It can be confused with *G. rogersii* but has shorter bracts and different corms.

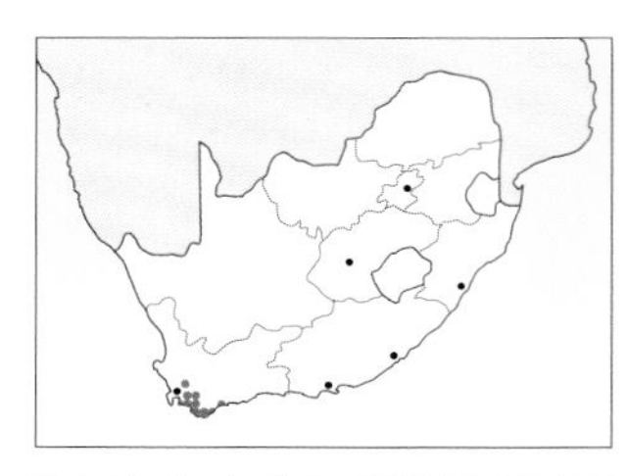

| J | F | M | A | M | J | J | A | S | O | N | D |
|---|---|---|---|---|---|---|---|---|---|---|---|

**DESCRIPTION** **Plant** 45–100cm. **Corm** ovoid, with woody tunics. **Cataphylls** upper up to 15cm above ground, green to purplish. **Leaves** 4, linear blade 2mm wide; midrib heavily thickened making the leaf oval in section with 2 narrow grooves on each surface. **Spike** unbranched, with 1 or 2 flowers. **Bracts** 3cm or longer with elongate tapering point, brownish purple to grey, prominently striated (ridged) above the veins. **Flowers** large, inflated and bell shaped, blue to violet; lower tepals lightly spotted with purple inside the flower, with yellow marks or a transverse band towards the tepal tips. Perianth tube up to 16mm. **Anthers** white to grey-blue. **Pollen** cream. **Capsules** ovoid and concealed in the bracts. **Seeds** ovate, pointed at one end, evenly winged. **Scent** unscented.

*The species is found only in the Caledon district, Western Cape.*

**DISTRIBUTION** Found only in the Caledon district in the Overberg region, from the Kogelberg and Houw Hoek in the west to Bredasdorp and Cape Infanta in the east.

**ECOLOGY & NOTES** Grows in sandstone- or occasionally limestone-derived soils among stones and rocks. Usually found in the hills and on mountain slopes, growing amid Ericaceae and Proteacae.

*Pale violet form; note the yellow markings and purple speckles.*

**POLLINATORS** Thought to be long-tongued bees.

**SIMILAR SPECIES** The closest relative is probably *G. rogersii*, whose flowers are similar in shape and colour.

| | Cataphylls | Floral bracts | Flowers |
|---|---|---|---|
| ***G. bullatus*** | green or purple | 30mm or longer, striated | 1 or 2 per spike |
| ***G. rogersii*** | sometimes mottled | 15–22mm, smooth | up to 6 per spike |

*Linear leaves have thickened midribs and appear oval in cross section.*

*Grey-purple bracts are ridged.*

*Perianth tubes curve abruptly below the bracts.*

*Violet form.*

*Pale lilac form.*

*Purple form.*

*Note transverse band on tepals.*

G. bullatus *flowering in Vogelgat near Hermanus, Western Cape.*

# Gladiolus blommesteinii

**Named after G van Blommestein, who made one of the first scientific collections of the species.**

Restricted to the southwestern Cape, *Gladiolus blommesteinii* has pale pink to mauve flowers with dark, streaked nectar guides. It flowers from August to October. It can be confused with *G. virgatus*, but the latter's flowers are unscented.

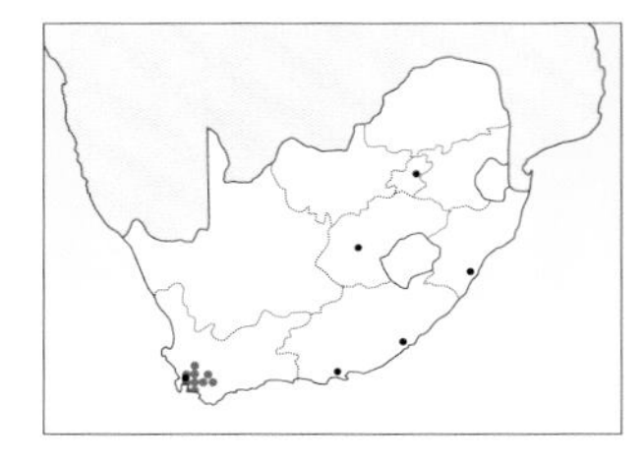

| J | F | M | A | M | J | J | A | S | O | N | D |
|---|---|---|---|---|---|---|---|---|---|---|---|

**DESCRIPTION** **Plant** 30–60cm. **Corm** ovoid, with woody tunics. **Cataphylls** above ground, uppermost green or purple. **Leaves** 4, lowest leaf is longest, narrow linear blade 1.5mm wide; margins and midrib lightly raised. **Spike** inclined, 1–4 flowers. **Bracts** 20–36mm, green to dark grey-green and lightly ridged. **Flowers** pale mauve-pink; tepals unequal and lanceolate, lower tepals yellow below, with streaks and lines of red, blue or purple and spots in the throat. Perianth tube 13–24mm. **Anthers** whitish. **Pollen** light grey. **Capsules** unknown. **Seeds** unknown. **Scent** lightly scented.

*Inclined spikes bear up to four flowers.*

**DISTRIBUTION** Found in the Overberg region, from the Hottentots Holland Mountains eastwards to Riviersonderend, in the Western Cape.

**ECOLOGY & NOTES** Grows in sandstone-derived soils, associated with mountain slopes. Although we found plants in flower in five-year-old vegetation, they flower best in the first two years after a fire. We saw a large number of plants in flower on Franschhoek Pass, but we found no seed capsules and can only surmise that the plants were not well pollinated. This site is extremely windy and it is possible that the pollinators struggle in the wind.

*Flower colour is variable; here deep mauve, Franschhoek.*

**POLLINATORS** Long-tongued bees, *Anthophora diversipes*, which are large and swift flying, and are attracted by the sweet scent.

**SIMILAR SPECIES** Similar vegetatively and in flower to *G. virgatus*, with which it could be confused, especially where their ranges overlap. Can also be confused with *G. hirsutus* (section *Linearifolii*).

| | Leaves | Perianth tube | Markings on lower tepals |
|---|---|---|---|
| ***G. blommesteinii*** | 4, 1.5mm wide, not hairy | 13–24mm | streaks, lines and spots of red, blue or purple |
| ***G. virgatus*** | 4, 2.5mm wide, not hairy | 22–27mm | spear-shaped whitish streak edged with dark pink |

*Bracts are ridged.*

*Pink form, Franschhoek.*

*Pale mauve form, Sir Lowry's Pass.*

*Blue form, Sir Lowry's Pass.*

*Pink form, Franschhoek.*

G. blommesteinii *flowering in the Franschhoek Pass, Western Cape.*

# Gladiolus virgatus

***virgatus* = willowy, named for the slender stem.**

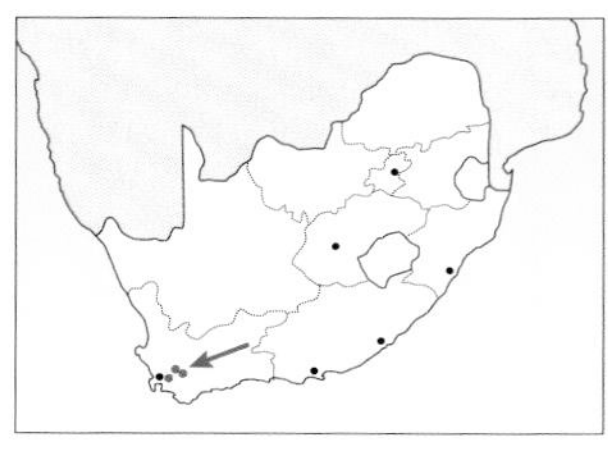

J F M A M J J A S O N D

*Gladiolus virgatus* is restricted to the coastal mountains of the southwestern Cape. Its large white, pale or dark pink flowers appear from mid-September to November and are pollinated by long-tongued flies.

**DESCRIPTION** **Plant** 40–50cm. **Corm** globose, with woody tunics. **Cataphylls** uppermost firm textured and green. **Leaves** 4, lower 2 basal, up to 2.5mm wide; margins and midrib lightly raised. **Spike** inclined, unbranched, 2–4 flowers. **Bracts** up to 5cm long, green or purplish, striated above the veins. **Flowers** pink to almost white; lower tepals have white spear-shaped marks surrounded by dark pink, V-shaped marks, speckled pink in the throat; tepal margins undulate. Perianth tube up to 27mm. **Anthers** light mauve. **Pollen** whitish. **Capsules** unknown. **Seeds** unknown. **Scent** unscented.

*Flowers are pale pink to almost white with V-shaped guides.*

**DISTRIBUTION** Known only from the coastal mountain ranges in the southwestern Cape, from Somerset West to north of Paarl, in the Du Toitskloof Mountains.

**ECOLOGY & NOTES** Grows on cool and well-watered mountain slopes. Seems to prefer the clay soils derived from shales. Tends to flower later at higher altitudes.

**POLLINATORS** Long-tongued flies such as the horse fly appear to pollinate these unscented flowers with reddish nectar guides.

*Flowers from the southern end of the range are larger than those in the north.*

**SIMILAR SPECIES** Similar vegetatively and in flower shape to *G. blommesteinii*, to which it is thought to be closely related.

| | Leaves | Perianth tube | Markings on lower tepals |
|---|---|---|---|
| ***G. virgatus*** | 4, 2.5mm wide, not hairy | 22–27mm | spear-shaped whitish streak edged with dark pink |
| ***G. blommesteinii*** | 4, 1.5mm wide, not hairy | 13–24mm | streaks, lines and spots of red, blue or purple |

*Narrow leaves have raised margins and midribs.*

*Spikes are slender and bracts are striated.*

G. virgatus *growing in the Helderberg Mountains, Western Cape.*

# Gladiolus debilis

***debilis* = delicate and feeble, for the delicate appearance of the flowers and the weak stem.**

Restricted to the mountains of the southwestern Cape, *Gladiolus debilis* flowers from September to October. White flowers have distinctive red markings, which vary by geographical location.

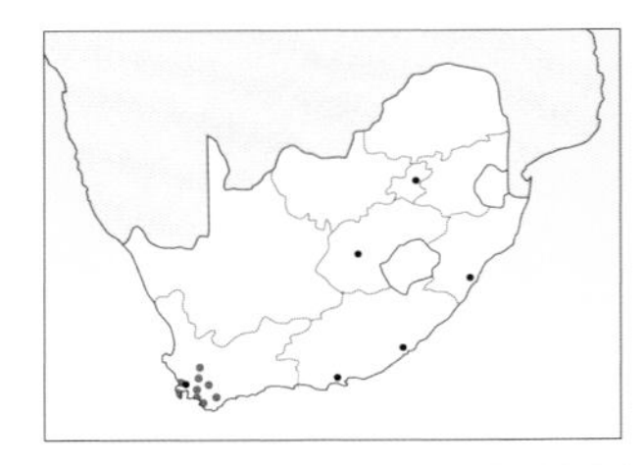

| J | F | M | A | M | J | J | A | S | O | N | D |
|---|---|---|---|---|---|---|---|---|---|---|---|

**DESCRIPTION** **Plant** 25–45cm. **Corm** wiry to woody tunics. **Cataphylls** above ground, green or purple. **Leaves** 4, plane blade, 1–2mm wide; thickened margins and midvein. **Spike** inclined, 1–3 flowers. **Bracts** light green to greyish, lightly striated above the veins. **Flowers** large, white to pale pink; distinctively shaped, with the tepals widely spread, lower 3 tepals marked with red spots, diamonds or streaks; throat speckled and streaked with red. Perianth tube long, up to 20mm. **Anthers** white above, purple below. **Pollen** cream. **Capsules** ellipsoid. **Seeds** dark brown, evenly winged. **Scent** unscented.

**DISTRIBUTION** Found from Bainskloof southwards to the Cape Peninsula and eastwards to Bredasdorp.

**ECOLOGY & NOTES** Occurs in mountain fynbos on rocky sandstone slopes. Like many other *Gladiolus* species in the southwestern Cape, they flower best in the first few years after a fire. Markings differ on each tepal and every flower, making this species unique in both the genus *Gladiolus* and in the family Iridaceae.

**POLLINATORS** Thought to be long-tongued flies.

**SIMILAR SPECIES** Only likely to be confused with *G. variegatus*.

| | Habitat | Leaves | Bracts | Markings |
|---|---|---|---|---|
| ***G. debilis*** | rocky sandstone in mountain fynbos | 4, 1–2mm wide, thickened midrib and margins | 15–28mm, ridged | diamond, spade or chevron shapes |
| ***G. variegatus*** | almost at sea level in limestone cracks or calcareous sands | 3, up to 2.5mm wide, margins not thickened | 30–45mm, smooth | deep red spots |

*Each spike bears up to three flowers.*

*Wide tepals give a distinctive shape to the flower.*

*Anthers are purple underneath.*

*Bracts are ridged, with hyaline margins.*

*There is variation in markings on lower tepals; red markings here are spear shaped.*

*These lower tepals have chevron markings.*

G. debilis *growing on sandstone at Honingklip near Bot River, Western Cape.*

# Gladiolus variegatus

***variegatus* = variegated, referring to the contrasting colour of the nectar guides.**

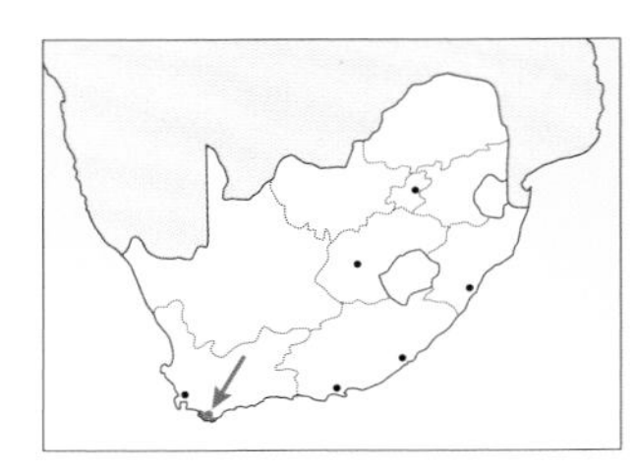

**STATUS** Vulnerable

*Gladiolus variegatus* is restricted to a small coastal area between Stanford and Agulhas in the Western Cape. Unscented, white to pale pink flowers, speckled with red on the lower tepals, appear from mid-September to mid-October.

**DESCRIPTION** **Plant** 20–40cm. **Corm** conical, with woody segmented tunics. **Cataphylls** uppermost green. **Leaves** 3, lower 2 basal and the longest, up to 2.5mm wide; margins not thickened, midrib raised. **Spike** unbranched, 1 or 2 flowers. **Bracts** dark green, smooth. **Flowers** white, throat and lower 3 tepals marked with red spots; dorsal tepal slightly inclined, other tepals spread widely. Perianth tube long, up to 35mm. **Anthers** white above, purple below. **Pollen** cream. **Capsules** unknown. **Seeds** unknown. **Scent** unscented.

*It has deep red spots on the lower tepals.*

**DISTRIBUTION** Known only from the southern Cape coast between Stanford and Cape Agulhas.

**ECOLOGY & NOTES** Recorded from sea level to an altitude of 250m. Plants are always in cracks in limestone or in calcareous sands. They flower best in the first few years after a fire.

**POLLINATORS** Probably long-tongued flies.

**SIMILAR SPECIES** Distinctive, can only be confused with *G. debilis*.

| | Habitat | Leaves | Bracts | Markings |
|---|---|---|---|---|
| ***G. variegatus*** | almost at sea level in limestone cracks or calcareous sands | 3, up to 2.5mm wide, margins not thickened | 30–45mm, smooth | deep red spots |
| ***G. debilis*** | rocky sandstone in mountain fynbos | 4, 1–2mm wide, thickened midrib and margins | 15–28mm, ridged | diamond, spade or chevron shapes |

*Smooth bracts are longer than the perianth tube.*

*Anthers are purple on the underside.*

G. variegatus *growing in limestone cracks near Stanford, Western Cape.*

# Gladiolus vigilans

**_vigilans_ = the watcher or sentinel, named for its type locality, which has views over the Atlantic Ocean and False Bay.**

*Gladiolus vigilans* is known only from the Cape Peninsula. Pale pink flowers have white markings with dark outlines. It flowers from October to November.

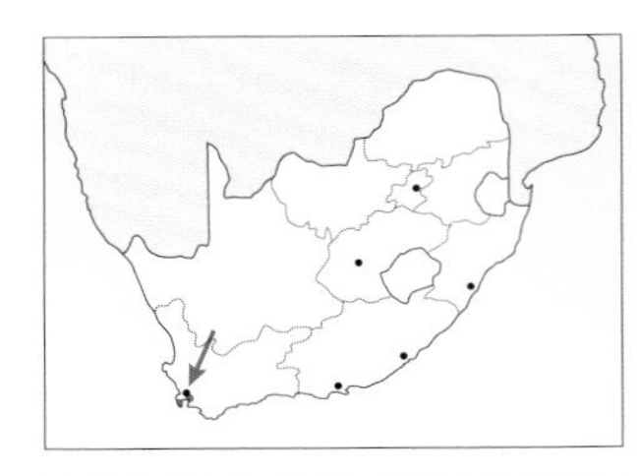

J F M A M J J A S O N D

**STATUS** Endangered

**DESCRIPTION** **Plant** 30–45cm. **Corm** woody, with segmented tunics. **Cataphylls** uppermost dark green or flushed with purple. **Leaves** 4, lower 2 basal and the longest, up to 1.5mm wide; margins and midrib thickened. **Spike** unbranched, 1–3 flowers. **Bracts** smooth, pale green. **Flowers** pale pink; dorsal tepal largest and inclined over the anthers, lower 3 tepals each have spade-shaped white marks outlined in dark red, with red marks and lines extending into the tube. Perianth tube up to 40mm, about twice as long as the bracts. **Anthers** dark purple. **Pollen** cream. **Capsules** ellipsoid. **Seeds** pale brown, evenly winged. **Scent** unscented.

*This species grows only on the Cape Peninsula.*

**DISTRIBUTION** Known only from a few hills at the southern tip of Cape Point in the Table Mountain National Park.

**ECOLOGY & NOTES** Grows among sandstone rocks in short fynbos. It seems to flower well even in unburned areas.

**POLLINATORS** Long-tongued flies, particularly *Philoliche rostrata* (a horse fly active in spring), are seen on the flowers.

*Flowers are usually pink but may be apricot coloured.*

**SIMILAR SPECIES** Originally described as a form of *G. carneus* (section *Blandi*), although they are not thought to be closely related. Related to *G. ornatus*, but they are unlikely to be confused.

| | Corm tunics | Leaves | Flowers |
|---|---|---|---|
| ***G. vigilans*** | woody | 4, up to 1.5mm wide, 2 basal and 2 sheathing the stem | pale pink with white nectar guides outlined in red, perianth tube up to 40mm long |
| ***G. carneus*** (sect. *Blandi*) | soft and papery | 4 or 5 in a fan, 6–14mm wide | white to deep pink with or without pink to red markings, perianth tube 38mm |
| ***G. ornatus*** | soft and membranous | 3, up to 2mm wide, 1 basal and 2 sheathing | dark pink with white nectar guides surrounded with darker pink, perianth tube up to 20mm |

*The dorsal tepal is the longest.*

*Tepals may be striped on the reverse.*

*Spear-shaped marks are outlined in red.*

*Smooth bracts cup long perianth tubes.*

G. vigilans *in old fynbos; Cape Point, Western Cape.*

# Gladiolus ornatus

***ornatus* = adorned, for the striking markings in the flower.**

*Gladiolus ornatus* has pale to deep pink flowers, appearing from September to October. It is restricted to the Cape Peninsula and nearby areas.

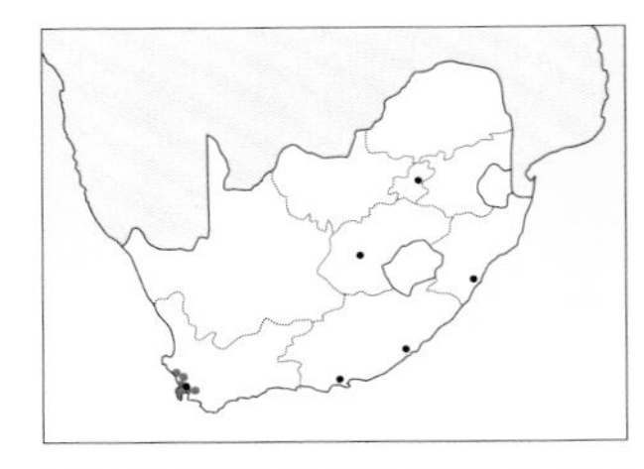



**DESCRIPTION** **Plant** 25–35cm. **Corm** globose, with soft membranous tunics. **Cataphylls** uppermost dark green or purplish. **Leaves** 3, 1–2mm wide; with thickened margins and midrib. **Spike** unbranched, 2 or 3 flowers. **Bracts** up to 30mm, dull green to purplish grey. **Flowers** pale to deep pink; lower tepals with white spear- to spade-shaped marks outlined in dark pink. Perianth tube up to 20mm. **Anthers** mauve. **Pollen** cream to mauve. **Capsules** ellipsoid, up to 20mm. **Seeds** pale brown, evenly winged. **Scent** unscented.

*Urban development has reduced the range of this species.*

**DISTRIBUTION** From the Mamre hills southwards to the Cape Peninsula and eastwards to the Jonkershoek Mountains of Stellenbosch.

**ECOLOGY & NOTES** Grows in damp to marshy areas in sandstone-derived soils. The species' habitat range has decreased substantially over the years owing to urban development, and it is now only commonly seen in protected areas such as Silvermine on the Cape Peninsula. The ones we photographed in Silvermine were flowering at the end of October, very late in the season.

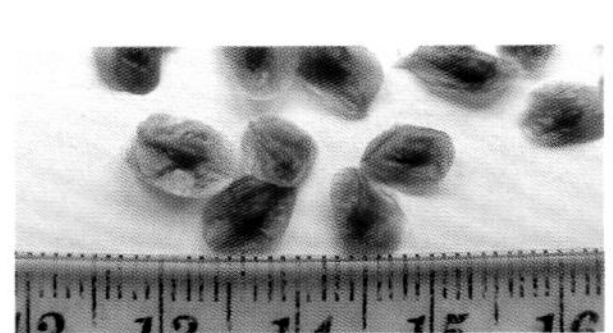

*Seeds are ±8mm, broadly and evenly winged.*

**POLLINATORS** Thought to be long-tongued bees.

**SIMILAR SPECIES** Most commonly seen in areas where *G. carneus* and *G. pappei* also occur. These species are thought not to be immediately allied to *G. ornatus* and are classified in section *Blandi*, but the flowers of all three species can be very similar and may lead to confusion.

| | Leaves | Perianth tube | Flowers |
|---|---|---|---|
| ***G. ornatus*** | 3, to 2mm wide, reaching to flower spike | 18–20mm | pink, lower tepals with spear- to spade-shaped white marks outlined in dark pink, white throat with pink streaks |
| ***G. carneus*** (sect. *Blandi*) | 4 or 5 in fan, up to 14mm wide, reaching to base of spike | 25–38mm | white to deep pink, lower tepals unmarked, or with pink to red spade or diamond markings |
| ***G. pappei*** (sect. *Blandi*) | 3 or 4, up to 3mm wide, reaching to base of spike | 30–35mm | pink, lower tepals with diamond- to heart-shaped white markings edged in red, base of throat red |

*Inclined spikes bear up to three flowers.*

*Bold, spear-shaped markings. Note the mauve anthers and pollen.*

G. ornatus *in one of the few sites left in its range at Silvermine on the Cape Peninsula, Western Cape.*

# Gladiolus inflexus

***inflexus* = bent abruptly, referring to the stem and spike.**

*Gladiolus inflexus* occurs in the southwestern Cape. The sweetly scented flowers are purple with cream tepals blotched with purple.

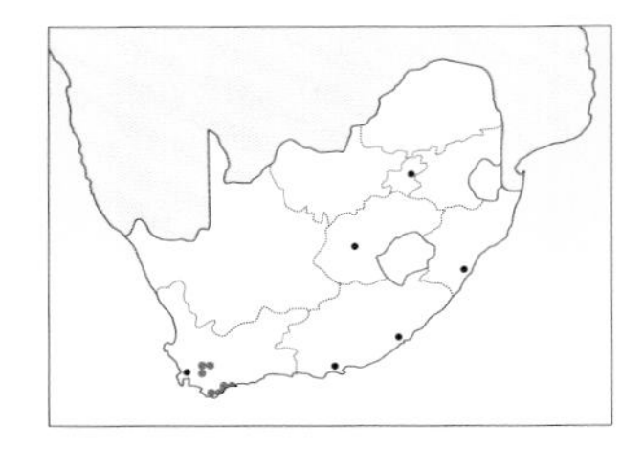

J F M A M J J A S O N D

**DESCRIPTION** **Plant** 20–35cm. **Corm** conic, with woody tunics. **Cataphylls** uppermost dark green or purple. **Leaves** 3, lowermost basal and the longest, up to 2mm wide, grey-green; midrib lightly raised and pale. **Spike** sharply flexed outwards in upper third above sheathing leaves, 1–3 flowers. **Bracts** up to 38mm, grey-green with transparent margins; tips long and sometimes twisted. **Flowers** purple with cream at base of tepals and throat; lower tepals speckled with purple, dorsal tepal arched. Perianth tube 20mm. **Anthers** purple-grey. **Pollen** cream. **Capsules** ellipsoid. **Seeds** yellowish brown, broadly winged. **Scent** sweetly scented.

**DISTRIBUTION** Known from the Breede River in the southwestern Cape, from the area around Worcester, from Potberg at the Breede River mouth and from Stilbaai.

**ECOLOGY & NOTES** Plants seem to grow in two very different habitats. Along the coast, they are found in soils derived from limestone, whereas in the Worcester area they grow in sandstone-derived soils among rounded pebbles. The flower colour also varies between these two populations: the coastal plants have paler flowers with violet streaks, whereas the Worcester population has purple and cream flowers with purple spots. Flowering is particularly responsive to fire.

**POLLINATORS** Long-tongued bees such as honeybees and *Anthophora* species.

**SIMILAR SPECIES** With its unusual purple and cream flowers, it is unlike any other species. It resembles *G. gracilis* in its corm tunics (which are hard and woody), flower shape and wiry flexuous stem, but the leaves and flower colour of *G. gracilis* are different.

*Flowers are in shades of purple. All the flowers on this spread are from populations near Worcester, Western Cape.*

*Dark mauve form.*

Immature seed capsules atop a flexed spike.

*Bracts are longer than perianth tube and may have twisted tips.*

*Maroon form.*

*Blue-purple form of* G. inflexus *growing in sandstone soils after a fire, near Worcester, Western Cape.*

# Gladiolus taubertianus

**Named after Paul Taubert (1867–1892), a German botanist.**

*Gladiolus taubertianus* has sweetly scented, pale mauve to blue flowers that appear in August and September. It is endemic to the mountainous regions of the Western Cape. Flowers are similar to *G. inflexus* but differ in markings.

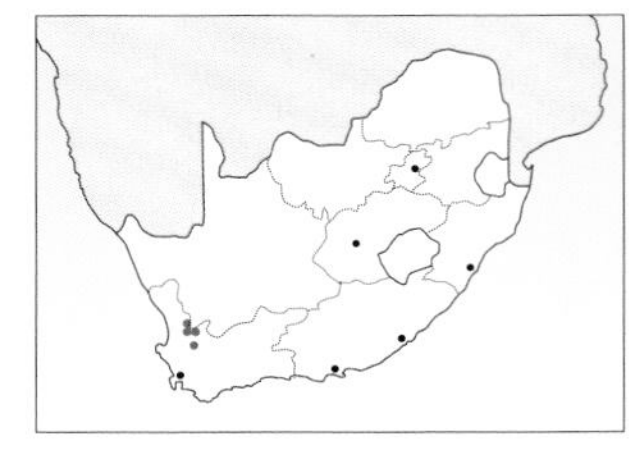

| J | F | M | A | M | J | J | A | S | O | N | D |
|---|---|---|---|---|---|---|---|---|---|---|---|

**STATUS** Rare

**DESCRIPTION** **Plant** 20–50cm. **Corm** globose, with hard layered tunics. **Cataphylls** uppermost purple and membranous. **Leaves** 3, narrow (up to 3mm wide), the lowermost as long as the spike or exceeding it; midrib raised and prominent. **Spike** flexed outwards above sheathing leaves, unbranched, 1–3 flowers. **Bracts** grey-green, inner bract twisted so that it lies under the outer bract. **Flowers** pale pinky mauve to blue with streaks of purple; lower tepals with pale yellow to cream transverse band outlined with darker mauve streaks. Perianth tube up to 14mm. **Anthers** pale lilac. **Pollen** cream. **Capsules** unknown. **Seeds** unknown. **Scent** lightly scented.

*There are streaks on all the tepals, with darker streaks on the lower ones.*

**DISTRIBUTION** Known only from the eastern and southern Cederberg in the southwestern Cape.

**ECOLOGY & NOTES** Grows among dry mountain fynbos in sandstone-derived soils, in areas that receive ±400–500mm of rain per annum. Often found among rocks, which presumably give them protection from porcupines and mole-rats. Although it is possible that they flower better after fires, we have seen them flowering well in mature fynbos on several occasions.

**POLLINATORS** Long-tongued bees.

**SIMILAR SPECIES** Closely related to *G. gracilis* and *G. inflatus* but in flower is unlikely to be confused with either.

| | Leaves | Flowers | Markings |
|---|---|---|---|
| ***G. taubertianus*** | blade linear, 2–3mm wide, midrib prominent | not bell shaped, pinky mauve | yellow transverse band with purple markings |
| ***G. inflexus*** | blade linear, 1–2mm wide, midrib raised | not bell shaped, shades of purple | purple streaks on a cream background |
| ***G. inflatus*** | blade oval to terete with 4 narrow longitudinal grooves | slightly bell shaped, pink to purple | spade- or spear-shaped markings on lower tepals |

*Very narrow leaves have a raised midrib.*

*Fine spikes bear up to three mauve flowers.*

*Dorsal tepals hood the stamens.*

*Mauve form.*

*Pale mauve form.*

*Pale pink form.*

*Pink form.*

G. taubertianus *grows in the dry sandstone soils of the Cederberg; in flower at Heuningvlei, Western Cape.*

# Gladiolus gracilis

***gracilis* = graceful; referring to its slender habit.**

Widespread in the southwestern Cape, *Gladiolus gracilis* has pale blue, grey or sometimes pink flowers with yellow lower tepals. It is sometimes confused with *G. caeruleus* but the latter is more floriferous and has different leaves.

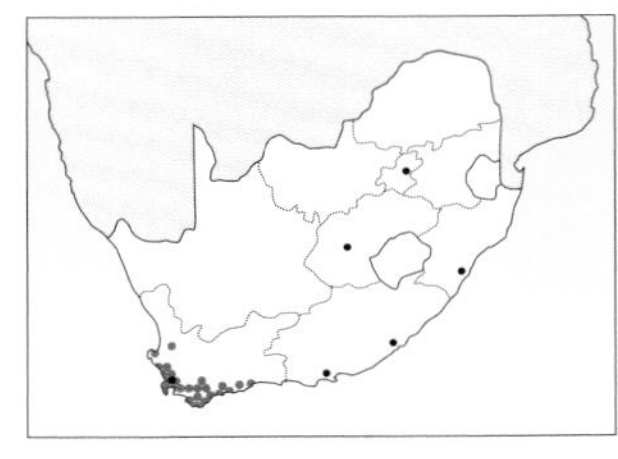

| J | F | M | A | M | J | J | A | S | O | N | D |
|---|---|---|---|---|---|---|---|---|---|---|---|

**DESCRIPTION** **Plant** 25–50cm. **Corm** tunics of hard concentric layers. **Cataphylls** uppermost green or flushed purple. **Leaves** 3 or 4, blade appears rounded but is linear, up to 2.5mm wide; margins raised into wings at right angles to the blade and arched inwards, margin edges usually rough or slightly hairy. **Spike** inclined at 45°, unbranched, 2–5 flowers. **Bracts** up to 40mm long, grey-green to dull purple, apices elongate and often twisted. **Flowers** very variable in colour from pale blue to grey to pink to dull yellow; lower tepals usually cream to yellow with dark purple streaks and spots, longer than upper tepals and extending downwards, often channelled; tepal margins undulate. Perianth tube up to 18mm. **Anthers** blue. **Pollen** cream or blue. **Capsules** oblong-ellipsoid. **Seeds** broadly and evenly winged, yellowish brown. **Scent** variable: unscented or sweetly scented.

*The plant has fine spikes bearing up to five delicate flowers with undulate tepals; flowers are very variable in colour.*

**DISTRIBUTION** A common species in the southwestern Cape, found from the Cape Peninsula eastwards to Albertinia and northwards to Darling and Hopefield.

**ECOLOGY & NOTES** Grows on heavier soils of clay or granite in renosterveld, but further north it is found on sandy soil in fynbos.

**POLLINATORS** Long-tongued bees such as honeybees and *Anthophora diversipes*.

**SIMILAR SPECIES** *G. gracilis* has similar flowers to *G. caeruleus*. When not in flower, it is vegetatively identical to *G. recurvus*.

| | Habitat | Leaves | Flowers |
|---|---|---|---|
| ***G. gracilis*** | usually in heavier clay or granitic soils | margins raised into wings which arch over leaf surface giving it a terete appearance | up to 5 flowers of various colours, funnel-shaped perianth tube to 18mm |
| ***G. caeruleus*** | sand in limestone outcrops or calcareous sands | margins raised into wings held at right angles to leaf surface | up to 14 pale blue flowers, marked with spots, lower tepals equal |
| ***G. recurvus*** | stony clay soils | as for *G. gracilis* above | 2 pearl to cream flowers, cylindric perianth tube up to 36mm, long narrow tepals with twisted apex |

*Brownish mauve form, Napier.*

*Orange-brown form, Napier.*

*Blue form, Napier.*

*Yellow-grey form, Bot River.*

*Mauve form, Napier.*

*Pink form, Stanford.*

# Gladiolus caeruleus

***caeruleus*** **= clear blue, named for the colour of the flower.**

*Gladiolus caeruleus* grows only on the West Coast of the Western Cape. It is scented and has many pale blue flowers on flexed spikes. It flowers from August to mid-September.

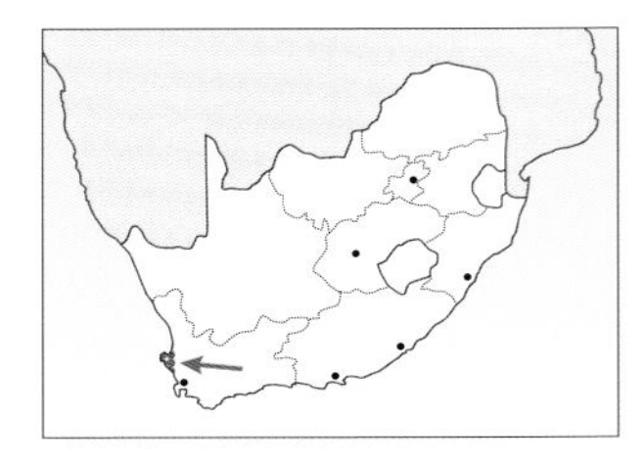

J F M A M J J A S O N D

**STATUS** Near Threatened

**DESCRIPTION** **Plant** 40–60cm. **Corm** tunics of hard concentric layers. **Cataphylls** uppermost green. **Leaves** 4, 3–7mm wide, margins extended at right angles to the surface, H-shaped in section. **Spike** flexed outwards at 45° above sheathing leaves, unbranched, 8–14 flowers. **Bracts** grey-green with membranous margins. **Flowers** pale blue with a darker line along the midline of each tepal; lower 3 tepals longer than upper and speckled with dark purple on a paler background. Perianth tube ±15mm. **Anthers** pale blue. **Pollen** cream or blue. **Capsules** ellipsoid. **Seeds** translucent light brown. **Scent** lightly scented.

**DISTRIBUTION** Known only from the West Coast of the southwestern Cape, from Yzerfontein in the south to St Helena Bay in the north.

**ECOLOGY & NOTES** All the known localities are close to the coast in sandy soils in outcrops of limestone, or in calcareous sands among coastal fynbos.

**POLLINATORS** Long-tongued bees.

*The plant has pale blue flowers.*

**SIMILAR SPECIES** Closely allied to *G. gracilis* and *G. recurvus*, both of which share its unusual leaf structure. The distribution of *G. caeruleus* overlaps with that of *G. gracilis*.

| | Habitat | Leaves | Flowers |
|---|---|---|---|
| ***G. caeruleus*** | sand in limestone outcrops or calcareous sands | margins raised into wings held at right angles to leaf surface | up to 14 pale blue flowers marked with spots, lower tepals equal |
| ***G. gracilis*** | usually in heavier clay or granitic soils | margins raised into wings which arch over leaf surface giving it a terete appearance | up to 5 flowers of various colours, funnel-shaped perianth tube up to 18mm |
| ***G. recurvus*** | stony clay soils | margins raised into wings which arch over leaf surface giving it a terete appearance | 2 pearl to cream flowers, cylindrical perianth tube up to 36mm, long narrow tepals with twisted apex |

*Leaves are H-shaped in cross section.*

*Bracts are long and grey-green.*

G. caeruleus *growing near Churchhaven in the West Coast National Park, Western Cape.*

# Gladiolus recurvus

***recurvus*** **= recurving, referring to the tips of the tepals, which are elongate, tapering and recurved.**

*Gladiolus recurvus* grows in the mountainous regions of the southwestern parts of the Western Cape, flowering from June to August. Flowers are pearly grey to yellowish and strongly scented.

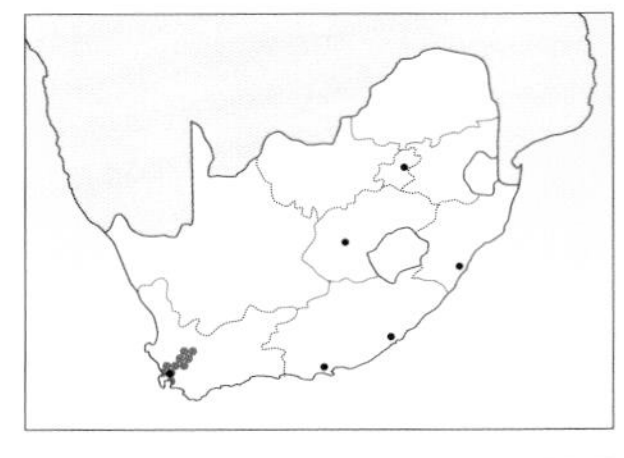

| J | F | M | A | M | J | J | A | S | O | N | D |
|---|---|---|---|---|---|---|---|---|---|---|---|

**STATUS** Vulnerable

*The pearl-coloured, strongly scented flowers are pollinated by moths.*

**DESCRIPTION** **Plant** 25–35cm. **Corm** globose, with tunics in hard, concentric layers. **Cataphylls** uppermost green. **Leaves** 4, the lowest leaf the longest, 2mm wide; margins are raised into wings extended at right angles to the blade and arching inwards. **Spike** flexed outwards above leaf sheaths, unbranched, with 1–4 flowers. **Bracts** 30–40mm long with apex twisted and in-rolled margins, dull green to greyish. **Flowers** pearl-grey to yellowish and cream; all tepals with purple spots along the midline, lower tepals irregularly spotted with purple or red, tepal margins undulating, with elongate tips that are twisted outwards and downwards. Perianth tube long, up to 36mm. **Anthers** dull grey to black. **Pollen** pale yellow. **Capsules** ovoid. **Seeds** golden brown, broadly winged. **Scent** strongly sweetly scented.

**DISTRIBUTION** A Vulnerable species owing to agricultural expansion and alien invasives. It is known from Citrusdal, southwards to the Witzenberg Mountains near Tulbagh and the Ceres Valley, and further south through the Breede River Valley to Somerset West.

**ECOLOGY & NOTES** Usually on lower slopes in heavy clay soils in renosterveld. Plants at higher elevations on the Kouebokkeveld flower later in spring; those at lower elevations flower in late winter and early spring.

**POLLINATORS** Moths pollinate these cream-coloured, long-tubed scented flowers.

**SIMILAR SPECIES** When in flower, can be mistaken for *G. liliaceus* and perhaps *G. maculatus*. When not in flower it is vegetatively identical to *G. gracilis*.

| | Leaves | Flowers |
|---|---|---|
| ***G. recurvus*** | winged margins arching inwards; H-shaped in section | cream, pearl-grey or pale lilac, tepal tips tapering, recurved, with undulate margins, perianth tube 27–36mm |
| ***G. maculatus*** | no thickened margins or midribs | dull yellow to lilac with brown speckles, perianth tube 23–35mm |
| ***G. liliaceus*** | margins and midrib strongly thickened and raised | translucent cream to rust coloured, tepal tips elongate, tapering, with undulate margins, perianth tube 40–53mm |
| ***G. gracilis*** | winged margins arching inwards; H-shaped in section, leaves appear terete | blue, grey, pink, dull yellow, tepal tips short, perianth tube 12–18mm |

*Long bracts have twisted tips.*

*Note the long bracts and perianth tubes.*

*The twisted, undulate, recurved tepals give the species its unusual shape, and name.*

*Dark anthers have yellow pollen.*

G. recurvus *growing in the Kouebokkeveld, Western Cape.*

# Gladiolus inflatus

***inflatus* = inflated, referring to the bell-shaped flowers.**

*Gladiolus inflatus* has pink to violet bell-shaped flowers with spear-shaped markings on lower tepals. It flowers from September to October, and is common in the mountain ranges of the Western Cape.

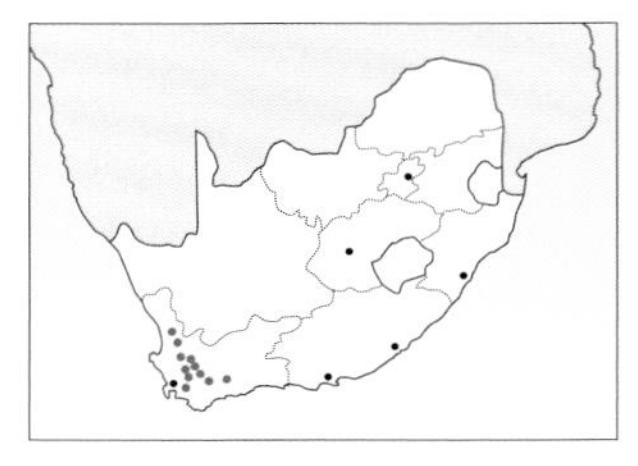

J F M A M J J A S O N D

**DESCRIPTION** **Plant** 20–45cm. **Corm** globose, with tunics of dark brown or blackish woody layers. **Cataphylls** pale or purple above ground, sometimes rough and may form a fibrous neck around stem. **Leaves** 3, lower 2 basal, sometimes slightly hairy, oval to terete with 4 hairline longitudinal grooves, 1mm diameter. **Spike** erect, unbranched, 1–3 flowers. **Bracts** pale green to greenish grey, with transparent veins. **Flowers** sometimes bell shaped, pink to shades of purple, rarely white; with dark spear- or spade-shaped markings on lower tepals. Perianth tube very variable in length, from 12–30mm. **Anthers** grey-purple. **Pollen** yellow. **Capsules** ellipsoid. **Seeds** broadly ovate, light brown, evenly winged. **Scent** unscented.

**DISTRIBUTION** Occurs from the Cederberg in the north, southwards to the Franschhoek Pass and the Riviersonderend Mountains, and eastwards to Matroosberg and the Langeberg.

**ECOLOGY & NOTES** Found in fynbos on stony sandstone soils, *G. inflatus* flowers particularly well in the first and second years after fire. However, it is also found flowering in mature vegetation. Plants at higher elevations flower later in the summer.

*Spikes are unbranched.*

**POLLINATORS** The anthophorid bee *Amegilla spilostoma* pollinates the short-tubed flowers, while the long-tubed flowers are thought to be pollinated by long-tongued flies.

**SIMILAR SPECIES** It can be difficult to identify this species because it is very variable in flower form and markings, and length of perianth tube. It can be mistaken for other bell-like flowers, such as *G. rogersii*, but it is most likely to be confused with *G. patersoniae* and, in the area around Ceres where a long-perianth form is found, with *G. cylindraceus*.

| | Leaves | Flowers | Perianth tube |
|---|---|---|---|
| ***G. inflatus*** | 1mm diameter, oval to terete, 4 hairline grooves | pink to purple, may be bell shaped, dark markings on lower tepals | 12–30mm |
| ***G. cylindraceus*** | 2mm, cross shaped in section, wide grooves | creamy pink to pink, long narrow tube with tepals at right angles, red markings | 25–52mm |
| ***G. patersoniae*** | 2mm, narrow grooves | blue to grey, bell shaped, yellow marks on lower tepals | 10–15mm |

*Note the bell-shaped flower.*

*Erect spikes bear up to three flowers, Waboomsberg.*

*Pink form, Franschhoek.*

*White form, Kouebokkeveld.*

*Pale purple form, Cederberg.*

*Mauve form, Cederberg.*

G. inflatus *growing high in the Matroosberg Mountains, Western Cape.*

# Gladiolus cylindraceus

***cylindraceus* = cylindrical, referring to the long slender shape of the perianth tube.**

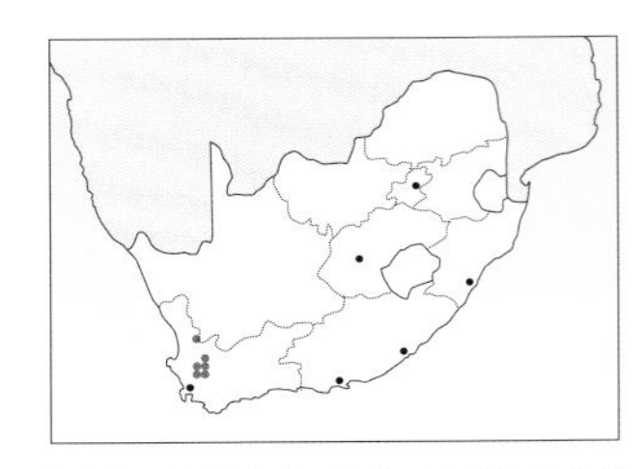

| J | F | M | A | M | J | J | A | S | O | N | D |
|---|---|---|---|---|---|---|---|---|---|---|---|

*Gladiolus cylindraceus* occurs high in the mountains of the Western Cape, where it flowers from mid-December to January. Flowers are pink to salmon, with reddish markings on the lower tepals.

**DESCRIPTION** **Plant** 30–50cm. **Corm** globose; tunics consist of light brown wiry fibre, with the previous season's corms below the current one. **Cataphylls** pale to purplish above ground, forming a fibrous neck at base of stem. **Leaves** 3, blade 2–3mm, cross shaped in section with wide grooves. **Spike** flexed at base and bent, with 5–8 flowers. **Bracts** pale greenish grey. **Flowers** pale creamy pink to salmon, tepals are unequal, wider at the top, with lower tepals longer in profile, tepals are straight and sometimes at right angles to the perianth tube, lower 3 tepals have a reddish diamond-shaped mark, some with white to yellow centre in lower midline. Perianth tube is very long, 35–52mm. **Anthers** white. **Pollen** yellow or cream. **Capsules** ovoid. **Seeds** ovoid, unevenly winged, 6mm. **Scent** unscented.

*Lower tepals have reddish diamond marks.*

**DISTRIBUTION** Localised and endemic to the Ceres district of the southwestern Cape, from Great Winterhoek along the Witzenberg and Skurweberg ranges to Mosterthoek peaks.

**ECOLOGY & NOTES** Strictly a montane species, which has not been recorded below 1,500m. It grows on rocky sandstone slopes and ridges in low fynbos vegetation.

*Pink form, Waboomsberg.*

**POLLINATORS** Long-proboscid flies.

**SIMILAR SPECIES** Distinguished from *G. inflatus* by the shape of the leaves, and from other similar species by the length of the perianth tube.

| | Leaves | Flowers | Perianth tube |
|---|---|---|---|
| ***G. cylindraceus*** | 2mm, cross shaped in section, wide grooves | creamy pink to pink, long narrow tube with tepals at right angles, red markings | 25–52mm |
| ***G. inflatus*** | 1mm diameter, oval to terete, 4 hairline grooves | pink to purple, may be bell shaped, dark markings on lower tepals | 12–30mm |
| ***G. patersoniae*** | 2mm, narrow grooves | blue to grey, bell shaped, yellow marks on lower tepals | 10–12mm |

*Leaves are narrow and cross shaped in section.*

*The cylindrical perianth tube gives the species its name.*

*Seeds are dark brown and unevenly winged.*

*These mature capsules have split, shedding seed.*

*Pale salmon form, Hansiesberg.*

*Yellow-apricot form, Witzenberg.*

G. cylindraceus *grows above 1,500m, here flowering at Waboomsberg beyond Ceres, Western Cape.*

# Gladiolus nigromontanus

***nigromontanus*** **= of the black mountain, refers to the Swartberg.**

Restricted to the Swartberg Mountains in the Western Cape, *Gladiolus nigromontanus* has white to pale pink flowers with red-streaked, channelled tepals. When in flower, from February to May, it can be confused with *G. engysiphon*.

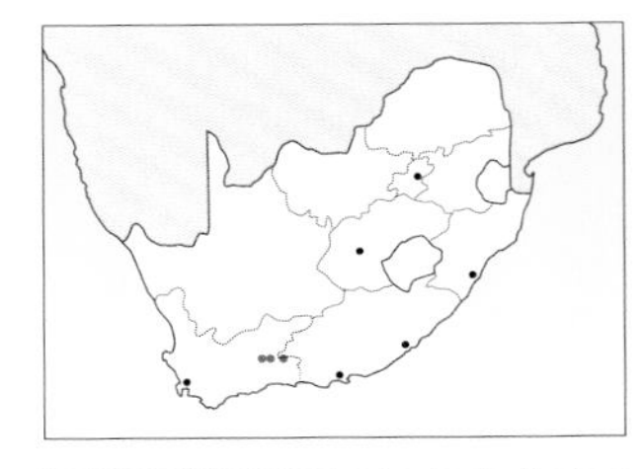

| J | F | M | A | M | J | J | A | S | O | N | D |
|---|---|---|---|---|---|---|---|---|---|---|---|

**STATUS** Rare

**DESCRIPTION** **Plant** 30–40cm high. **Corm** globose, with coarse, netted tunics. **Cataphylls** membranous, uppermost green and extending just above ground. Upper cataphylls and corm tunics form a neck around the stem. **Leaves** 3, the lower basal is the longest leaf, sheathing the lower half of the stem; the other two leaves sheath the stem almost completely. **Spike** inclined, more or less straight, with 3–6 flowers. **Bracts** 14–22mm, pale green or purple with brown tips. **Flowers** white to pink; with unequal tepals, top tepals larger, inclined to hooded over stamens, upper lateral tepals are lanceolate, extended forwards then outwards and upwards, lower tepals narrow and curving downwards; the lower 3 tepals have dark red longitudinal streaks, widest at the base of the tepal. Funnel-shaped perianth tube, 20–25mm. **Anthers** dark purple. **Pollen** cream. **Capsules** obovoid with rounded apices. **Seeds** pale brown, small, evenly winged. **Scent** unscented.

*The unscented flowers suggest pollination by long-tongued flies.*

**DISTRIBUTION** Endemic to the Swartberg range in the Western Cape.

**ECOLOGY & NOTES** Flowers poorly in the first season after fire, then strongly thereafter for two to three seasons until fynbos growth shades out the plants and flowering is poor. We photographed plants on south-facing slopes.

**POLLINATORS** Unknown, possibly long-tongued flies

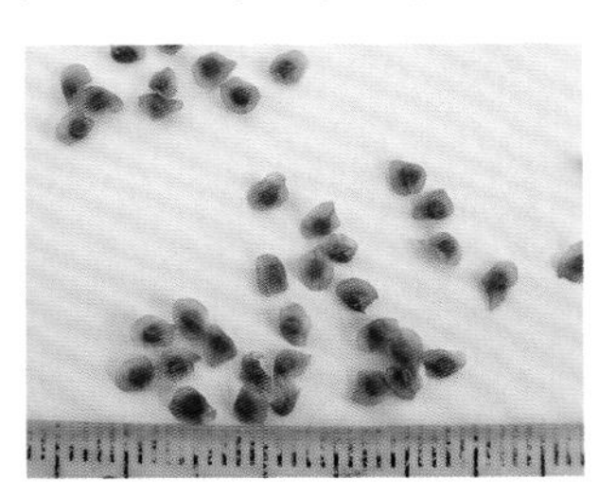

*Small seeds are evenly winged.*

**SIMILAR SPECIES** When not in flower it can be confused with *G. mutabilis*, from which it is distinguished by leaf shape. Has similar flowers to *G. engysiphon*, but *G. nigromontanus* can be distinguished by its shorter perianth tube and longer leaves.

| | Leaves | Flowers |
|---|---|---|
| ***G. nigromontanus*** | 3, sheathing, midrib and margins heavily thickened | white to pale pink flowers with red-streaked, channelled tepals, perianth 20–25mm |
| ***G. engysiphon*** | 3, sheathing, blades shorter than *G. nigromontanus* | white to cream, sometimes flushed with mauve or pink, perianth long, 40–60mm |
| ***G. mutabilis*** | 4, fleshy, no raised midrib or margins | blue, mauve to brown and cream, perianth enclosed in bracts |

*Lower tepals are clawed.*

*Perianth tubes are much longer than the bracts.*

G. nigromontanus *growing on south-facing slopes on the Swartberg Mountains, Western Cape.*

# Gladiolus engysiphon

***engysiphon* = elongate tube, referring to the shape of the flowers.**

*Gladiolus engysiphon* occurs in the Langeberg of the southern Cape. It flowers from March to April. Flowers are white to cream with a red median streak.

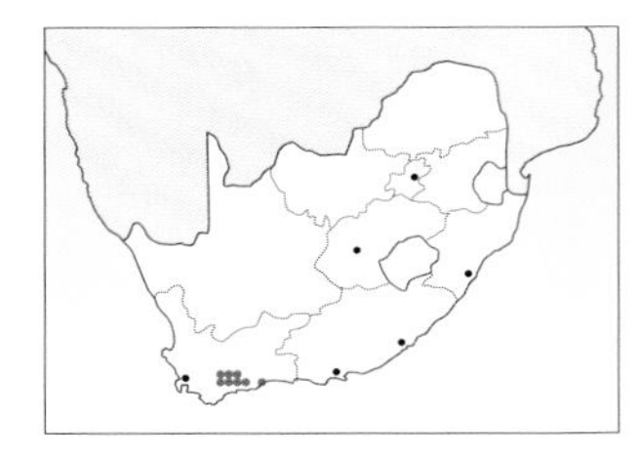

| J | F | M | A | M | J | J | A | S | O | N | D |
|---|---|---|---|---|---|---|---|---|---|---|---|

**STATUS** Vulnerable

**DESCRIPTION** **Plant** 35–45cm. **Corm** ovoid, with light brown, papery, layered tunics. **Cataphylls** pale purplish and membranous above ground, older plants may have a neck around the stem. **Leaves** 3, sheathing the stem, with basal leaf sheathing the lowest two-thirds of the stem. Non-flowering plants have a single terete leaf. **Spike** flexed at base, with 2–6 flowers. **Bracts** dull green-grey to brown. **Flowers** white or cream, sometimes with mauve flush, may have pink on outside of perianth tube; tepals are unequal and narrowly lanceolate with dorsal the largest, upper lateral and lower 3 tepals all have a red streak in lower midline. Perianth tube is nearly cylindrical and very long, 40–60mm. **Anthers** purple. **Pollen** cream. **Capsules** obovoid. **Seeds** ovate, evenly winged. **Scent** unscented.

**DISTRIBUTION** Endemic to the winter-rainfall areas at the foot of the Langeberg Mountains between Swellendam and Riversdale. The species is Vulnerable, owing to habitat loss from farming.

**ECOLOGY & NOTES** Occurring in clay soils or clay loam of the southern Cape coastal plain. Flowers in autumn, before winter rains begin, and continues growing through the wet season.

**POLLINATORS** Long-tongued fly, *Prosoeca longipennis*.

**SIMILAR SPECIES** Closely related to *G. nigromontanus*. The perianth tube of *G. engysiphon* is longer and the leaf blades are shorter than in *G. nigromontanus*. *G. bilineatus* (section *Blandi*) grows in the same area and has a similar flower.

*Spikes bear up to six flowers.*

*Lower tepals have a red mark.*

*The perianth tube is cylindrical and very long, giving the flower a distinctive shape.*

*Purple anthers have cream pollen.*

G. engysiphon *flowers in autumn; seen here at Suurbraak near Swellendam, Western Cape.*

# Gladiolus patersoniae

**Named after Florence Paterson (1869–1936), who made the type collection in the Baviaanskloof Mountains in 1910.**

*Gladiolus patersoniae* is widespread in the winter-rainfall areas. Sweet-scented, bell-like blue flowers appear from August to September at lower altitudes and October to November at higher ones.

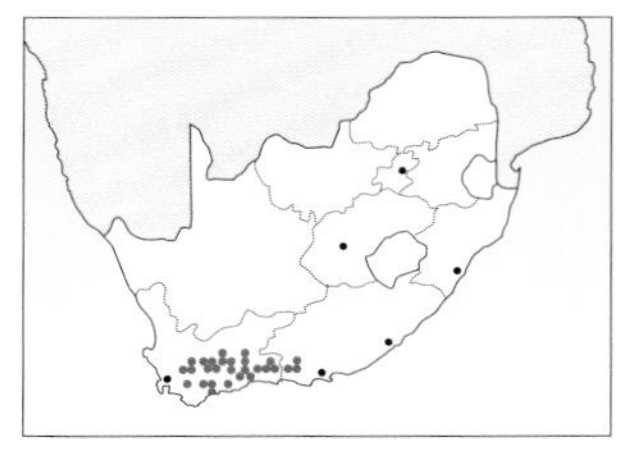

J F M A M J J A S O N D

*Lower tepals are marked with a yellow band.*

**DESCRIPTION** **Plant** 30–60cm. **Corm** globose, with tunics of hard, coarse fibres. **Cataphylls** membranous, upper cataphylls often concealed by corm fibres accumulated around the stem. **Leaves** 3, a distinguishing feature, lowest leaf the longest, 2mm wide. **Spike** inclined, flexuous, with 2–4 and occasionally 5 flowers. **Bracts** grey-green, often flushed with light purple. **Flowers** bell-like, ranging in colour from purple or deep blue through slate-grey to cream; tepals are unequal, the dorsal tepal (hooded over the stamens) is usually more intensely coloured, lower 3 tepals with yellow band across distal third. Perianth tube is 10–12mm. **Anthers** whitish. **Pollen** whitish. **Capsules** obovoid-ellipsoid. **Seeds** ovate. **Scent** with strong apple and carnation scent.

**DISTRIBUTION** Widespread in winter-rainfall areas, from Worcester in the southwestern Cape eastwards to the Baviaanskloof Mountains in the Eastern Cape. The wide range is unusual; plants occur both in the mountains and near the coast.

**ECOLOGY & NOTES** Exposed rocky habitats in interior ranges. Also at the coast near Cape Infanta and on stony alluvial flats near Worcester. Associated with soils derived from Cape sandstones.

**POLLINATORS** Long-tongued bees, *Anthophora diversipes*, which are large and swift flying, and are attracted by the sweet scent.

**SIMILAR SPECIES** Can be confused with other bell-like flowers. It is most closely related to *G. inflatus*, which also has terete leaves, but whose flowers differ in colour.

| | Leaves | Flowers | Perianth tube |
|---|---|---|---|
| ***G. patersoniae*** | 2mm, narrow grooves | blue to grey, bell shaped, yellow marks on lower tepals | 10–12mm |
| ***G. inflatus*** | 1mm diameter, oval to terete, 4 hairline grooves | pink to purple, may be bell shaped, dark markings on lower tepals | 12–30mm |

*The lowest leaf is longest and sheaths the stem base.*

*Spikes usually bear two to four flowers.*

*Pink form.*

*Purple form.*

*Cream form.*

*Pale blue form.*

*Mauve form.*

G. patersoniae *flowering near McGregor, Western Cape.*

# *Gladiolus subcaeruleus*

***subcaeruleus*** **= nearly blue, describing the pale blue flowers.**

*Gladiolus subcaeruleus* is confined to the southern areas of the Western Cape. Pale blue flowers with yellow median marks appear in March and April.

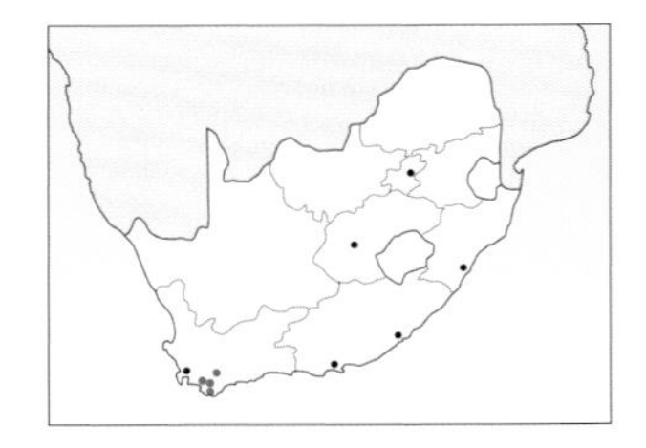

J F M A M J J A S O N D

**STATUS** Near Threatened

**DESCRIPTION** **Plant** 15–30cm **Corm** conic, with tunics of firm, papery layers. **Cataphylls** pale and membranous, uppermost is 2–5cm above ground, purple or dry and brown. **Leaves** 3, rarely 4, 60–80mm long; lowermost the longest. After flowering, a single foliage leaf is produced from a separate shoot. This remains green through winter. **Spike** slightly inclined, unbranched, with 3–5 flowers. **Bracts** grey-green. **Flowers** pale blue; lower lateral tepals have yellow transverse or spear-shaped median markings outlined in dark purple. Perianth tube 15–17mm. **Anthers** pale mauve. **Pollen** cream. **Capsules** obovoid, rounded apically. **Seeds** 7mm, ovate, broadly and evenly winged. **Scent** usually unscented, but may have a slightly metallic odour.

**DISTRIBUTION** Endemic to a narrow belt in the winter-rainfall region, from Bot River eastwards along the southern slopes of the Swartberg and Riviersonderend ranges to the town of Riviersonderend. Some populations found near Elim and Napier. The species is classified as Near Threatened owing to the encroachment of farming activities on its habitat.

**ECOLOGY & NOTES** Usually found in heavy clay soils near the interface of shale and sandstone strata.

**POLLINATORS** Long-tongued bees.

**SIMILAR SPECIES** *G. mutabilis* is sometimes mistaken for *G. subcaeruleus* but their distributions differ. It can be mistaken for *G. vaginatus*, whose distribution overlaps, but *G. subcaeruleus* lacks the dark blue streak that characterises *G. vaginatus*.

*This species usually grows on clay soils but is sometimes found in sandy soil; here in flower at Albertinia.*

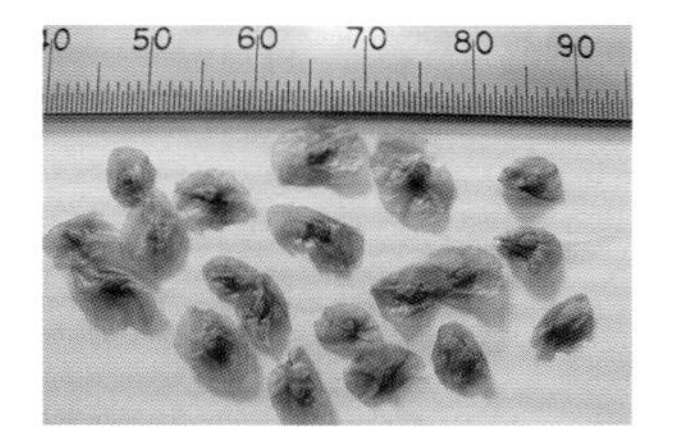


*Seeds are evenly winged.*

*Bracts are as long as the perianth tube.*

*The species is endemic to a narrow belt in the winter-rainfall region.*

*Lower tepals are distinguished by their yellow transverse bands outlined in darker purple.*

G. subcaeruleus *near Napier, Western Cape.*

# Gladiolus martleyi

**Named after JF Martley, who brought this species to the attention of botanists in the 1930s.**

*Gladiolus martleyi* has whitish or pale to deep pink to mauve flowers that appear in March and April. It is widespread in the Western Cape and extends just into the Northern Cape.

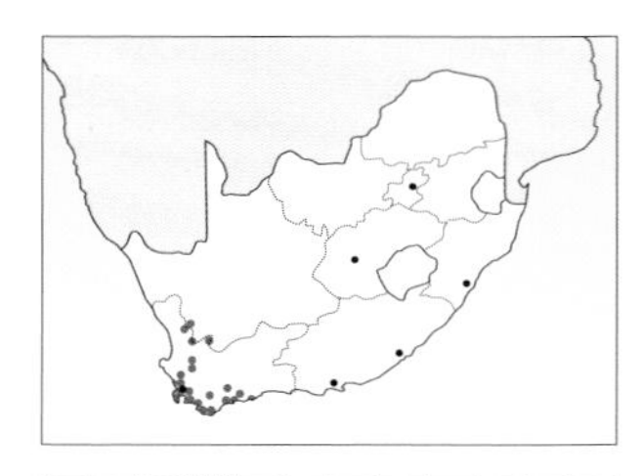


**DESCRIPTION** **Plant** 30–60cm. **Corm** globose, with papery tunics. **Cataphylls** extending up to 5cm above ground, green to dull purple, closely resembling leaves of flowering stem. **Leaves** 2 or 3, evenly spaced, sheathing the flowering stem entirely; 1 or 2 long foliage leaves are produced in autumn from a separate shoot after the flowers have faded. **Spike** lightly inclined, unbranched, 5–11 flowers. **Bracts** greyish to purplish green. **Flowers** pale to deep pink or lilac to mauve; lower lateral tepals have a spear-shaped yellow mark edged with darker pink or purple in upper half. Perianth tube 11–12mm. **Anthers** dark purplish blue. **Pollen** pale blue. **Capsules** narrowly ellipsoid. **Seeds** dark brown, ovate, with light brown and broad even wings. **Scent** often strongly sweetly scented, although may be unscented.

*The flowers usually have bold nectar guides.*

**DISTRIBUTION** Widespread but not common in winter-rainfall areas from the southwestern Cape to just into the Northern Cape. Extends from Bokkeveld escarpment southwards to the Cape Peninsula and eastwards to Albertinia.

**ECOLOGY & NOTES** Found in deep and sandy soils, usually in fairly rocky sites, but sometimes in heavier soils along interface of sandstone and shale strata of the Cape system.

*Pink form, near Riversdale.*

**POLLINATORS** Assumed to be pollinated by bees of genus *Amegilla*.

**SIMILAR SPECIES** In flower it is similar to *G. brevifolius* (section *Linearifolii*) and may also be confused with *G. jonquilliodorus*.

| | Corm | Leaves | Scent |
|---|---|---|---|
| ***G. martleyi*** | fleshy | 1 or 2 foliage leaves after flowering, needle-like with 4 fine grooves | usually scented |
| ***G. brevifolius*** (sect. *Linearifolii*) | hard, leathery | 1 foliage leaf after flowering, sword shaped, plane, hairy | seldom scented |
| ***G. jonquilliodorus*** | membranous | 2 or 3 foliage leaves after flowering, needle-like with 4 fine grooves | strongly scented |

*Inclined stems bear up to 11 small flowers.*

*The short perianth tube suggests bee pollination.*

*White form, Suurbraak.*

G. martleyi *in flower near Riversdale, Western Cape.*

# Gladiolus jonquilliodorus

***jonquilliodorus* = describing the strong, sweet scent, like a jonquil.**

The sweet-scented *Gladiolus jonquilliodorus* is found on the Cape Peninsula and West Coast, where it flowers in December and January. Flowers are cream to yellow to grey, with a transverse yellow band on the lower tepals.

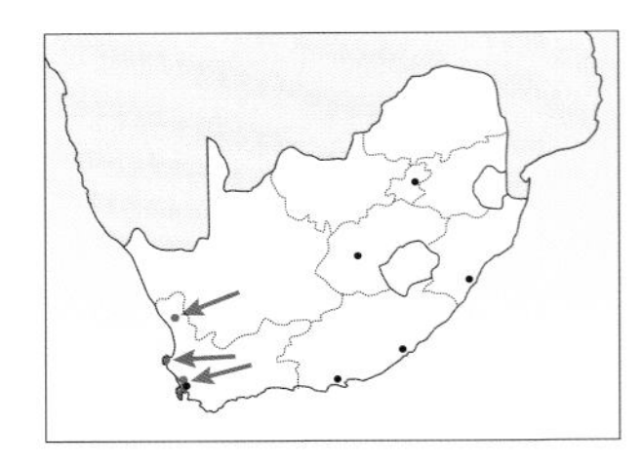

J F M A M J J A S O N D

**STATUS** Endangered

**DESCRIPTION** **Plant** 40–70cm. **Corm** globose, with membranous to papery tunics. **Cataphylls** up to 10cm above ground, often dry and brown. **Leaves** 2 or 3 on flowering stem, entirely sheathing; 2 or 3 terete foliage leaves of 30–50cm are produced after flowering. **Spike** unbranched, inclined, with 7–14 flowers. **Bracts** yellow-green to grey. **Flowers** cream, pale yellow or pearl grey, sometimes flushed with pink or light mauve; lower 3 tepals with transverse yellow band, often streaked with narrow purple lines below. Perianth tube 8–9mm long, enclosed in bracts. **Anthers** mauve. **Pollen** pale yellow. **Capsules** ellipsoid. **Seeds** ovate, light brown even wings. **Scent** sweetly scented by day.

*Spikes bear up to 14 small, funnel-shaped flowers.*

**DISTRIBUTION** A rare and fairly localised endemic of the extreme southwestern Cape, occurring on the Cape Peninsula and along the West Coast as far as Saldanha and Vredendal. Botanical surveys have demonstrated an extended range and new sites but it remains Endangered owing to coastal development and alien plant invasion.

**ECOLOGY & NOTES** Usually grows in deep alluvial sand among clumps of Restionaceae. Preference for habitat is not consistent; plants may grow on well-drained stony hill slopes.

*Lower tepals are clawed.*

**POLLINATORS** Long-tongued bees: *Apis mellifera, Amegilla spilostoma* and other small halictid (sweat) bees.

**SIMILAR SPECIES** Can be confused with *G. martleyi* and to a lesser extent with *G. brevifolius* (section *Linearifolii*).

| | Corm | Leaves | Scent |
|---|---|---|---|
| ***G. jonquilliodorus*** | membranous | 2 or 3 foliage leaves after flowering, needle-like with 4 fine grooves | strongly scented |
| ***G. martleyi*** | fleshy | 1 or 2 foliage leaves after flowering, needle-like with 4 fine grooves | usually scented |
| ***G. brevifolius*** (sect. *Linearifolii*) | hard, leathery | 1 foliage leaf after flowering, sword shaped, plane, hairy | seldom scented |

*Flowering plants do not have foliage leaves.*

*Bracts extend beyond the perianth tube.*

*Sweetly scented flowers are pollinated by bees.*

G. jonquilliodorus *growing in the mountains above Kommetjie on the Cape Peninsula, Western Cape.*

# *Gladiolus trichonemifolius*

***trichonemifolius* = with leaves like a *Trichonema* (a synonym for *Romulea*).**

*Gladiolus trichonemifolius* is found in the southwestern parts of the Western Cape. It bears cream to yellow zygomorphic flowers from August to September.

**DESCRIPTION** **Plant** 12–25cm. **Corm** globose, with woody to leathery layered tunics. **Cataphylls** uppermost barely above ground, dark purple. **Leaves** 3, lowermost the longest, sheathing in lower third. **Spike** erect, unbranched, flexuous, with 1–3 sometimes 4 flowers. **Bracts** dark green, veins almost transparent. **Flowers** cream to bright yellow; with outer tepals often flushed purple on reverse, upper 3 tepals are usually larger than lower, lower 3 tepals have paired brown lines on lower half, deeper yellow across the middle, and sometimes purplish at apices. Perianth tube 16–20mm long, tube mouth is edged with a purple star-shaped mark. **Anthers** pale yellow. **Pollen** cream. **Capsules** ovoid-ellipsoid. **Seeds** rich brown, ovate, broadly and evenly winged. **Scent** strongly sweetly freesia scented, although may be unscented.

**DISTRIBUTION** The range originally extended from Bredasdorp to Hopefield, but agriculture has considerably lessened its distribution and the species is now considered Vulnerable. There are abundant colonies on wetlands near Darling. We also found populations in Bot River and Sir Lowry's Pass.

**ECOLOGY & NOTES** Grows mostly at low altitudes but does occur in marshy lands in mountainous areas near Ceres. The plants we photographed in Bot River and Sir Lowry's Pass were growing in low-lying marsh lands.

**POLLINATORS** Long-tongued bees especially *Apis mellifera*.

**SIMILAR SPECIES** *G. sufflavus, G. pritzelii* and *G. delpierrei* all have yellow flowers but the zygomorphic flowers of *G. trichonemifolius* make it unlikely to be mistaken for any of them.

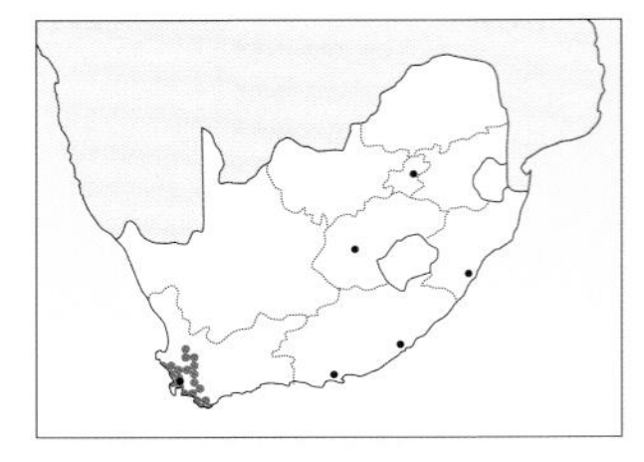

| J | F | M | A | M | J | J | A | S | O | N | D |
|---|---|---|---|---|---|---|---|---|---|---|---|

**STATUS** Vulnerable

*The habitat of this species has been disrupted by agriculture.*

*Flowers of this species are bilaterally symmetrical.*

*Visible cataphylls are purple.*

*Flowers may be white.*

*Green bracts have almost transparent veins.*

*The reverse side of the tepals is often flushed purple.*

*Flowers are often sweetly scented.*

G. trichonemifolius *flowering in marshlands near Sir Lowry's Pass, Western Cape.*

# Gladiolus sufflavus

***sufflavus* = pale yellow, named for the colour of the flowers.**

*Gladiolus sufflavus* is restricted to the Bokkeveld escarpment of the Northern Cape. Its pale yellow to greenish flowers are sweetly scented and appear from August to September.

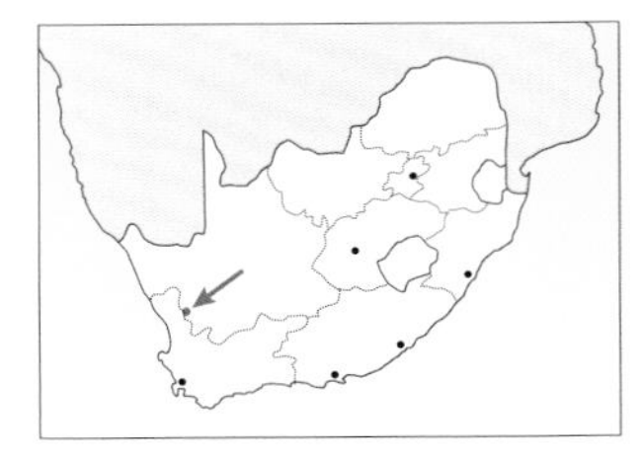

| J | F | M | A | M | J | J | A | S | O | N | D |
|---|---|---|---|---|---|---|---|---|---|---|---|

**STATUS** Vulnerable

**DESCRIPTION** **Plant** 45–70cm. **Corm** leathery tunic becoming fibrous with age. **Cataphylls** upper 2 are purple and reaching above ground. **Leaves** usually 4, lower 2 basal, lowermost is the longest, reaching between base and mid-spike, with a terete blade. Other leaves are hairy. **Spike** inclined, unbranched, with 4–6 flowers. **Bracts** olive-green. **Flowers** pale yellow to greenish; tepals form a bell, dorsal tepal is the largest, and hooded over stamens, lower tepals have obscure, dark yellow, greenish or brown markings. Perianth tube 14–15mm. **Anthers** purple. **Pollen** cream. **Capsules** ovoid-ellipsoid. **Seeds** ovate, broadly and evenly winged. **Scent** sweet, lily-like fragrance.

**DISTRIBUTION** Restricted to Bokkeveld escarpment in the Northern Cape.

**ECOLOGY & NOTES** Occurs in seasonally waterlogged soils and marshy soils. Most commonly seen after fires but does flower in old growth. Despite its unusual colouring, it can be difficult to spot as it easily blends into the surroundings.

**POLLINATORS** Long-tongued bees. The anthophorid bee *Anthophora diversipes* is thought to be the main pollinator.

**SIMILAR SPECIES** Could be confused with *G. pritzelii* but flowers differ in colour and markings.

*The yellow-green flowers are unmistakable.*

*Spikes bear up to six flowers.*

*Basal leaf is terete, other leaves are hairy.*

*Bracts encase the perianth tube.*

*Flowers alternate on spikes.*

*Flowers have a bell-like shape.*

G. sufflavus *flowering in old growth at Nieuwoudtville, Northern Cape.*

# *Gladiolus pritzelii*

**Named after EG Pritzel (1875–1946), a 19th-century botanist.**

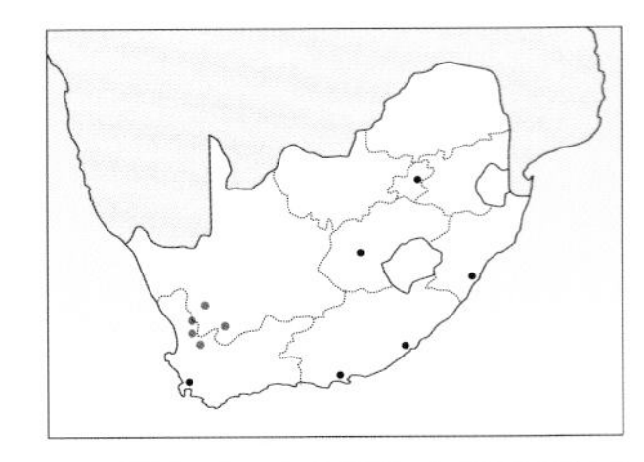

| J | F | M | A | M | J | J | A | S | O | N | D |
|---|---|---|---|---|---|---|---|---|---|---|---|

Occurring in the Bokkeveld of the Northern Cape, and the Cederberg and western Karoo regions of the Western Cape, *Gladiolus pritzelii* flowers in August and September. Its yellow bell-like flowers have red markings. In the Cape mountains, they tend to flower after fires, while in the Karoo, flowering is rainfall dependent.

**DESCRIPTION** **Plant** 20–60cm. **Corm** globose, woody to leathery layered tunics fragmenting into coarse teeth or fibres. **Cataphylls** green or purple above ground, often dry at apex. **Leaves** 3, lowermost basal and reaching base of spike. **Spike** lightly inclined, flexuous, erect to inclined and unbranched, 1–3 flowers. **Bracts** pale to olive-green. **Flowers** yellow; unequal tepals form a bell, upper tepals are ovate, with the large dorsal arched over stamens, lower lateral tepals have a bright yellow transverse band outlined in red, purple or brown, the lower median with a spade-shaped yellow mark outlined in red or purple. Perianth tube 11–12mm. **Anthers** reddish brown. **Pollen** yellow. **Capsules** unknown. **Seeds** ovoid, broadly and evenly winged, 10mm. **Scent** sweetly scented.

*Yellow flowers are inflated and bell shaped.*

**DISTRIBUTION** Southern parts of the Northern Cape and northern mountain ranges of the Western Cape, extending from the Bokkeveld escarpment down to the southern Cederberg. It also occurs on Hantamsberg near Calvinia and in the Roggeveld escarpment in the western Karoo, Northern Cape.

**ECOLOGY & NOTES** Grows in rocky habitats in thin sandstone soils; flowers best after fires, except in the western Karoo. There, plants grow on Beaufort sandstone soils and their flowering is related to rainfall, not fire. Despite these differences, *G. pritzelii* shows no sign of local differentiation.

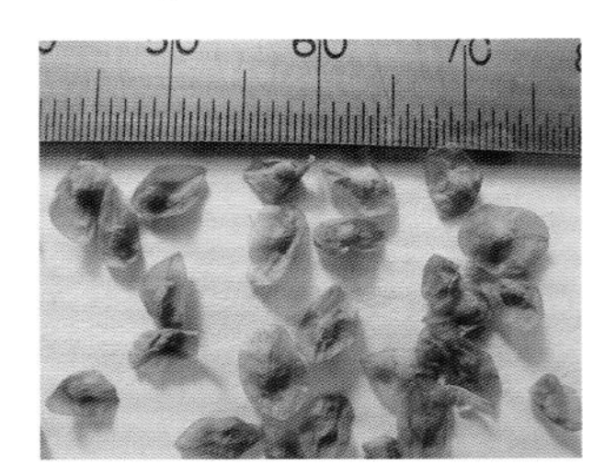

*Seeds are large and evenly winged.*

**POLLINATORS** Has not been studied but presumed to be pollinated by anthophorid bees.

**SIMILAR SPECIES** Although thought to be closely related to *G. delpierrei*, it is unlikely to be confused with it because both species have distinctive flower markings.

*Bracts are considerably longer than the perianth tube.*

*Short-tubed flowers are presumed to be bee pollinated.*

*Flowers may lack the usual markings on lower tepals.*

*Karoo plants depend on rainfall rather than fire.*

# Gladiolus delpierrei

**Named after George Delpierre, professor of biochemistry and a bulb enthusiast who co-authored *The Winter-growing Gladioli of South Africa* with N du Plessis in 1974.**

*Gladiolus delpierrei* is known only from the high regions of the Cederberg, where it flowers in December and early January. Flowers are creamy yellow.

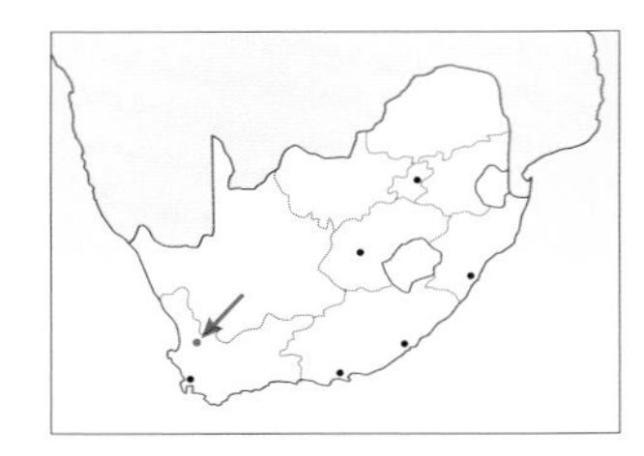

J F M A M J J A S O N D

**STATUS** Rare

**DESCRIPTION** **Plant** 55–65cm. **Corm** globose, with finely fibred tunics. **Cataphylls** uppermost reaching to 8cm above the ground. **Leaves** 3, scabrid, plane; with thickened margins and midrib. **Spike** inclined, unbranched, with 5–7 flowers. **Bracts** pale grey-green or flushed light purple on upper sides. **Flowers** short tubed, creamy yellow; with unequal tepals, dorsal the largest, inclined over stamens, lower tepals are dark yellow in lower third with a pair of fine red lines at base, upper laterals have narrow red-streaked midline. Perianth tube up to 8mm, mouth of tube is dark red. **Anthers** light mauve. **Pollen** cream. **Capsules** ellipsoid-obovoid. **Seeds** ovoid, 7mm. **Scent** thought to be lightly scented.

**DISTRIBUTION** A narrow endemic of the Cederberg range in the Western Cape. The known only populations are at 1,300m on Sneeuberg and adjacent peaks.

**ECOLOGY & NOTES** Grows in leached sandstone soils in seepage areas that remain moist until midsummer, when plants flower.

**POLLINATORS** Unknown; thought to be long-tongued bees.

**SIMILAR SPECIES** Although closely related to *G. pritzelii*, it is unlikely to be confused with it.

*The species grows only in the Cederberg Mountains.*

*The basal leaf has a ridged appearance.*

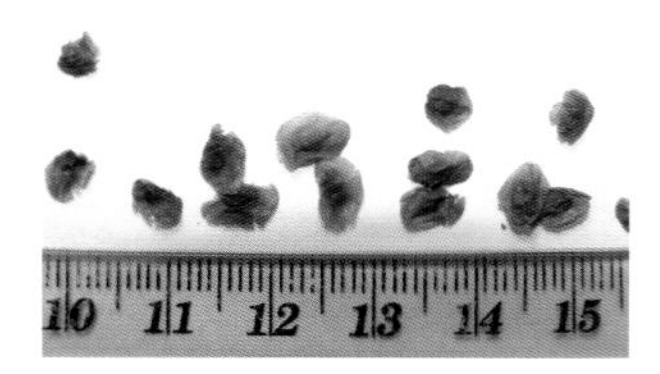


*Seeds are ovoid.*

*Flowers alternate on spikes.*

*Bracts completely enclose the short perianth tube.*

*Mature capsules have split.*

G. delpierrei *grows above 1,300m in the Cederberg, Western Cape.*

# Gladiolus hyalinus

***hyalinus*** **= translucent, referring to the edges of the upper tepals.**

*Gladiolus hyalinus* is widespread in the winter-rainfall area, where it flowers from June to mid-August. Cream and brown flowers are moth pollinated.

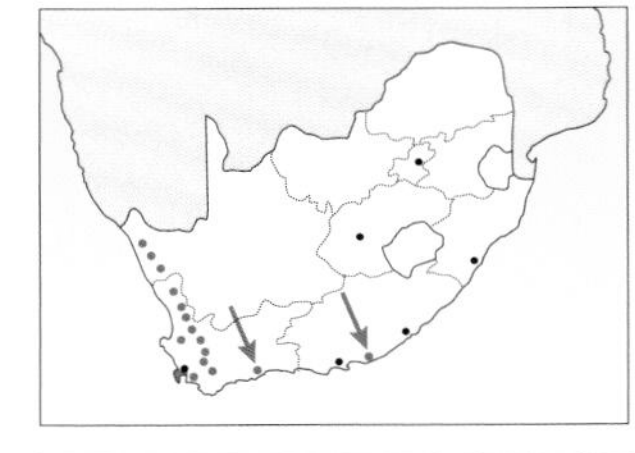

| J | F | M | A | M | J | J | A | S | O | N | D |
|---|---|---|---|---|---|---|---|---|---|---|---|

**DESCRIPTION** **Plant** 30–60cm. **Corm** globose-conic, with woody tunics. **Cataphylls** uppermost extending 3–5cm above ground, membranous and pale purple. **Leaves** 3, reaching to base of spike. **Spike** inclined, unbranched, with 1–3 or more flowers. **Bracts** pale green to grey. **Flowers** light to reddish brown on pale cream background; tepals usually darker along midlines, lower half of dorsal tepal is transparent as is the lower part of the throat. Perianth tube 25–36mm. **Anthers** long, cream. **Pollen** yellow. **Capsules** elongate-ellipsoid. **Seeds** ovate, broadly and evenly winged. **Scent** unscented, rarely sweetly scented in evenings.

*The species is widespread in the winter-rainfall region.*

**DISTRIBUTION** One of the most widespread species in the winter-rainfall region of South Africa. Most common in the southwestern part of the Western Cape, extending northwards to Nieuwoudtville in the Bokkeveld. Also occurs in Kamiesberg in central Namaqualand and near Steinkopf in the Richtersveld. It is poorly documented beyond the Riviersonderend Mountains although populations have been recorded in Plettenberg Bay and near Gqeberha (Port Elizabeth).

**ECOLOGY & NOTES** At higher elevations the plants continue to flower through September. Occurring in heavy soils in fynbos–renosterveld transition, between sandstone and shale. Populations also occur on granite or sandstone.

*Tepals are usually darker along the midlines.*

**POLLINATORS** Pollinated by moths.

**SIMILAR SPECIES** Most closely resembles *G. liliaceus*. The flowers of *G. maculatus* are similar but the leaves differ.

| | Flowers | Leaves |
|---|---|---|
| ***G. hyalinus*** | brown and cream with transparent tepal edges, odourless or sometimes scented | 3, linear, 1.7–2.5mm wide, margins and midrib thickened |
| ***G. liliaceus*** | large, light brown, pink-red, purple; tepal and sutures transparent; changes colour to blue-mauve at night, scented at night | 3, 2–4mm wide, plane with scabrid edges |
| ***G. maculatus*** | dull yellow to lilac, dorsal tepal transparent, scented day and night | 3 or 4, plane with fleshy blades |

*Leaves are grooved.*

*Perianth tubes bend abruptly.*

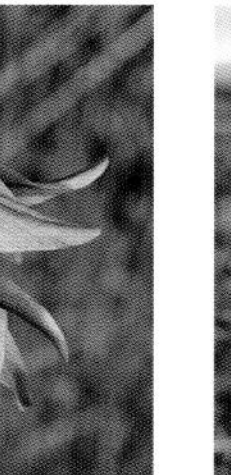

*Flowers are moth pollinated.*

*Dorsal tepal is partly translucent.*

*Lateral tepals wing outwards.*

*Nieuwoudtville, Northern Cape.*

G. hyalinus *flowering in the Kouebokkeveld, Western Cape.*

# Gladiolus liliaceus

***liliaceus* = lily-like, a reference to the large flower.**

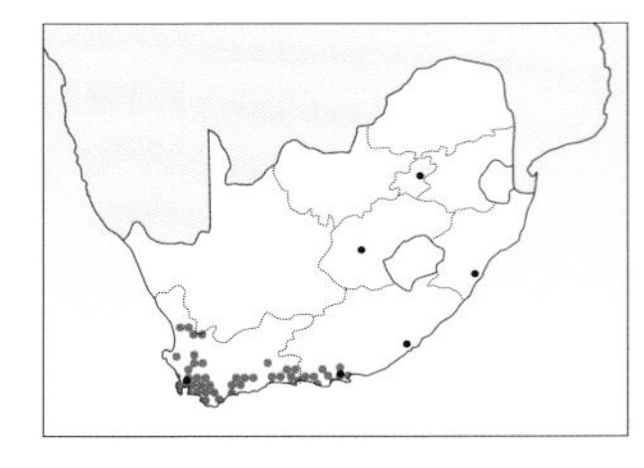

J F M A M J J A S O N D

*Gladiolus liliaceus* is widespread in the winter-rainfall areas. Flowers vary in colour from cream to brown, purple or greenish yellow. At night, tepals become mauve and the flowers become very sweetly scented, an adaptation for moth pollination.

**DESCRIPTION** **Plant** 30–80cm. **Corm** globose, with woody-leathery layered tunics. **Cataphylls** green and leathery above ground. **Leaves** 3, with lowermost the longest. **Spike** inclined, unbranched, with 1–4 flowers, sometimes more. **Bracts** pale green to dull grey. **Flowers** in a wide range of shades from light brown to dull pinkish red, purplish or cream to greenish yellow; throat is speckled with dark brown; tepals are darker along midlines and lower tepals are pale yellow to cream in lower third, lower edges of dorsal tepals and sutures between tepals are transparent, changing colour after nightfall and becoming blueish mauve and scented. Perianth tube 40–45mm. **Anthers** long and almost horizontal, brown or purple. **Pollen** pale yellow to cream. **Capsules** oblong-ellipsoid. **Seeds** ovate, broadly and evenly winged. **Scent** strongly sweet smelling and clove scented.

*Large flowers open fully in the evening, releasing a strong scent.*

**DISTRIBUTION** Restricted to winter-rainfall regions where it is relatively common in the southwestern and southern parts of the Western Cape. The range extends from the Cederberg Mountains to Gqeberha (Port Elizabeth), favouring heavy clay soils and lowland habitats.

**ECOLOGY & NOTES** Flowers mostly in August and September, but at higher elevations, flowering occasionally continues beyond September; one form flowers between November and December. Flowering physiology is unusual: tepals are partly closed and translucent brown by day, changing at sunset to translucent mauve, with tepals opening widely and flowers releasing a strong clove-like scent after sunset. This is considered an adaptation for moth pollination.

*Flowers are strongly scented at night, attracting moths.*

**POLLINATORS** Moths, including owlet moths (family Noctuidae) and hawk moths (family Sphingidae).

**SIMILAR SPECIES** Closely related to *G. hyalinus*, but the latter has smaller flowers and plane leaves.

*Leaf margins and midribs are thickened.*

*Tepals change colour to mauve in the evenings.*

*Inclined spikes bear up to four or more flowers.*

*Flowers are moth pollinated.*

G. liliaceus *on Van der Stel Pass in the Overberg, Western Cape.*

# Gladiolus tristis

***tristis* = sad, alluding to the pale colour of the flowers.**

*Gladiolus tristis* is widely spread across the winter-rainfall region. It only opens fully in the late afternoon to evening. The sutures of its yellow to greenish flowers are transparent. Weakly scented in the day, the flowers, which appear from September to November, become strongly fragrant at night and are pollinated by moths.

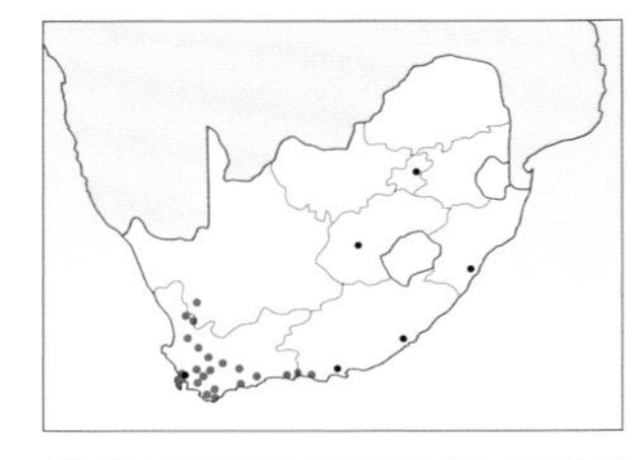

| J | F | M | A | M | J | J | A | S | O | N | D |
|---|---|---|---|---|---|---|---|---|---|---|---|

*It is widespread and easily recognised by its large flowers.*

**DESCRIPTION** **Plant** 40–120cm. **Corm** papery to woody tunics. **Cataphylls** up to 4cm above ground, purple to brown. **Leaves** 3. **Spike** more or less straight, with 2–4 flowers, sometimes more. **Bracts** pale to dark green, sometimes flushed with grey. **Flowers** pale yellow to greenish yellow or cream; tepals are darker on midline and may have purple to reddish median streaks; back of tepals is usually dark grey, with purple to reddish on midline and sometimes tips. Perianth tube long, 40–63mm. **Anthers** long, pale yellow to light purple. **Pollen** pale yellow. **Capsules** oblong-ellipsoid. **Seeds** ovate, broadly and evenly winged. **Scent** variable: weakly scented by day, with strong carnation and clove scent at night.

**DISTRIBUTION** Widespread in the winter-rainfall region, occurring from sea level to high elevations, spanning an area from Gqeberha (Port Elizabeth) to the Bokkeveld plateau near Nieuwoudtville. It is particularly common near Bredasdorp and Riversdale, where it sometimes forms dense colonies in damp flats.

**ECOLOGY & NOTES** Soil conditions seem relatively unimportant, but it is always found in wet habitats. It produces large quantities of sweet nectar.

**POLLINATORS** Moths, including owlet moths (family Noctuidae) and hawk moths (family Sphingidae).

**SIMILAR SPECIES** *G. tristis* has similar flowers to *G. longicollis* subsp. *longicollis*. Their distributions overlap slightly. In those areas, if subsp. *longicollis* has a shorter perianth tube than the species norm, it can be confused with *G. tristis*.

*Pale form.*

*The perianth tube is long and funnel shaped.*

*Tepals are usually marked on the reverse.*

*Colour variation is thought to be partly linked to moisture and soil conditions.*

*Flowers are pollinated by moths.*

G. tristis *flowering near Caledon, Western Cape.*

# Gladiolus longicollis

***longicollis* = long-necked, referring to the perianth tube.**

*Gladiolus longicollis* is widespread in the summer-rainfall area. The pale flowers have very long perianth tubes and are pollinated by moths. There are two subspecies.

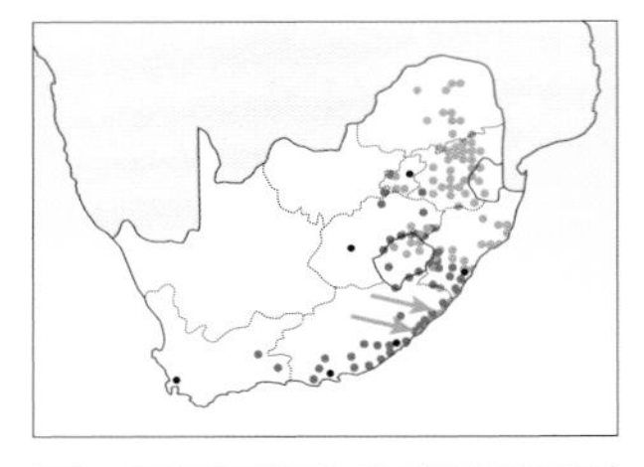

J F M A M J J A S **O N D**

**DESCRIPTION** **Plant** 30–50cm. **Corm** conic, with fine- to medium- to coarse-fibred tunics. **Cataphylls** reaching just above the ground, green and leather-like. **Leaves** 3, lowest with well-developed blade. **Spike** erect, unbranched, 1–3 flowers, sometimes more. **Bracts** grey-green. **Flowers** white to pale yellow, usually uniform but sometimes mottled with brown, opening in the evening and closing early morning; tepals unequal, with outer tepals flushed and purplish-brown veined or green on reverse, lower margins of tepals usually transparent. Perianth tube very long, 50–110mm. **Anthers** green to purplish. **Pollen** cream to yellow. **Capsules** elongate-ellipsoid. **Seeds** ovate, broadly and evenly winged. **Scent** with sweet carnation and clove scent.

*Subsp.* longicollis. *Cream-yellow flowers have unequal tepals.*

**SUBSPECIES** *G. longicollis* has two subspecies.

***G. longicollis* subsp. *longicollis*** (green on map) Can be distinguished from subsp. *platypetalus* by its shorter perianth tube and transparent upper tepals; flowers October to November.

***G. longicollis* subsp. *platypetalus*** (orange on map) ***platypetalus* = broad petals.** Has a much longer perianth tube and broader petals than subsp. *longicollis*; may flower into February.

*Subsp.* platypetalus. *Outer-whorl tepals are longer than those of the inner whorl.*

**DISTRIBUTION** Widespread in the summer-rainfall areas, occurring from the Swartberg and Kammanassie mountains in the southern Cape, northeast through the Eastern Cape and Free State, to Limpopo. Subsp. *longicollis* is found around Oudtshoorn in the Western Cape, through the Eastern Cape and Free State and into Lesotho and Gauteng, usually on moist, south-facing slopes in low, grassy fynbos or rocky grassland. Subsp. *platypetalus* occurs from central KwaZulu-Natal to Mpumalanga and Limpopo, extending into Eswatini. It grows in rocky grasslands and while common in elevations above 1,000m, may also occur along the coast.

**ECOLOGY & NOTES** Usually seen in low grassland in late spring and early summer, but some populations (especially in Mpumalanga) flower in midsummer. Plants in the Western and Eastern Cape have a shorter tube length than those of northern parts, and the latter have little to no speckling on the perianth tube.

**POLLINATORS** Adapted for pollination by night-flying moths such as the Convolvulus hawk moth (*Agrius convolvuli*).

**SIMILAR SPECIES** *G. longicollis* subsp. *longicollis* has similar flowers to *G. tristis*. Their distributions overlap slightly; if subsp. *longicollis* has a shorter perianth than the species norm, it can be confused with *G. tristis*.

*Flowers may resemble those of* G. tristis. *Here, subsp.* longicollis *is mottled with brown.*

*Subsp.* platypetalus. *Flowers are adapted for pollination by moths.*

*A spider exploring subsp.* longicollis.

*Subsp.* platypetalus *has a very long perianth tube.*

*Subsp.* platypetalus *in grasslands at Ntendeka, KwaZulu-Natal.*

# Gladiolus symonsii

**Named after RE Symons (1884–1972), who made the first scientific collection of this plant.**

*Gladiolus symonsii* is found high in the Drakensberg and is seldom seen. Its small, pale to bright pink flowers appear in December and January.

*Editor's note: Rod and Rachel Saunders' initial searches using existing site data were unsuccessful; they did not find a single plant in any of the six recorded locations. It took them three years and directions from multiple networks to find a single, small population. This was the last species they found before their deaths. Their photographic archive includes 17 images of what appear to be the same three or four plants; we are unable to infer from it how large the population might be.*

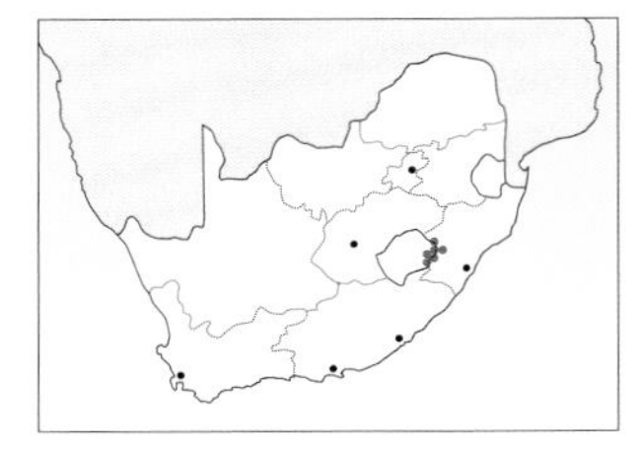

| J | F | M | A | M | J | J | A | S | O | N | D |
|---|---|---|---|---|---|---|---|---|---|---|---|

**STATUS** Rare

**DESCRIPTION** **Plant** 25–45cm. **Corm** globose, with fine- to medium-textured tunics. **Cataphylls** barely reaching above ground, purplish. **Leaves** 3, lowermost the longest and sheaths the lower stem; all leaf blades terete. **Spike** inclined, unbranched, with 2–4 flowers. **Bracts** soft textured, usually dull green. **Flowers** small, pale rose to bright pink with a whitish throat; tepals are subequal and lanceolate, dorsal tepals streaked or spotted at the base. Perianth tube 8–12mm. **Anthers** yellow. **Pollen** yellow. **Capsules** unknown. **Seeds** unknown. **Scent** unscented.

**DISTRIBUTION** A narrow distribution in the main Drakensberg range of KwaZulu-Natal and eastern Lesotho, above 1,800m. Although assessed as not threatened, sightings have become increasingly rare.

**ECOLOGY & NOTES** Grows in montane habitats. Collections were made from 2,000m and above in short grasslands on rocky pavement of sandstone or basalt.

**POLLINATORS** Unknown; presumed to be unspecialised.

**SIMILAR SPECIES** None.

*The pink flowers are almost radially symmetrical.*

*Bracts are usually green but may be purplish.*

*Leaves are cross shaped in section.*

*The reverse side of the tepals may be marked.*

G. symonsii *has a restricted range; encountered here near Cobham, KwaZulu-Natal.*

# Gladiolus watsonius

***watsonius* = watsonia-like, referring to the similarity of its flowers to those of *Watsonia* species.**

*Gladiolus watsonius* occurs in the western parts of the Western Cape. Its orange-scarlet flowers are unscented. Flowering time is August to mid-September. It is assumed to be pollinated by sunbirds.

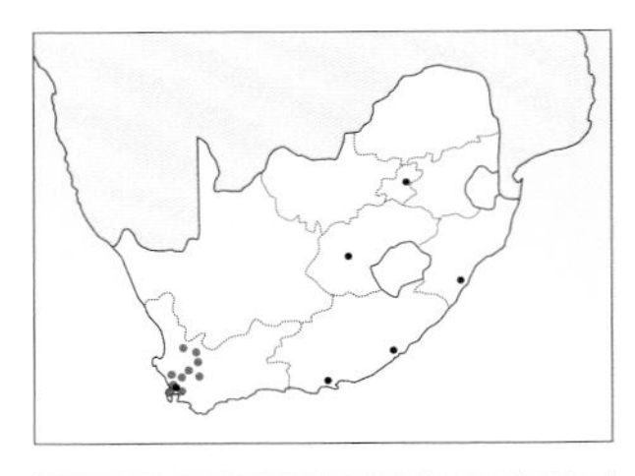

| J | F | M | A | M | J | J | A | S | O | N | D |
|---|---|---|---|---|---|---|---|---|---|---|---|

**STATUS** Near Threatened

**DESCRIPTION** **Plant** 30–70cm. **Corm** globose, often with numerous small cormlets at base, woody to leathery layered tunics. **Cataphylls** up to 18cm above ground, greenish to purple. **Leaves** 3, sheathing stem; margins and midrib thickened, forming wings. **Spike** erect, unbranched, usually with 4–6 flowers. **Bracts** pale green or flushed light brown. **Flowers** orange-red to scarlet, with unequal tepals; lower tepals often translucent, occasionally yellowish below and speckled dark red near base. Perianth tube up to 55mm. **Anthers** dark purple. **Pollen** cream to pale yellow. **Capsules** elongate-ellipsoid. **Seeds** ovate, broadly and evenly winged. **Scent** unscented.

*Erect spikes bear up to six flowers.*

**DISTRIBUTION** Found in the western part of the Western Cape from Piketberg to the Cape Peninsula and no further east than the Upper Breede River Valley, west of Worcester.

**ECOLOGY & NOTES** Favours heavier soils, particularly clay soils derived from Malmesbury and Bokkeveld shales. Along the West Coast, it grows on sandy soils derived from decomposed granite. The species is Near Threatened because of the intrusion of urbanisation and intensive agriculture on its original habitat.

**POLLINATORS** Assumed to be adapted for pollination by sunbirds of the genus *Nectarinia*.

**SIMILAR SPECIES** Can be confused with *G. priorii*, which used to be included in this species. Closely related to *G. teretifolius*. Flowers may be mistaken for *G. quadrangularis*, but that plant has softer corm tunics and leaf blades that are cross shaped in transverse section.

| | Flowering time | Leaves | Flowers |
|---|---|---|---|
| ***G. watsonius*** | Aug–Sep | 4–6mm wide, midrib thickened, margins strongly thickened forming wings | scarlet, purple anthers |
| ***G. priorii*** | May–Jun | slightly fleshy, no venation | scarlet, yellow on lower tepals, yellow anthers |
| ***G. teretifolius*** | Jul–Aug | 1–2mm wide, oval to nearly terete with margins and midrib thickened | scarlet, purple anthers |

*Leaf margins are thickened, forming 'wings'.*

*Long, bent perianths carry the flowers horizontally.*

*Orange-red flowers are likely to attract sunbirds.*

G. watsonius *flowering after fire in Elandsberg Reserve, Western Cape.*

# Gladiolus teretifolius

***teretifolius* = alluding to the rounded leaf shape in transverse section.**

*Gladiolus teretifolius* is restricted to the region between Caledon and Riversdale in the Western Cape. Scarlet flowers appear from mid-July to August. It may be mistaken for *G. watsonius* but is more slender with finer leaves.

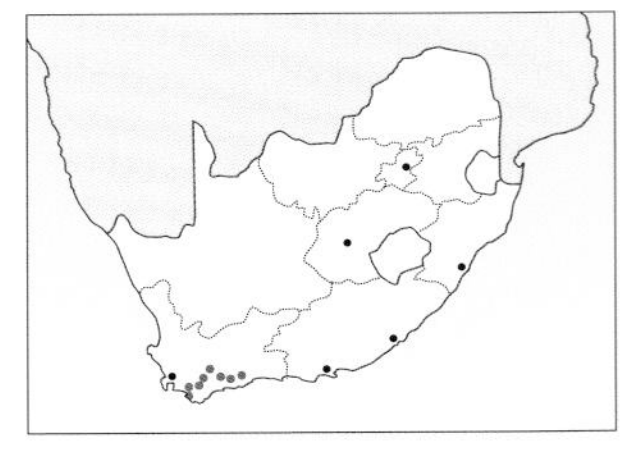

| J | F | M | A | M | J | J | A | S | O | N | D |
|---|---|---|---|---|---|---|---|---|---|---|---|

**STATUS** Near Threatened

**DESCRIPTION** **Plant** 40–70cm high. **Corm** globose, with woody to leathery layered tunics, often with many small cormlets around base. **Cataphylls** reaching 5–7cm above ground, bright green. **Leaves** 3, the lowermost as long as the spike. **Spike** nearly straight, unbranched, with 1–5 flowers, sometimes more. **Bracts** grey-green, sometimes flushed with purple. **Flowers** scarlet, and rarely cream; tepals are unequal, dorsal tepal the largest and translucent along lower edges. Perianth tube is bright red, 30–45mm. **Anthers** dark violet. **Pollen** yellow. **Capsules** ovoid-ellipsoid, 20mm. **Seeds** ovate, broadly and evenly winged. **Scent** unscented.

*Erect spikes can bear up to five flowers.*

**DISTRIBUTION** Local endemic of the Western Cape, restricted to the Caledon, Napier, Montagu, Swellendam and Riversdale districts of the southern Cape. Its habitat has been diminished by wheat farming.

**ECOLOGY & NOTES** Occurs in renosterveld at low elevations in heavy soils, either clay or clay loam. Now rarely seen and confined to small islands of undisturbed vegetation in pasture and grain fields.

**POLLINATORS** Adapted for pollination by sunbirds.

**SIMILAR SPECIES:** Distinguished from *G. watsonius* by a more slender habit; the smaller and narrower leaf blades are oval or round in transverse section. Has similar flower colour to red-flowered *G. priorii*. Flowers may also be mistaken for *G. quadrangularis*, but that plant has softer corm tunics and leaf blades that are cross shaped in transverse section.

| | Flowering time | Leaves | Flowers |
|---|---|---|---|
| ***G. teretifolius*** | Jul–Aug | 1–2mm wide, oval to nearly terete with margins and midrib thickened | scarlet, purple anthers |
| ***G. watsonius*** | Aug–Sep | 4–6mm wide, midrib thickened, margins strongly thickened forming wings | scarlet, purple anthers |
| ***G. priorii*** | May–Jun | slightly fleshy, no venation | scarlet, yellow on lower tepals, yellow anthers |

*Plants have three fine leaves.*

*Red flowers are adapted for pollination by sunbirds.*

*Dorsal tepal margins are translucent.*

G. teretifolius *in the Overberg, Western Cape.*

# Gladiolus quadrangularis

***quadrangularis*** **= referring to the leaf, which is cross shaped in transverse section.**

The unscented red flowers of *Gladiolus quadrangularis* appear in late September and October. The species is restricted to the Ceres region of the Western Cape.

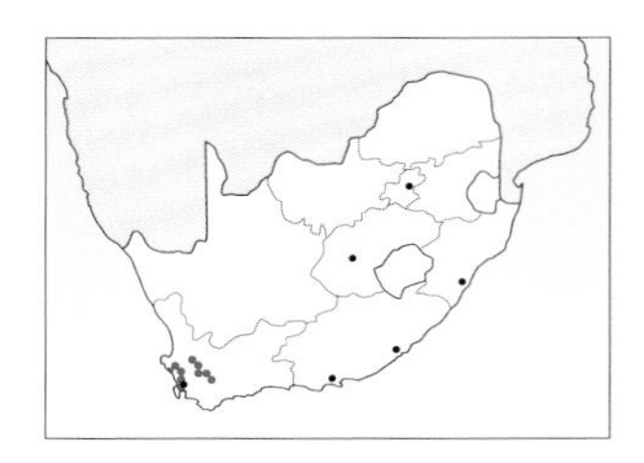

| J | F | M | A | M | J | J | A | S | O | N | D |
|---|---|---|---|---|---|---|---|---|---|---|---|

**DESCRIPTION** **Plant** 35–70 cm. **Corm** globose-conic, with papery to leathery layered tunics. **Cataphylls** greenish purple above ground. **Leaves** 3, lowermost the longest, cross shaped in section. **Spike** erect, unbranched, with 4–8 flowers, sometimes more. **Bracts** green. **Flowers** scarlet to brick-red, rarely pink; dorsal and lower lateral tepals become transparent along the edges. Perianth tube long, up to 55mm. **Anthers** long, yellow. **Pollen** yellow. **Capsules** ellipsoid. **Seeds** ovoid, 7mm. **Scent** unscented.

**DISTRIBUTION** Narrow endemic of winter-rainfall areas restricted to the Ceres district of the Western Cape.

**ECOLOGY & NOTES** Favours heavier clay and loamy soils, areas of which are now heavily farmed. Flowers until late October at high elevations.

**POLLINATORS** Sunbirds.

**SIMILAR SPECIES** Distinguished from *G. watsonius* and *G. teretifolius* by softer corm tunics and leaf blades that are cross shaped in transverse section.

*Flowers are scarlet to red on erect spikes.*

*Upper lateral tepals spread outwards; anthers are yellow.*

*Seeds are ovoid.*

*Leaves are cross shaped in transverse section.*

*Note the erect spikes and long perianth tubes.*

*Dorsal and lower tepals become translucent on the edges.*

G. quadrangularis *flowering on the road verge, Kouebokkeveld, Western Cape.*

# *Gladiolus huttonii*

**Named after H Hutton, a 19th-century collector.**

*Gladiolus huttonii* grows along the southern coastal plain of the Eastern Cape. It has large red-orange flowers with yellow markings, appearing in August and September. It may be confused with *G. fourcadei*, but the latter has smaller flowers with shorter lower tepals.

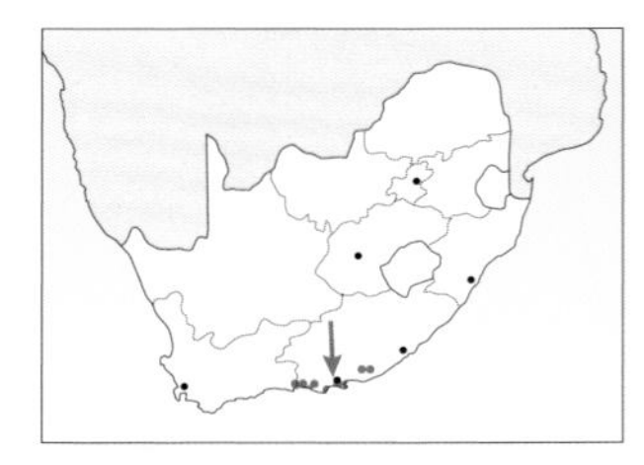

| J | F | M | A | M | J | J | A | S | O | N | D |
|---|---|---|---|---|---|---|---|---|---|---|---|

**STATUS** Vulnerable

**DESCRIPTION** **Plant** 20–45cm. **Corm** globose, with cartilaginous tunics. **Cataphylls** green above ground. **Leaves** 3, lowermost sheathing only near ground, cross shaped in section. **Spike** erect or slightly inclined, unbranched, with 3 or 4 flowers, sometimes more. **Bracts** green or flushed purple. **Flowers** orange-red; dorsal tepals light orange to yellow in lower half with broad translucent edges. Perianth tube is long, ±50mm. **Anthers** long, orange. **Pollen** yellow. **Capsules** elongate-ellipsoid. **Seeds** ovate, broadly and evenly winged. **Scent** unscented.

**DISTRIBUTION** The only *Gladiolus* species that occurs in the Eastern Cape with red, funnel-shaped flowers and a narrow basal leaf that is cross shaped in section. The range extends along the southern coastal strip from Storms River eastwards to East London and ranges inland a short distance. Afforestation and urban and agricultural expansion threaten its habitat.

**ECOLOGY & NOTES** Grows in fertile soils, often in sandy loam and light clay.

**POLLINATORS** Probably sunbirds.

**SIMILAR SPECIES** May be confused with *G. fourcadei* but is distinguished by bigger flowers with larger lower dorsal tepals.

*The large orange-red flowers have yellow markings.*

*Leaves are cross shaped in section.*

*Flowers make a bright splash at Storms River.*

*Bracts may be flushed purple.*

*Perianth tubes are longer than the bracts.*

*Lower tepals are orange-yellow and smaller than the upper tepals.*

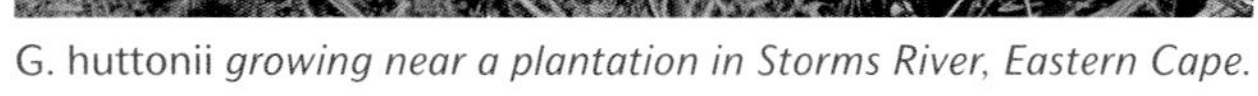

G. huttonii *growing near a plantation in Storms River, Eastern Cape.*

# Gladiolus fourcadei

**Named after HG Fourcade (1865–1948), a botanist who surveyed the entire flora of the southern Cape.**

*Gladiolus fourcadei* grows in the Uniondale district of the southern Cape, flowering between October and November. Yellowish-green flowers are veined with red.

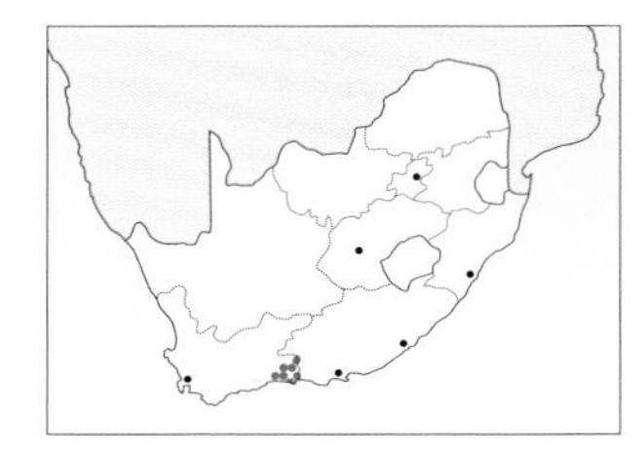

J F M A M J J A S O N D

**STATUS** Endangered

**DESCRIPTION** **Plant** 40–60cm. **Corm** globose-obconic, with cartilaginous tunics. **Cataphylls** uppermost 5–7cm above ground, green or flushed with purple. **Leaves** 3, lowermost the longest, cross shaped in section. **Spike** erect and more or less straight, unbranched, with 3–5 flowers. **Bracts** pale brownish green, sometimes lightly flushed with purple. **Flowers** yellowish green on tepals; upper 3 tepals flushed and veined with dusky red, reverse of tepals and tube red, lower 3 tepals unicoloured or speckled with tiny red dots. Perianth tube long, up to 46mm. **Anthers** long, light purple. **Pollen** yellow. **Capsules** unknown. **Seeds** unknown. **Scent** unscented.

*It has unusually coloured, tubular flowers.*

**DISTRIBUTION** Rare and poorly collected, *G. fourcadei* is considered Endangered because of agricultural and urban encroachment on its habitat. It is now mainly centred in the Uniondale region, in the Upper Langkloof, Kammanassie and southern Outeniqua mountains, among small remnants of fynbos.

**ECOLOGY & NOTES** Favours heavy soils (clay to clay loam) at the sandstone–shale transition; grows in renosterveld or transitional fynbos. This species' habitat is threatened by agricultural expansion and poor land management.

**POLLINATORS** Probably sunbirds.

**SIMILAR SPECIES** Distinguished by the unusual coloration, lacking the red of the other species that have similar tubular flowers and a slender basal leaf.

*A green form, Kammanassie.*

*Long perianth tubes bend abruptly from the bracts.*

*Upper and lower tepals may be markedly different in colour.*

*Sometimes tepals are all the same colour.*

G. fourcadei *has a striking form; here seen on the Kammanassie Pass, Western Cape.*

# Gladiolus abbreviatus

***abbreviatus* = abbreviated, referring to the very short lower tepals.**

*Gladiolus abbreviatus* grows in the southern parts of the Western Cape and flowers from mid-June to August. Its unusual orange-green flowers are horizontal, with elongate upper tepals and minute lower tepals. It is probably pollinated by sunbirds.

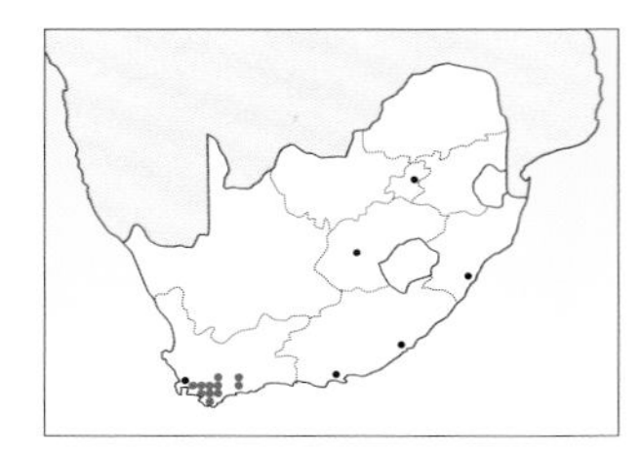

| J | F | M | A | M | J | J | A | S | O | N | D |
|---|---|---|---|---|---|---|---|---|---|---|---|

**STATUS** Vulnerable

**DESCRIPTION** **Plant** 35–55cm. **Corm** globose-obconic, with cartilaginous tunics. **Cataphylls** reaching 12cm above ground, dark purple. **Leaves** 3, with lowest reaching or exceeding top of spike, cross shaped in transverse section. **Spike** erect, unbranched, with 4–6 flowers. **Bracts** the only species in South Africa with coloured bracts: greenish grey flushed with dark red, or olive-green flushed with brown or orange. **Flowers** dark red to orange; tepals uneven, dorsal tepal flushed brownish and pale to transparent on lower edges, upper lateral and lower tepals either darker red, brown or green, lower tepals very short. Perianth tube long, up to 52mm. **Anthers** long, dull red. **Pollen** yellow. **Capsules** elongate-elliptic. **Seeds** ovate, usually tapering at one end, broadly and evenly winged. **Scent** unscented.

*The species has unusual orange-green flowers on an erect spike.*

**DISTRIBUTION** Occurs in a restricted area of the southern portion of the Western Cape, south of Riviersonderend from Caledon to Stormsvlei. It is considered Vulnerable owing to habitat loss caused by agricultural expansion and alien invasive species.

**ECOLOGY & NOTES** Grows on clay and shale banks and slopes in renosterveld, and is most common on wetter, south-facing slopes. It also grows in grass when native flora has been eliminated, but is less common in such environments.

**POLLINATORS** Large quantities of nectar suggest that it is adapted for sunbird pollination.

**SIMILAR SPECIES** None.

*Lower tepals are markedly shortened; note the transparent lower edges on dorsal tepal.*

*Bracts are coloured and leaf midribs and margins are thickened.*

*Dorsal tepals extend horizontally, giving a flattened appearance.*

G. abbreviatus *flowering near Bot River, Western Cape.*

# GLOSSARY OF TERMS

**actinomorphic** radially symmetrical, star-shaped
**acuminate** reaching a tapered point
**anther** the part of a flower that produces pollen
**anthophorid** a member of the family Anthophoridae (mostly solitary bees)
**apiculus** acute point or tip
**attenuate** elongate
**bract** modified leaf associated with a flower
**capsule** simple fruit, often dry, containing seed
**cataphyll** modified leaf without a blade; produced at the base of a plant, often underground
**cauline** relating to the stem
**clade** a natural group
**coriaceous** leathery
**corm** underground storage organ
**deciduous** losing its leaves during the dry part of a cycle
**dehiscent** of a capsule, splitting open and releasing seed at maturity
**depressed** slightly concave
**distal** distant from an originating point
**distichous** arranged alternately in two opposing rows/ranks (see secund)
**dorsal** upper, back
**evergreen** retaining leaves throughout the cycle
**falcate** sickle-shaped
**filament** stalk bearing the anther
**foliage leaf** main photosynthetic leaf
**glabrous** smooth, free of hair or down
**glaucous** dull greyish-green or blue, powdery colour
**globose** globe-shaped, spherical
**hyaline** glassy, translucent
**hybrid** plants that have been cross-bred from two different species
**intermediate** when natural hybrids of some species cross with each other and parents (backcrossing), what appears to be a transition between species is produced
**internode** part of the stem between leaf nodes
**keel** sharp ridge or angle along the midline
**lanceolate** oval and pointed at each end; lance-like
**nectar guides** markings on tepals that guide pollinators to nectar
**obconic** shaped like an inverted cone
**obovoid** egg-shaped, with widest end at the top
**ovate** shaped like an egg, with pointed end at the top
**papillae** small bumps
**perianth** outer, non-reproductive part of a flower
**plane** completely flat, level
**pollen** powdery substance produced by anthers and carrying male gametes
**pseudostem** overlapping leaf sheaths that form a stem-like structure
**pubescent** softly hairy
**scabrid** rough-textured
**secund** arranged on single side (see distichous)
**spathulate** broad and rounded at the ends, tapering to the base
**speciation** the formation of new species as a result of evolution
**spike** collection of flowers attached directly to the stem
**stellate** like a star
**stolon** a shoot from the base of a plant, producing roots, buds or corms
**sub-equal** slightly unequal
**suture** fusion of tepal edges
**synonym** (botany) different name for the same plant
**tepal** outer part of a flower, used when petals and sepals cannot be differentiated easily
**terete** round or cylindrical in cross section
**type** a specimen with the essential characteristics of its species
**variant** a stable difference within a species or subspecies
**vestigial** remaining or underdeveloped
**zygomorphic** bilaterally symmetrical

*actinomorphic flowers*

*bract*

*hyaline tepal*

*cataphyll*

*glabrous leaf, hyaline margin*

*internode*

*keel*

*nectar guides*

*distichous flowers*

*secund flowers*

# BIBLIOGRAPHY


Almond, J. 2018. Growing southern African *Gladiolus*. *The Plantsman*. March 10–17.

Delpierre, GR. & Du Plessis, NM. 1974. *The Winter-growing Gladioli of South Africa*. Tafelberg. Cape Town.

Engle, M. 2010. *Summer Birds: The Butterflies of Maria Merian*. Henry Holt and Co. New York.

Fraser, M. & Fraser, L. 2011. *The Smallest Kingdom: Plants and Plant Collectors at the Cape of Good Hope*. Royal Botanic Gardens. Kew.

Gericke, N. 2014. Ethnobotanical records from a corporate expedition in South Africa in 1685. *Herbalgram: The Journal of the American Botanical Council* 102: 48–61.

Goldblatt, P. & Manning, J. 1998. *Gladiolus in Southern Africa*. Fernwood Press. Cape Town.

Goldblatt, P. & Manning, J. 1999. New species of *Sparaxis* and *Ixia* (Iridaceae: Ixioideae) from Western Cape, South Africa, and taxonomic notes on *Ixia* and *Gladiolus*. *Bothalia* 29(1): 59–63.

Goldblatt, P. & Manning, J. 2000. The long-proboscid fly pollination system in Southern Africa. *Annals of the Missouri Botanical Garden* 87(2): 146–170.

Goldblatt, P. & Manning, J. 2002. Evidence for Moth and Butterfly Pollination in *Gladiolus* (Iridaceae–Crocoideae). *Annals of the Missouri Botanical Garden* 89(1): 110–124.

Goldblatt, P. & Manning, J. 2019. Iridaceae of southern Africa. *Strelitzia* 42. South African National Biodiversity Institute. Pretoria.

Goldblatt, P. & Vlok, JHJ. 1989. New species of *Gladiolus* (Iridaceae) from the southern Cape and the status of *G. lewisiae*. *South African Journal of Botany* 55(2): 259–264.

Goossens, E., Goossens, R. & Dold, T. 2020. Hide and Seek in Baviaanskloof. *Veld & Flora*. 106(4): 36–39.

Hahn, N. & Roux, H. 2014. *Gladiolus filiformis*, a poorly known species from North West Province, South Africa. *Bothalia* 44(1): Article 174.

Lewis, GJ. & Obermeyer, AA., with Barnard, TT. 1972. *Gladiolus: A Revision of the South African Species*. Purnell Press. Cape Town.

Library of Parliament. 1973. *Francois Le Vaillant: Traveller in South Africa and his collection of 165 water-colour paintings 1781–1784*, Volume II. Library of Parliament. Cape Town.

Manning, JC. & Goldblatt, P. 2009. Three new species of *Gladiolus* (Iridaceae) from South Africa, a major range extension for *G. rubellus* and taxonomic notes for the genus in southern and tropical Africa. *Bothalia* 39(1): 37–45.

Manning, JC., Goldblatt, P. & Winter, PJD. 1999. Two new species of *Gladiolus* (Iridaceae: Ixioideae) from South Africa and notes on long-proboscid fly pollination in the genus. *Bothalia* 29(2): 217–223.

Polcha, E. 2019. Breeding insects and reproducing white supremacy in Maria Sibylla Merian's ecology of dispossession. *Lady Science*. 57 (online).

Raimondo, D., von Staden, L., Foden, W., Victor, JE., Helme, NA., Turner, RC., Kamundi, DA. & Manyama, PA. (eds). 2009. *Red List of South African Plants 2009*. Strelitzia 25. South African National Biodiversity Institute, Pretoria.

Schiebinger, L. 2008. The art of medicine. *The Lancet*. 371: 718–719.

South African National Biodiversity Institute (SANBI). 2012. Red List of South African Plants. Available at: http://redlist.sanbi.org. Accessed on: 2021/03/30.

Valente, LM., Manning, JC., Goldblatt, P. & Vargas, P. 2012. Did pollination shifts drive diversification in Southern African *Gladiolus*? Evaluating the model of pollinator-driven speciation. *The American Naturalist* 180(1): 83–98.

# SOURCES AND PHOTOGRAPHS

- The key sources for descriptions are Peter Goldblatt and John Manning's *Gladiolus in Southern Africa*, published in 1998, augmented by additional publications and Rod and Rachel Saunders' notes and observations, such as we have been able to retrieve.
- Maps were compiled from the Saunders' research notes, SANBI datasets, Goldblatt and Manning's text, and other publications.
- The images that accompany the species descriptions are drawn from the Saunders' extensive archive and, to the best of our knowledge, all were taken by Rod and Rachel Saunders, barring those of *G. uitenhagensis*, which was found after their deaths and photographed by Tony Dold. We are grateful for his permission to use the photographs here. Most of the Saunders' photographs are of known populations, although in some cases they were able to identify new populations or to record the demise of previously recorded ones. They were also able to collect information that had been missing from some scientific descriptions. These data are preserved in their archive.

*A robust, dark pink form of* G. ferrugineus *grows in rock crevices in the Steenkampsberg, Mpumalanga.*

# INDEX TO *GLADIOLUS* SPECIES

210
212
214
216
218
220
232
234
236
238
240
242
256
258
260
262
264
266
282
284
286
288
290
292
306
308
310
312
314
316
330
332
334
336
338
340
354
356
358
360
362
364